Millet Proteins
Nutritional, Structural and Functional Attributes

Food Chemistry, Function and Analysis

For a list of titles in this series see: rsc.li/food-chem

How to obtain future titles on publication:
A standing order plan is available for this series. A standing order will bring delivery of each new volume immediately on publication.

For further information please contact:
Book Sales Department, Royal Society of Chemistry, Thomas Graham House, Science Park, Milton Road, Cambridge, CB4 0WF, UK
Telephone: +44 (0)1223 420066, Fax: +44 (0)1223 420247
Email: booksales@rsc.org
Visit our website at books.rsc.org

Millet Proteins
Nutritional, Structural and Functional Attributes

Edited by

Shweta Suri
University of Petroleum and Energy Studies (UPES), Dehradun, India
Email: shwetasuri94@gmail.com

Anuradha Dutta
Govind Ballabh Pant University of Agriculture and Technology, India
Email: adutta.gbpuat@gmail.com

and

Tarun Belwal
Texas A & M University, USA
Email: tarungbpihed@gmail.com

ROYAL SOCIETY
OF CHEMISTRY

Food Chemistry, Function and Analysis No. 48

Hardback ISBN: 978-1-83767-739-9
PDF ISBN: 978-1-83767-740-5
EPUB ISBN: 978-1-83767-741-2
Print ISSN: 2398-0656
Electronic ISSN: 2398-0664

A catalogue record for this book is available from the British Library

The Royal Society of Chemistry is a charity, registered in England and Wales, Number 207890, and a company incorporated in England by Royal Charter (Registered No. RC000524), registered office: Burlington House, Piccadilly, London W1J 0BA, UK, Telephone: +44 (0)20 7437 8656.

For further information see our website at www.rsc.org

For general enquiries, please contact books@rsc.org

For EU product safety enquiries, please email books@rsc.org or contact Royal Society of Chemistry Worldwide (Germany) GmbH, Römischer Hof, Unter den Linden 10, 10117 Berlin.

Printed in the United Kingdom by CPI Group (UK) Ltd, Croydon, CR0 4YY, UK

Preface

In recent years, there has been a notable shift in global dietary patterns, driven by an increasing awareness of health, environmental sustainability, and ethical considerations. A growing number of consumers are actively seeking alternatives to animal-based proteins, which are often associated with adverse environmental impacts such as greenhouse gas emissions, excessive land and water use, and waste generation. In this context, millet protein has emerged as a promising, sustainable, and nutritionally rich alternative. Millets, a group of ancient cereal grains, have traditionally been cultivated and consumed in regions of Africa and Asia. They are known for their resilience to adverse climatic conditions and minimal input requirements, making them an environmentally sustainable crop. Among their nutritional benefits, protein stands out as the second most abundant macronutrient in millets, with content ranging from 7% to 13%—a value that exceeds many other staple cereals. Moreover, millet protein contains essential amino acids required for human health, positioning it as a valuable component in plant-based diets. In addition to their nutritional profile, millet-derived bioactive peptides have demonstrated a range of functional properties, including antioxidant, antimicrobial, antihypertensive, anticancer, and hypoglycemic effects. These attributes further enhance the potential of millets as a functional food ingredient capable of supporting preventive health and overall well-being.

Despite these advantages, the scientific literature on the extraction, characterization, and application of millet proteins remains fragmented. There is an urgent need for a consolidated resource that brings together current research findings, technological advancements, and practical insights in this field. This book aims to fill that gap by providing a comprehensive and systematic overview of millet protein, with particular emphasis on both thermal and non-thermal processing techniques, including

Food Chemistry, Function and Analysis No. 48
Millet Proteins: Nutritional, Structural and Functional Attributes
Edited by Shweta Suri, Anuradha Dutta and Tarun Belwal

Published by the Royal Society of Chemistry, www.rsc.org

ultrasound, pulsed electric fields, and high-pressure processing. The contents of this book cover a wide spectrum of topics, including the classification and structure of millet storage proteins, amino acid composition, emerging bioactive peptides, digestibility, functional and therapeutic benefits, and recent innovations in extraction and processing methods. Furthermore, the book explores the prospects and challenges associated with the integration of millet protein into food systems, highlighting its potential to serve as a viable alternative to animal-derived proteins.

Chapter 1 provides an overview of millets, including their species, types, and nutritional value. It emphasizes the agronomic aspects and breeding conditions necessary for cultivating millets and highlights the health benefits of incorporating millet into the daily diet. Chapter 2 offers a comprehensive review of millet protein, focusing on its distribution, composition, and the effects of processing on millets. Chapter 3 delves into novel millet peptides, also known as millet-derived bioactive peptides. It discusses their functional role, applications, and the challenges associated with them. This chapter explores the potential of millet-derived bioactive peptides in functional foods, nutraceuticals, and therapeutic interventions. Chapter 4 examines the nutritional value and digestibility of millet protein, as well as its impact on the gut microbiome. Chapter 5 explores the functional, structural, and engineering attributes of millet protein. It explains the structural properties of millet proteins, including how various processing methods affect these structures. This chapter also highlights the health benefits of millet proteins beyond nutrition, such as their antioxidant activity and support for digestive health. Chapter 6 discusses the medicinal properties of millet proteins, thoroughly examining millet protein-based bioactive peptides and their therapeutic potential, which include antibacterial, antioxidant, antihypertensive, anticancer, and antidiabetic properties. Chapter 7 focuses on the extraction, isolation, and purification of millet proteins. It features various protein purification and separation techniques, including dialysis, ultrafiltration, chromatography, and electrophoresis. These techniques enhance the functional, physical, and digestible properties of millet protein, making it suitable for use as a functional food item in the bakery and confectionery industries. Chapter 8 explores the bioactive properties of millet protein, such as phenolic acids, flavonoids, and antimicrobial peptides, which contribute to its antioxidant, antimicrobial, antihypertensive, antidiabetic, and anti-inflammatory effects. Chapter 9 discusses thermal and non-thermal methods of processing millet protein, including their effects on the functionality and structure of these proteins. Chapter 10 describes the applications of millet protein in both food and non-food sectors, highlighting its importance in the context of global food security. Chapter 11 examines advancements, market value, and the future outlook of millet protein. It provides a thorough understanding of millet protein's role in the evolving food industry by analyzing its developments and potential future applications.

This book serves as a valuable resource for food scientists, researchers, academicians, principal investigators, agriculturists, laboratory scientists, graduate students, and policy makers. The editors have made a deliberate effort to select contributors with recognized expertise in millets and their processing fractions. We extend our sincere gratitude to all the authors for their valuable contributions. Special thanks are also due to our readers and the editorial staff at the Royal Society of Chemistry for their support and insightful suggestions.

Shweta Suri, Anuradha Dutta and Tarun Belwal

Acknowledgement

Kṛṣṇārpaṇam astu (May this be offered to Krishna). I express my sincere gratitude to Lord Krishna for his divine guidance and blessings throughout the process of editing and compiling this book. I am deeply thankful to my husband and family for their constant support, encouragement, and understanding during this endeavor. I also extend my appreciation to my co-editors for their valuable cooperation and contributions, and I acknowledge all those who contributed directly or indirectly to this publication.

Shweta Suri

Food Chemistry, Function and Analysis No. 48
Millet Proteins: Nutritional, Structural and Functional Attributes
Edited by Shweta Suri, Anuradha Dutta and Tarun Belwal

Published by the Royal Society of Chemistry, www.rsc.org

Contents

Food Chemistry, Function and Analysis No. 48
Millet Proteins: Nutritional, Structural and Functional Attributes
Edited by Shweta Suri, Anuradha Dutta and Tarun Belwal

Published by the Royal Society of Chemistry, www.rsc.org

CHAPTER 1

Millets: Species, Types and Nutritional Value

MANISHA BHANDARI, RAJAN SHARMA* AND HANUMAN BOBADE

Department of Food Science and Technology, Punjab Agricultural University, Ludhiana, India
*Email: rajan-fst@pau.edu

1.1 Introduction

Food is one of the most crucial factors in the maintenance of human health and physical well-being. The driving force for the development and maximization of human genetic potential is the nutritional quality of food. For maintenance of fitness and human health, the dietary quality of food must be taken into consideration to solve the problem of malnutrition. Therefore, diversification of food, accompanied by increasing yields, is encouraged at both household and national levels. Many crops are not used as food in households due to a lack of awareness amongst people. One such crop is millets, which are used either as animal feed or as bird feed. Millets, being important crops at the global level, economically impact the developing countries. They are reported to possess several nutritious and medicinal functions. Millets are neglected and underutilized crops as many people have little knowledge of them and they possess some other problems like lower cooking quality, bioavailability and taste.[1]

Millets belong to the *Poaceae* family (true grass) and have grains of small to medium size. They are grown throughout the tropical and subtropical regions of the world. These crops are generally distributed in countries like India, China, Australia, Malaysia and Sri Lanka. There are seven types of

Food Chemistry, Function and Analysis No. 48
Millet Proteins: Nutritional, Structural and Functional Attributes
Edited by Shweta Suri, Anuradha Dutta and Tarun Belwal

Published by the Royal Society of Chemistry, www.rsc.org

millets, differentiated by their colors, sizes, cultivation areas and shapes.[2] Millets are further classified into two types – major millets (pearl millet, foxtail millet, proso millet and finger millet) and minor millets (barnyard millet, kodo millet and little millet),[3] as shown in Figure 1.1.

Millets comprise carbohydrates (60–70%), proteins (7–11%), fat (1.5–5%), crude fiber, vitamins and minerals. They contain high amounts of B vitamins, antioxidants, magnesium, manganese, iron and phosphorus. Millets have sufficient amounts of essential amino acids (except lysine and threonine) and have large reserves of amino acids containing sulfur (methionine and cysteine). They are considered suitable sources of proteins in our diet, owing to their gluten-free nature and hypoallergenic properties. They tend to have an amino acid profile in a balanced form. After starch, protein is the second most abundant nutrient found in millets, *i.e.*, approximately 10% by weight, as the protein content in various millet seeds ranges from 7.52% to 12.1%, comparable to other cereal grains. The amount, quality and composition of protein depends on the cultivar, growing conditions, agricultural practices (water content, growth temperatures, soil quality and chemicals given to the crops) and species of millet.[4]

Large amounts of essential fatty acids are also present in free (palmitic, linoleic and oleic acids) and bound (monogalactosul, digalactosyl diacylglycerols, phosphatidyl serinediacyl glycerols, phosphatidyl ethanolamine and phosphatidyl choline) forms. Oils extracted from millets can be a good source of linoleic acid and tocopherols. Millets play a key role in the synthesis of energy in the body due to the presence of vitamin B (niacin, riboflavin, thiamine and folacin).[1] Millets are rich sources of micronutrients such as vitamins, β-carotene, *etc.*, that are consumed as pharmaceutical pills in today's generation. Currently, millets are used as superfoods and are claimed to be a solution for malnutrition and obesity in the majority of the Indian population.[5]

Millets are drought and pest-resistant grains, imparting them with significant advantages. Millets can grow in shallow and low-fertility soils with a pH ranging from acidic (4.5) to basic (8.0). They are capable of good growth in high-salinity soil. A lower amount of water is required for the maintenance and growth of millet groups when compared to other cereal crops. Millets are generally grown in places where there is a risk of famine. They are cheaper and safer and are an important aspect for food safety. The physiological, biochemical and anatomical features of millets make them an important industrial crop. They can grow in stress-related conditions due to their short stature, dense root system, thick cell walls, adaptive nature and leaf area. Crops of millets behave as eco-friendly purifiers contributing to the mitigation of environmental issues.[6] Millets are a part of a group of C4 cereals. These plants absorb carbon dioxide from the atmosphere and convert it into oxygen. Highly efficient water use makes millets environmentally friendly crops. Thereby, millets are grown in conditions that have climatic uncertainties and they contribute to a reduction in atmospheric carbon dioxide and help in climate change mitigation.[7] Growing millets in fields

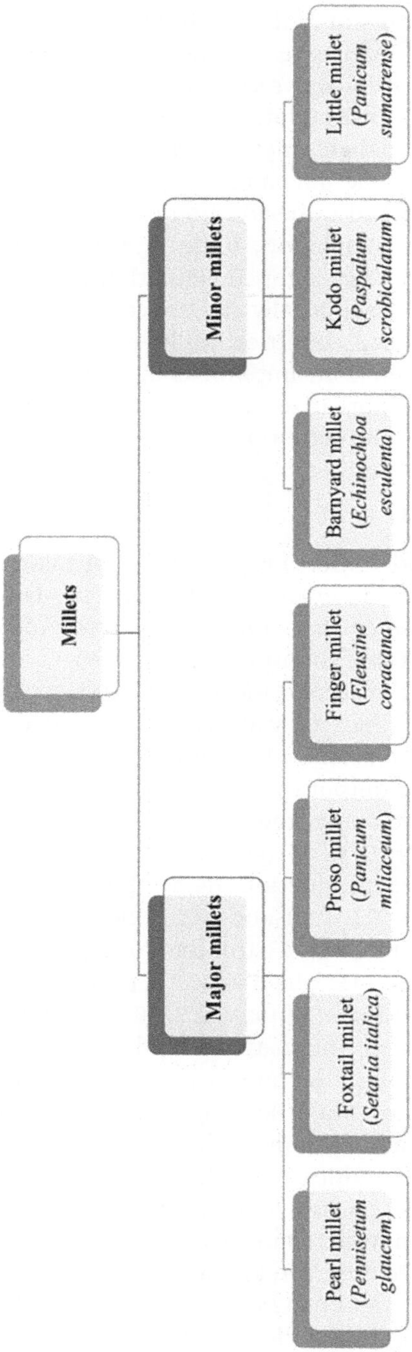

Figure 1.1 Millets classified as major and minor millets.

solves the problem of pollution and helps in the attainment of a sustainable agriculture system. All these features of millets cause them to act as a "next generation crop". Benefits of millets include food security (sustainable food sources, resistant to climatic stress, pests and diseases), nutritional security (rich in micronutrients, bioactive compounds with a balanced amino acid profile), safety from diseases (gluten-free, low glycemic index, prevents cardiovascular problems, diabetes mellitus, *etc.*) and economic security (climate resilient crop, sustainable income source for farmers, low investment needed for production and value-addition can lead to economic gains).[6]

With an increasing population, there is an immense need to explore various plants that can be used for consumption purposes. One such crop is millets, which can be grown locally and are consumed by humans in places like India and other Asian countries. Millet-based foods and beverages are now becoming common consumption choices of consumers worldwide.[8] The present chapter summarizes various types of millets, their breeding conditions, agro-economic conditions and their nutritional composition.

1.2 Production

Millets are a vital food source and are grown in many parts of the world, including Asia, South America and Africa. The Food and Agriculture Organization (FAO) states that millets are grown in over 100 countries, among which the top three countries are India, Niger and China. In African countries, millets are the second most widely grown cereal crops, generally in the semi-arid areas where they are grown as rainfed crops.[9] In 2021, the total amount of millet produced worldwide was 30.1 million tonnes, of which 41% of the total output was from India, with Niger also producing a sizeable amount. Developing countries in Asia and Africa produce and consume approximately 97% of the world's total millet.[10]

According to millet production data, the worldwide production of millet in 2024 was 31 103 thousand MT, amongst which India produced 13 500 thousand MT, which was 43% of the world production of millets. Niger produced 3400 thousand MT, which was 11% of the total world production of millets. The total production of China was 2700 thousand MT, which was 9% of the total millet production (International Production Assessment Division).

India is known as one of the largest producers of millets. Millet is cultivated in 21 Indian states. Major leading states in the production of millets are Rajasthan, Karnataka, Tamil Nadu, Telangana, Jharkhand, Haryana, Gujarat, Maharashtra, Andhra Pradesh, Kerala, Uttarakhand and Madhya Pradesh. India produces 26.60% of the global production of millets and 83% of the production from Asian countries. In India, millet is grown in an area of over 13.8 million hectares, producing a total yield of 17.3 million tonnes and having an average yield of 1247 $kg\,ha^{-1}$. India is the topmost producer of kodo millet (100%), little millet (100%), barnyard millet (99.9%), finger millet (53.3%) and pearl millet (44.50%) and produces around 12.46 million metric tonnes in an area of 8.87 million ha. The highest millet cultivating area is Rajasthan, followed by Maharashtra, Karnataka, Uttar Pradesh, Madhya Pradesh, Gujarat and Tamil Nadu having cultivation around 29.05%, 20.67%, 13.46%, 8.06%, 6.11%, 3.94% and 3.74% respectively. Gujarat and Madhya Pradesh have comparatively improved the areas under which millets are cultivated over the years. The highest yields of millet are recorded in Andhra Pradesh, Tamil Nadu, Haryana, Gujarat and Madhya Pradesh, having yields of 2626.58 kg ha^{-1}, 2153.22 kg ha^{-1}, 1906.78 kg ha^{-1}, 1762.05 kg ha^{-1} and 1729.70 kg ha^{-1}, respectively. Better levels of productivity are observed in Gujarat and Andhra Pradesh compared to others.[11]

1.3 Species, Types and Nutritional Value of Millets (Including Agronomic Aspects and Breeding Conditions)

1.3.1 Major Millets

1.3.1.1 Pearl Millet

Pearl millet is the sixth most important cereal grain in the world and is generally grown in arid and semi-arid areas of Asia and Africa. Pearl millet is grown as a food and forage crop in India, and as a forage crop in Africa. India is the largest producer of pearl millet and has seven major pearl millet-producing states. The progenitor of cultivated pearl millet is *P. glaucum* subsp. *monodii*, which is found in the Sahelian region of Africa. The domestication of pearl millet occurred through either single or multiple events, and evidence of the presence of millet at the earliest archaeological site was recorded 3000 years ago in northern Ghana.[12]

1.3.1.1.1 Nutritional Composition. Pearl millet [*Pennisetum glaucum* (L.)] grains are similar or even superior to other cereals in terms of composition of energy, minerals, proteins and vitamins. Also, they are rich in dietary fiber, micronutrients and phytochemicals. Pearl millet is a rich source of energy, giving 361 kcal per 100 g, which, compared to other cereals, *i.e.*, wheat (346 kcal per 100 g), maize (325 kcal per 100 g), sorghum (349 kcal per 100 g) and rice (345 kcal per 100 g), is high. In pearl millet, the protein content ranges from 8% to 14% depending upon the varieties, while the fat content ranges from 6% to 10%, and the protein digestibility ranges from 53% to 68%. The amino acid content and quality of pearl millet is comparable to wheat, rice and barley, and better than sorghum and maize. It contains a high content of vitamin B (niacin, riboflavin, thiamine), vitamin E, vitamin A, potassium, magnesium, zinc, manganese, phosphorus, iron and copper.[13] Protein and amino acid contents are amongst the major considerations for assessing the quality of the food grains. The protein content of pearl millet is more than or comparable to that of cereal grains but less than that of legumes. The amino acid profiles of all the cereals are almost similar except for the amounts of lysine, leucine, tryptophan, tyrosine and isoleucine.

The average amino acid content in pearl millet is 0.67%, but the amount of non-essential amino acids is 1.22%. The amounts of essential amino acids, *i.e.*, methionine (0.31%), lysine (0.43%) and histidine (0.39%), are less than those of leucine (1.7%) and phenylalanine (0.71%). The limiting amino acid, *i.e.* lysine, is comparatively higher than that in sorghum and maize. In pearl millet, the level of lysine ranges from 0.43 to 0.55 g per 100 g proteins, and the ratio of isoleucine/leucine is lower than maize and sorghum but comparable to other grains like rice, wheat and barley. The balance of amino acids in pearl millet makes it a nutritious source of protein for human consumption.[14] Nutritional compositions of millets are given in Table 1.1.

Pearl millet flour tends to retain the alkaline properties that are beneficial for people suffering from wheat allergy, hence they are gluten-free cereal grains.[15] The amylase content of starch, starch content and soluble sugars in pearl millet range from 2.86% to 21.5%, 62.8% to 70.5% and 1.2% to 2.6%, respectively. The free sugars (glucose, sucrose, raffinose and fructose) of pearl millet range from 1.2% to 2.5%. The non-starch polysaccharide fraction is found as monosaccharide (xylose, uronic acids, arabinose and glucose) of pearl millets.[16]

1.3.1.1.2 Agronomic Conditions. Pearl millet grains are cultivated in rainfed conditions. The major constraints are drought, drought-induced heat stress, low soil fertility and low phosphorus content in soil. Phosphorus deficiency in soil limits the productivity of the crop and the application of nitrogen.[17] Pearl millet possesses reproductive, physiological and climate-smart vegetative features that make this crop suitable for growth in harsh conditions such as low fertility of soil, high saturation of aluminium in soil, high temperature, limited rainfall, high pH of soil, low moisture in soil and high soil salinity. Pearl millets can be grown in areas where the annual precipitation is as low as 250 mm. These crops can tolerate temperatures over 42 °C in their reproductive phase, suggesting their cultivation in hot summers using proper irrigation.[18]

1.3.1.1.3 Breeding. A prerequisite of crop breeding is to explore the genetic variability that exists in the crop for yield and other related traits. This results in the development of better varieties, capable of addressing the growing demands of an increasing population. Genotypes and the surrounding environment (season) are governed by characteristics like micronutrients and grain yield at the time of grain filling.[19]

Breeding objectives depend primarily upon whether the plant is used for grain production or for forage production. Various traits that are important for grain production include seed size, seed color, floret compactness, pest resistance, days to seed maturity, plant height, seed weight, inflorescence length and diameter, stalk strength and tillering ability. The primary traits for forage production include dry matter, response to day length, tillering ability, pest resistance, dry matter digestibility and regrowth potential. Other characteristics of plants include their variety, disease, seed and ear.[20]

Table 1.1 Nutritional composition of millets.

	Pearl millet[41,85,86]	Finger millet[41,86]	Foxtail millet[41,86–88]	Proso millet[41,86,87]	Little millet[41,89]	Barnyard millet[41,88,90]	Kodo millet[41]	Sorghum[91]
Protein (%)	10.97–11.65	7.16	12.30	12.50	8.42–11.20	6.2	8.3	8.27 (IVPD: 13.62%, Lysine: 1.21 g per 16 g of N)
Fat (%)	5.14–5.96	1.92	4.30	1.10	1.8–2.10	9.8	1.4	1.87
Crude fiber (%)	2.07–2.63		20.8 Soluble dietary fiber: 1.1% Insoluble dietary fiber: 19.7%		3.20 Dietary fiber: 12.51%	2.2–5.3	9	3.34
Ash (%)	2.05–2.72		2.4		1.75–2.3			1.73
Carbohydrates (%)	66.49–68.85	68.82	60.09 Total starch content: 18.5% Amylose content: 4%	70.04	72–74.75	65.5 Total starch: 12.6% Amylose content: 8%	65.9	84.77
Minerals	Calcium: 40.07–42.67 mg per 100 g; Phosphorus: 255.67–327.82 mg per 100 g; Iron: 5.08–8.12 mg per 100 g; Calcium: 10–80 mg per 100 g	2.7 g per 100 g; Iron: 3.6–6.8 mg per 100 g; Phosphorus: 283 mg per 100 g; Calcium: 376–515 mg per 100 g	3.3 g per 100 g; Sodium: 8 mg per 100 g; Potassium: 252 mg per 100 g; Iron: 2.8–3.7 mg per 100 g; Calcium: 31–40 mg per 100 g; Phosphorus: 238 mg per 100 g	1.9 g per 100 g; Iron: 0.8 mg per 100 g; Phosphorus: 206 mg per 100 g; Calcium: 14 mg per 100 g	4.5 g per 100 g; Iron: 3.98–5.12 mg per 100 g; Zinc: 27.9–41.8 mg per 100 g; Phosphorus: 220 mg per 100 g; Calcium: 17 mg per 100 g	4.4 g per 100 g; Calcium: 20–60 mg per 100 g; Phosphorus: 252–280 mg per 100 g; Iron: 5–15.2 mg per 100 g; Sodium: 18 mg per 100 g	2.6 g per 100 g; Iron: 0.5 mg per 100 g; Phosphorus: 188 mg per 100 g; Calcium: 27 mg per 100 g	

Table 1.1 (*Continued*)

	Pearl millet[41,85,86]	Finger millet[41,86]	Foxtail millet[41,86–88]	Proso millet[41,86,87]	Little millet[41,89]	Barnyard millet[41,88,90]	Kodo millet[41]	Sorghum[91]
Vitamins	Thiamine: 0.38 mg per 100 g; Riboflavin: 0.21 mg per 100 g; Niacin: 2.8 mg per 100 g	Thiamine: 0.42 mg per 100 g; Riboflavin: 0.19 mg per 100 g; Niacin: 1.1 mg per 100 g	Thiamine: 0.59 mg per 100 g; Riboflavin: 0.11 mg per 100 g; Niacin: 3.2 mg per 100 g	Thiamine: 0.41 mg per 100 g; Riboflavin: 0.28 mg per 100 g; Niacin: 4.5 mg per 100 g	Thiamine: 0.30 mg per 100 g; Riboflavin: 0.09 mg per 100 g; Niacin: 0.2 mg per 100 g	Thiamine: 0.30 mg per 100 g; Riboflavin: 0.1 mg per 100 g; Niacin: 4.2 mg per 100 g	Thiamine: 0.33 mg per 100 g; Riboflavin: 0.09 mg per 100 g; Niacin: 0.2 mg per 100 g	
Bioactive	Polyphenols: 403.59–445.38 mg per 100 g; *p*-coumaric acid: 1350 μg g^{-1}; Gallic acid: 15.3 μg g^{-1}; Ferulic acid: 199 μg g^{-1}; Syringic acid: 7.4 μg g^{-1}		Total phenolic content: 1.7–1.83 mg g^{-1}; Total flavonoid content: 0.71–0.78 mg g^{-1}; Proanthocyanid: 0.59–0.79 mg g^{-1}					Total phenols: 92.62 mg GAE per 100 g; Tannin: 8.46 mg TAE per 100 g

Improved productivity and reliability of the cultivars depend upon the various approaches used for population improvement. The maximum potential of the pearl millet can be improved by utilization of hybrid vigor. This was utilized in the mid 1940s, where two hybrids were developed and released, increasing the yield by 45% in comparison to local types. Hybridization in India increased in the mid-1960s, although the region of plantation remained static. Hybridization of pearl millet increased because it gave rise to superior inbreds. Generally, the inbreds of pearl millet are vigorous, having general properties of the species and specific properties that are a combination of both species. The use of hybrids increased due to the improved seed sets and yields, disease resistance, plant height, plant vigor and standability. A major constraint in hybridization is the lack of infrastructure that is required for the distribution of seeds.[21]

1.3.1.2 Foxtail Millet

Foxtail millet (*Setaria italica*) is an important staple crop that is grown throughout the world, predominantly in arid and semi-arid areas of China, India and Japan. South and North America grow foxtail millet for silage and hay. Other names of foxtail millet are Kangni (Hindi); Navane (Kannada); Kaon dana (Bengali); Kavalai and Tenai (Tamil); Kang (Gujrati); Kangam (Oriya); Su, Xiaomi, Shao-mi, Kou wei tsao in China; Mohar in Russia; Panico, Milho panico, and Milho panico de Itálica in Portugal; Kimanga in Kenya; Awa in Japan; Millet des oiseaux and millet d'Italie in France; Siberian millet and Dawa in Indonesia. Foxtail millet is the most cultivated millet after pearl millet.[22] Foxtail millet is the sixth highest-yielding grain worldwide.

Foxtail millet belongs to the genus *Setaria*, family *Poaceae* and subfamily *Panicoideae*. Foxtail millet is known as one of the world's oldest cultivated crops, and its archaeological remains are primarily found in northern China dating back 7400–7935 years. Several remains were found in Europe that date to 4000 years ago. Presently, foxtail millet is cultivated in 26 countries and is the second most widely produced millet worldwide. In terms of yield, foxtail millet ranks fourth amongst millets. A few more alternative names of foxtail millet include Italian millet, German millet, foxtail bristle grass, Italian foxtail and Siberian millet.[23]

1.3.1.2.1 Nutritional Composition. Foxtail millet is a coarse grain possessing nutritional compounds in applicable amounts, *i.e.* starch, vitamins, proteins and minerals. Due to its coarse nature, it constitutes 79% of the digestible fiber, while the remaining 21% is undigestible fiber, and there is a presence of some anti-nutritional compounds. Its richness in dietary fiber makes it a compound that helps in the digestive process and improves the bowel movements, thereby producing a laxative effect that is an essential requirement for a healthy digestive system.[24] The biological value of foxtail millet protein is better than that of rice and wheat, and it contains 7 of the 8 essential amino acids that the human body cannot synthesize. It is 2.5 times higher in edible fiber than rice, thereby it is an edible source of dietary fiber that plays an important role in stomach and intestinal health. Bran of foxtail millet carries around 8–10% crude oil that is a rich source of linoleic (66.5%) and oleic acid (13.0%).[22]

1.3.1.2.2 Agronomic Conditions. Foxtail millet grows in regions that are cooler and drier when compared to other millets and grows up to 2–5 feet tall. It is usually grown in sandy and loamy soils. It requires approximately one-third more moisture content than corn. Foxtail millet can grow across a wide altitudinal range, from sea level to 1500 metres. Once planted, it can be harvested after 75–90 days and it is an economical crop in the semi-arid tropical areas. Foxtail millet has the potential to grow under saline conditions and can be grown in drought-resistant areas that have low rainfall. Glasshouses with densities of up to 100 plants per square metre are suitable for the growth of foxtail millet, making it suitable for research purposes. Foxtail millet is generally cultivated in China, India, the USA, Africa, Russia and many parts of Europe. Foxtail millet is a C4 photosynthetic plant and has high nitrogen use efficiency and water uptake efficiency, along with several other morpho-physiological traits, including deep and dense root systems, thickening of cell walls and small leaf area, that enhance the tolerance of cells to a wide range of abiotic stresses, *i.e.* heat, salinity and drought. For germination, foxtail millet requires an amount of water equivalent to 26% of the grain weight, while other major cereals (rice, maize and wheat) require an amount of water equivalent to at least 45% of their grain weight. Similarly, 257 g of water is required for the production of 1 g of biomass, while other major cereals, *i.e.* wheat and maize, require 470 and 510 g of water, respectively. Foxtail millet is known to possess a novel stress-responsive gene that participates in stress tolerance. Heat, salt stress and drought stresses are the major abiotic stresses that occur as an impact of sudden changes in the climatic conditions.[22]

1.3.1.2.3 Breeding. Breeding of foxtail millet is carried out to achieve a higher quality of nutrition and results in improved yields, resistance to disease and tolerance of stress. Utilization of germplasm majorly depends upon two factors: (i) identification of the useful gene sources, and (ii) evaluation and characterization. The hybrids are generally screened for physiological, quality, agronomic and pathological parameters. Breeding values of the germplasm are assessed *via* many field trials, followed by the germplasm catalogue. Efficiency of germplasm utilization can be enhanced by forming subsets and further making them available to breeders, and the constraint to breeding is that the area of cultivation of foxtail millet is neglected. Therefore, the genetic potential of foxtail millet is lower compared to other dryland crops. Breeders must focus on the above limitations to improve the resilience to adverse climates and the productivity of the crop. The major threat to foxtail millet is the blast pathogen.[25]

1.3.1.3 Proso Millet

Proso millet is known by different names such as Broomcorn millet (China), Barri (India), Kashfi millet, Red millet, White millet, Brown millet, Chinese millet, Kibi, Mijo (Spain), Gijang (Korea), Common millet (USA), Broomtail millet and Panic millet (France). Domestication of proso millet started in Northern China. It is widely grown throughout the world due to its wide adaptability to different climatic zones. India, Nepal, Russia, Belarus, Turkey, China, Africa, Ukraine, the Middle East and Romania are the countries where proso millet is generally grown. German-Russian immigrants were responsible for the introduction of proso millet in North America in 1875, and the plantation was done along the eastern Atlantic coast.[26] Developing countries derive one-third of their total energy and protein from proso millet, yet it remains an underutilized crop. Despite possessing several benefits (gluten-free, high protein and fiber content and low glycemic index), proso millet still contributes majorly to birdseed industries and struggles to enter retail consumer markets. However, the recent increase in demand for gluten-free food has caused an increase in the requirement of these ancient grains. Different food sectors in which proso millet can be utilized are breakfast cereals, fermented products, brewing and bakery products.[27]

1.3.1.3.1 Nutritional Composition. Proso millet has several nutritional and health benefits, and when incorporated into our daily diets, it can help overcome several nutritional deficiencies. Proso millet is a rich source of protein, vitamins, micronutrients (iron, copper, manganese and zinc), phytochemicals, minerals and several other non-nutritional components that protect the plant. The protein content of proso millet is 11.6%, within the range of protein content in wheat, *i.e.* 8–15%, and it has a higher content of leucine, methionine and isoleucine. Out of the total carbohydrate content (68.2%), 52% is the starch content. The dietary fiber content ranges from 8.9% to 12.5%. Proso millet also contains hemicellulose (generally made up of arabinose, xylose, glucose and uronic acid and is considered a dietary fiber). The major phenolic acids in proso millet are ferulic, ferulic acid dehydrodimers, soluble chlorogenic acids and *p*-coumaric acids; however, the phenolic content (both soluble and bound) of proso millet is at least compared to kodo, foxtail, little, finger, pearl and barnyard millet.[28]

1.3.1.3.2 Agronomic Conditions. Proso millet can be best grown in those agricultural areas where the rainfall is less than 100 mm. It has a short growing season of 60–90 days and has a shallow root system (90–120 cm), which makes it a suitable dryland crop. Proso millet is an important rotational crop in Central Great Plains, where it is rotated with corn (*Zea mays* L.) and sunflower (*Helianthus annuus* L.), ascribing to its short growing season and high-water use efficiency. The short growing season of proso millet favors the sowing and harvesting of millets before winter crops.[27]

Proso millet is an annual grass grown in summers as a late-seeded crop of summer and completes its life cycle in roughly 60–100 days. Proso millet is named "broom corn" millet due to the presence of a compact panicle that droops at the top like an old broom. The grains of proso millet are enclosed *via* smooth hull, and the grains are 3 mm in length and 2 mm in width. The grain is present in various colors, *i.e.* creamy white, red, yellow and white, along with grey, black and brown. The U.S. is generally known for the white-seeded varieties followed by the red seeds. The plant of proso millet is 30–100 cm tall, having few tillers and an adventitious root system.[29]

Proso millet is more prone to grow in plateau and hilly areas that are arid lands and receive an annual precipitation of as low as 200–500 mm, and have the potential for production at 350 mm of annual rainfall. Proso millet is a warm-season crop that requires warm temperatures (20 to 30 °C) for germination and production and is sensitive to frosts. It is a C4 crop that has a low transpiration ratio and can efficiently fix carbon under conditions of high temperature, limited nitrogen, carbon dioxide and drought. It has the potential to avoid drought sensitivity by maturing rapidly. Once the temperature increases above 30 °C, the vegetative growth stops, and the flowering ceases. Also, it maintains its primary stem at a shorter height to resist dry conditions.

The soils in which proso millet can grow are sandy loam, saline, low-fertility soils and slightly acidic. The growth of proso millet is poor in waterlogged soil conditions and in sandy and coarse soils. They also thrive in low pH soils, and the germination occurs mostly in the pH range of 5.5 to 6.5. Proso millets that grow in soil with a pH over 7.8 exhibit symptoms of iron chlorosis.[29]

1.3.1.3.3 Breeding. Generally, proso millet is self-pollinating, but cross-pollination may exceed 10%. This leads to the creation of artificial crosses that generate intervarietal hybrids. However, breeding of proso millet improvement is challenging due to disease and pest tolerance, flowering time and floral morphology, grain shattering and lodging, phenotyping and high-throughput phenotyping, wide hybridization, genome editing, mutation breeding, *etc.* Plant breeding has many methods for the creation, selection and establishment of superior phenotypes that have the goal of improved cultivars that have enhanced yield, better biotic and abiotic stress tolerance and enriched nutritional qualities. Breeding efforts of proso millet are carried out in several countries around the world, but only slight success was achieved due to a lack of sufficient genetic characterization. The major contributors to proso millet breeding are the USA, China, Russia, Kenya and India. The effort of proso millet improvement in India was led by All India Coordinated Small Millets Improvement Project (AICSMIP). Germplasm was collected from many sources and made available to the breeders, and AICSMIP have access to 939 accessions of proso millet. Since this project started, 16 different types of proso millet cultivars have been released, and these were developed using backcrossing, pedigree breeding and pure-line selection. The first cultivar was Ram Cheena, which was a pure line selection, cultivated in 1960, and has an improved yield of 2000–2500 $kg\,ha^{-1}$. The leader of variety development of proso millet is Tamil Nadu Agricultural University (TNAU), as it has developed 10 cultivars from 1980 to 2011. K2 was the first cultivar of proso millet that was developed by screening and selection of local lines by TNAU. TNAU released two cultivars (TNAU 145 and 151) in 2009 that were resistant to rust and shoot fly.[30]

1.3.1.4 Finger Millet

Finger millet is the oldest crop of India. It was once referred to as "*nrtta-kondaka*", which meant "dancing grain", in the ancient time when the basic language for literature was Sanskrit. It was also called "*rajika*" or "*markataka*". Finger millet is also recognized by several names throughout the world: ragi and mandua (India); fingerhirse (Germany); bulo (Uganda); wimbi and mugimbi (Kenya); finger millet, African millet and koracan (England); kambale, lupoko, mawele, amale and bule (Zambia); petit mil, eleusine cultivee, coracan and koracan (France); poho, rapoko, zviyo, njera and mazhovole (Zimbabwe); dagussa, tokuso and barankiya (Ethopia); and kaddo (Nepal). Finger millet is a staple crop of certain parts of eastern Africa, Central Africa and India. It was a domesticated crop in certain parts of India and was popular by the name *nachni* in Maharashtra, which meant dancer, and as *umi* in Bihar. Grains can be roasted, sprouted and dried, ground as flour and sieved. The pinkish flour of finger millet was used as a ball, sweetened and salted. It is also popularly used as a weaning food. According to ancient Tamil literature, red finger millet is called "*Kelavaragu*". Words like "*Purananuru*" refer to the drying, husking and cooking of the finger millet grains.[31]

1.3.1.4.1 Nutritional Composition. In general, finger millet is a rich source of calcium, fiber and many other minerals. The carbohydrate content of finger millet ranges from 72% to 79.5%. The major component of carbohydrates is starch, and its quantity ranges from 59.0% to 70.2%. Starch granules of finger millet possess a polygonal rhombic shape. Out of the total starch content, 80% to 85% is amylopectin, and 15% to 20% is amylose. The hypoglycemic effect of the millet is ascribed to high fiber content, and it contains complex carbohydrates that are digested slowly and absorbed, causing a reduction in postprandial glucose. After carbohydrates, the major nutritional component in millet is protein (7%); however, it varies from 5.6% to 12.70%. The protein content of brown-seeded finger millet is higher than white-seeded finger millet. Out of the total protein, the major fraction is prolamin, ranging from 24.6% to 36.2%. The quality of protein depends upon the type and quality of essential amino acids present.[32] Out of the total amino acid content of finger millet, 44.7% is the essential amino acid content, which is significantly higher than the total amount of essential amino acids present in the reference protein, *i.e.* 33.90%. The ratio of essential amino acids to total amino acids is good in finger millet. The limiting amino acid in finger millet is lysine, while tryptophan is the second most deficient amino acid. However, the content of threonine is fairly good when compared to rice, sorghum and wheat grains and, among all the millets, finger millet is the most balanced in essential amino acids as it contains significantly better amounts of lysine, valine and threonine. Methionine and lysine content are inversely correlated to the finger millets' protein content. Albumin and globulin fractions of protein contain better amounts of essential amino acids, while prolamine has significantly larger amounts of proline, isoleucine, phenylalanine, glutamic acid, valine and lysine, but has lower content of glycine and arginine. The ratio of leucine to isoleucine is roughly 2, which is similar to that of rice and wheat. The amount of sulfur-containing amino acids in finger millet is equal to the amount of sulfur-containing amino acids in milk. The protein efficiency ratio (PER) and protein digestibility content of finger millet flour are 1.40% and 76%, respectively. However, protein digestibility can range from 55.4% to 88.1% and is in negative correlation to phytates, tannins, phenols and antipyretic activity.[33]

1.3.1.4.2 Agronomic Conditions. Finger millet grows in tropical and subtropical areas of the world and is an annual hardy crop. It can withstand severe drought conditions, thereby having the potential to grow in adverse weather conditions, including fair growth in poor and marginal soils. It is cultivated in large amounts in West Africa (Senegal, Niger, and Nigeria), Madagascar, China, tropical Africa (Ethiopia, Eritrea, Mozambique, Zimbabwe, and Namibia), Japan and Malaysia. The major states of India that have high finger millet production are Karnataka, Odisha, Tamil Nadu and Andhra Pradesh. These crops are generally grown in the areas of heavy rainfall in the Eastern Ghats and are grown in areas dominated by tribal communities, where the soil has low fertility and water absorption capacity. The Eastern Ghats consist of non-uniform mountain ranges, plateaus, stretches and hills spread across Chhattisgarh, Tamil Nadu, Karnataka, Odisha and Andhra Pradesh. Finger millet in Orissa grows in rainfed areas during the rainy season, which is from June to October, and it comprises 20% of total cropped area.[34]

Finger millet is termed a drought-tolerant crop and predominantly cultivated during the monsoon season, where drought stress (DS) for 25 to 30 days invariably occurs at one or the other crop growth stages. DS for 25 days during the grand growth phase to ear head emergence decreased the grain yield up to 18% and to the extent of 109.8% with prolonged DS from 28 days until crop maturity. DS additionally increases micro-climate temperature by 4 to 5 °C and induces disease infestation (*Helminthosporium*). The most critical crop growth stage for DS response in finger millet is the ear head emergence stage. Therefore, selection of a specific physiological or agronomic trait from amongst released stable varieties could be more useful for further yield improvement under DS. In addition, queries are being raised by the farming community for a drought-tolerant variety of finger millet.[35]

1.3.1.4.3 Breeding. Finger millet is a self-pollinated plant, and the cross-fertilization by insects or wind is less than 1%. Finger millet has 4 principal growth stages termed as S1, S2, S3 and S4. S1 is the stage when one fourth of the inflorescence has emerged and is also known as the booting stage. The inflorescence is light green, and only the spikelets are visible. The S2 stage is named the anthesis stage, and the inflorescence at this stage appears yellow. The anthers emerge during this stage, and florets are also visible. S3 is the grain filling stage, where the inflorescence and developing grains are greener in color and the developing grains, which remain covered with lemma and palea, are swollen. In the S4 stage, the spikelets are 50% ripened, and the caryopsis is hardened, making it is difficult to separate *via* thumbnail. The inflorescence is dried, and the color changes to brownish yellow and the grains are visible as gaped florets. Variations in the shape of grains and head are useful to distinguish closely related species. Finger millet is known to have been first domesticated in the African highlands, from where it moved to the southern lowlands, followed by its movement to India. The flow of genes from wild African countries and cultivated subpopulations indicates the natural hybridization among the sympatric subspecies. Hybridization can be done using the contact method, hot water treatment, cold water and plastic bag method. Mutation breeding is a cost-effective, quick and robust method of hybridization. More than 210 plant species have been hybridized using this technique from over 70 countries, among which 2 cultivars are of finger millet. The most common method of mutation breeding is the treatment of seeds with physical, chemical or combination mutagens and from the subsequent population desirable mutants are selected that are superior to their parents.[36]

1.3.2 Minor Millets

1.3.2.1 Barnyard Millet

The oldest domesticated small millet is barnyard millet, which has two main species – *Echinochloa esculenta* (Japanese) and *Echinochloa frumentacea* (Indian). The Indian barnyard millet is also known as Billion Dollar Grass. Barnyard millet belongs to Poaceae family and is a self-pollinating crop. It is one of the hardest millet and is known by various common names, *i.e.* Sanwa and Jhangora (Hindi), Oodalu (Kannada), Swank (Punjabi), Shyama (Sanskrit), Shamula (Bengali), Kuthiravaali (Tamil), Sama (Gujarati), Udalu and Kodisama (Telugu) and Shamul (Marathi). Barnyard millet is cultivated in many countries few of which are India, Japan, East Indies, United States of America, China, Malaysia and Africa. The major states of India where these are cultivated are Orissa, Madhya Pradesh, Bihar, Gujarat, Maharashtra, Tamil Nadu, Punjab and the hills of Uttarakhand.[7]

1.3.2.1.1 Nutritional Composition. Barnyard millets are rich in dietary fiber, zinc, protein, fat, iron, calcium, magnesium, vitamins and some essential amino acids. The carbohydrate content varies from 59.5% to 62%, which is significantly lower than other millets, while the crude fiber content ranges from 8.1% to 16.3%, which is higher than other millets. The ratio of carbohydrate to dietary fiber is high, which signifies slow release of sugars in the blood, thereby maintaining blood sugar levels. The presence of resistant starch is known to lower the levels of glucose in blood, cholesterol in serum and triglycerides in rats. A high content of protein is present, ranging from 11.2% to 12.7%, while the amounts of total minerals, fat, amino acids and ash are comparable to other cereal grains. Barnyard millet possesses iron content ranging from 15.6 to 18.6 mg per 100 g, which is more than other grains, thereby making it a good choice of consumption for anemic patients. Barnyard millet contains twice the amount of polyphenols and carotenoids compared to that of finger millet. The presence of alkaloids, carbohydrates, tannins, flavonoids, steroids, glycosides and phenols results in ethno-medical properties like anti-carcinogenic, anti-microbial, biliousness, alleviating constipation associated diseases, anti-oxidant and anti-inflammatory activities.[37]

The nutritional composition of 100 grams of barnyard millet is protein (10.10%), fat (3.90%), carbohydrates (68.80%), moisture (8.70%), crude fiber (6.70%) and 398 kcal per 100 g energy. The dietary fiber content is around 12.50%, which includes both soluble (4.2%) and insoluble (8.40%) fractions. It is a multi-season crop that contains an appropriate amount of absorbable protein and is an excellent source of dietary fiber. The carbohydrate content is low and is slowly digestible, making barnyard millet an excellent crop for humankind. The essential fatty acids present in barnyard millet are linoleic and oleic acids. It has comparatively high amounts of resistant starch since it has a high level of amylase retrogradation. Thereby, it is suitable for consumption by people who are suffering from diabetes mellitus and cardiovascular sickness.[7]

1.3.2.1.2 Agronomic Conditions. Barnyard millet is a fast-growing annual summer crop grown under natural precipitation and is used for food and fodder. It is an important crop for famine areas as it can grow in marginal conditions, is drought resistant and has rapid growth. Temperature plays an important role in the growth of barnyard millet, as low temperature conditions (10/15 °C day/night) tend to cease the growth, and it also affects the plant development, especially in the dry weight and leaf area. The optimum temperature ranges for growth are 27–33 °C and 15–22 °C during the day and night, respectively. If the conditions for growth are favorable, in the first 3 weeks plants show profuse growth, while after 3 weeks the transition from vegetative growth to inflorescence occurs very rapidly within 40 days of the emergence. In long-day conditions, the plants grown are robust and have higher seed output, while in short-day conditions, the flowering of the plants is quicker, and they are smaller but abundant. Barnyard millet has a C4 photosynthetic pathway and has competitive advantages when grown with C3 crops. It also performs better than rice in terms of water stress conditions, as it has high water-use efficiency and 60% greater carbon efficiency ratio.[38]

1.3.2.1.3 Breeding. Barnyard millet possesses the potential for autogamy, and the cross-pollination rate is sufficient for gene exchange in the population. However, artificial hybridization and emasculation are difficult due to pollen's short viability, non-availability of pollen grains, smaller size of flowers, slight flower opening for a short time period and early time periods of flowering. Generally, breeding of barnyard millet is carried out in Tamil Nadu and Uttarakhand. The major breeding opportunities for crop improvement are mass selection and pure line selection, along with a few varieties that use the pedigree method of selection for development.[38]

1.3.2.2 Kodo Millet

Kodo millet (Araka or *Paspalum scrobiculatum* L.) originated in India, and its domestication is known to have taken place about 3000 years ago. The best-suited conditions for the growth of kodo millet are tropical and sub-tropical conditions. Kodo millet is a member of the *Poaceae* family, having other names such as cow grass or ditch grass. It is mainly cultivated in India, and Madhya Pradesh is the largest producer of kodo millet in the country, which produces 35% of the total produce in India and 50% of the cultivating area. The Government of MP emphasizes several schemes and practices to promote the cultivation and marketing of kodo millet. It is a small-seeded grain and is a monocot crop having a length of 2 mm and a width of 1.5 mm. The color ranges from light brown to dark grey and constitutes a type of covered caryopsis. Among all the minor millets, kodo millet is known as the highest drought resistant plant and has a good production yield with a short growing period of 80–135 days. The name of kodo millet varies from region to region, such as Kodo in Bengali; Kodon in Hindi; Harka in Kannada; Koduain in Odia; Varagu in Tamil; Kodra in Gujarati, Punjabi and Marathi; and Arikelu, Arika in Telugu. As estimated, kodo millet is grown in an area of approximately 907 800 ha, and the annual production in tonnes is about 310 710.[39]

1.3.2.2.1 Nutritional Composition. Kodo millet grain consists of carbohydrate (65.65%), protein (8.35%), ash (2.95%) and fat (1.45%). They are rich in dietary fiber, protein, magnesium, calcium, phosphorus and polyphenols. Magnesium can reduce the occurrence and minimize the effect of heart attacks and migraines, while phosphorus is an important constituent in adenosine triphosphate (ATP), which is an energy precursor in the human body. Kodo millet is rich in phytochemicals that contain sulfur. It also contains a sufficient amount of essential amino acids, *i.e.* lysine, threonine and valine, and the ratio of leucine and isoleucine is 2.0; however, it is deficient in tryptophan. The major vitamins present in kodo millet are vitamin B6, folic acid and vitamin B3, and minerals present include magnesium, calcium, zinc and potassium. The protein content of kodo millet is 8.30%, of which the major proportion is glutelin. The crude fiber content is high, *i.e.* 9%, when compared to wheat, *i.e.* 1.2%. The iron content in kodo millet ranges from 25.86 ppm to 39.60 ppm. Millets are rich in bioactive compounds and possess several functional properties. The major groups of phenolic compounds found in kodo millet are gallic acid, gentisic acid, caffeic acid, syringic acid, *para*-coumaric acid, 5-*n*-alkylresorcinols, tannins, protocatechuic acid, vanillic acid, ferulic acid and *trans*-cinnamic acid. Phenolic compounds have the potential to scavenge free radicals, leading to the prevention and treatment of diseases caused by free radicals. Dietary fiber content is high due to the presence of cellulose, hemicellulose and pectin in the seed coat that are not easily hydrolyzable by digestive enzymes. This affects the metabolic and digestive processes in terms of cholesterol levels and glucose absorption. Kodo millet and little millet have 37% and 38% dietary fiber, respectively, which are the highest among the cereals. They are low in fat content; however, the polyunsaturated fatty acid (PUFA) content is high. Kodo millet possesses the highest radical scavenging activity, followed by sorghum and finger millet. It protects from oxidative stress and type 2 diabetes by regulating glucose concentrations. Food products that have high phenolic content impart antiglycemic, antioxidant and antimutagenic characteristics that can be used for the development of health products. In addition, the grains of millets are gluten-free in nature, a rich source of phytochemicals and are easy to digest.[40,41]

1.3.2.2.2 Agronomic Conditions. Kodo millet growth is favored even in poor soil conditions of arid and semi-arid areas of India and African countries. It is a significant small-seeded crop in the Deccan Plateau. The common areas in India where kodo millet is grown are Gujarat, Chhattisgarh, parts of Tamil Nadu, Karnataka and Eastern Madhya Pradesh. It is an annual plant that grows up to a height of 90 cm. The grain color of kodo millet varies from light red to darker shades of grey and grains are bound in tough husk and are very difficult to separate. The husk of kodo millet grain is persistent, hard and corneous. The root system of kodo millet is shallow, which favors intercropping.[42]

1.3.2.2.3 Breeding. The flower opening type of kodo millet is cleistogamous, where only 15% to 20% of the flower opening is noticed, making it a self-pollinated crop. Due to the cleistogamous pattern of flowering, the recombination breeding in kodo millet is limited. However, a few varieties are released for the purpose of commercial cultivation. These are generally developed from selected landraces and superior germplasm accessions. The outcome from selected germplasm accessions was the GK 2, KMV 20 and APK 1 varieties. For the development of superior-yielding cultivars, pure line selection was carried out. To generate variability in kodo millet, induced mutagenesis is also practiced. A variety recognized as JK 76 was developed by gamma irradiation with a 5-Kr dose, leading to the development of a protogynous mutant with two rows of spikelets on the rachis. These successful variants in millet breeding endeavors will aid recombinant breeding.[43]

1.3.2.3 Little Millet

Little millet (*Panicum sumatrense*) is a minor millet that is native to the Indian subcontinent. The ancestral crop *Panicum psilopodium* was present throughout the country and is usually confused with proso millet (*Panicum miliaceum*) in terms of panicle and size, and it is smaller than little millet. The best regions for the growth of little millet are hilly tracts in India, along with a few regions of West Burma and Nepal. The major areas where finger millet is cultivated in India are Madhya Pradesh, Tamil Nadu, Andhra Pradesh, Chhattisgarh, Jharkhand, Karnataka, Bihar and Orissa. Grains of little millet can be stored for 10 years without major losses caused by deterioration, ascribing to its nature of a reserve food crop.[44]

Little millet is a reliable catch crop as it grows early and is resistant to several adverse agro-climatic conditions. It is generally used as an alternative to rice crop and in any recipe where rice is used as a basic material. Under poor conditions, it can be used as grain as well as fodder. Farmers of the dry zone find little millet promising and popular as it is adaptable to moisture-stress conditions and sowing time of crops.[45]

The lower productivity of little millets could be due to imbalanced use of fertilizers, weed control measures and non-adoption of crop varieties by the farmers. The soils of arid and semi-arid areas are nitrogen and carbon deficient, as the decomposition rate of organic material is higher due to the high temperatures of soil.[46]

With harsh atmospheric conditions and low soil fertility, effective nutrient management is of considerable importance to overcome limited yields in these areas. Nutrient management could be a key issue in achieving higher biomass of any crop plant and also maintaining soil fertility. Nitrogen is the primary nutrient which determines the growth and yield of a crop. Further, nitrogen is important in little millet for the manufacturing of proteins.[46]

1.3.2.3.1 Nutritional Composition. Little millet comprises fat (4.64 to 4.70 g per 100 g), ash (4.74 to 5.75 g per 100 g), protein (6.87 to 7.26 g per 100 g), carbohydrates (69.70 to 78.53 g per 100 g) and energy (374 to 374 kcal per 100 g) (Srilekha *et al.* 2019). It has a low glycemic index and exhibits a hypoglycemic effect due to the presence of a high amount of dietary fiber. Little millet is highly nutritious, thereby playing a significant role in providing nutraceutical components such as tannins, phenols and phytates, along with macronutrients and micronutrients. It exhibits hypo-lipidemic effects and fecal bulking effects.[47]

1.3.2.3.2 Agronomic Conditions. Little millet (*Panicum sumatrense* Roth. Ex. Roemer and Schultes) is also known as 'poor man's crop' and is grown in soils which are unsuitable for other crops. It has the potential to be grown in both drought and waterlogged conditions. It is also known as a famine-resistant food crop as it is produced under drought conditions where all other grains fail to produce. However, the yield of little millet is limited as it is grown in marginal or sub-marginal lands that have an imbalanced amount of nutrition, and also the cultivation practices are ignored.[48] Little millet is grown in tropical and sub-tropical areas and even in areas with an altitude of 7000 feet above mean sea level.

1.3.2.3.3 Breeding. Little millet can be genetically improved by recombinant breeding, resulting in increased yield, improved quality and tolerance against stress, especially from fly resistance. Hybridization of little millet can be done by contact methods, *i.e.* crossing and artificial hybridization with the help of manual emasculation or by hot water emasculation. When regular crossing cannot achieve genetic variety, then mutual breeding is carried out, particularly in selfed crops. Variants of little millet with high iron and high zinc content are being produced. Approximately 20 varieties of little millets are produced through various methods of breeding, *i.e.* mutual breeding, pedigree breeding and conventional selection. For the generation of crops with high iron and zinc, a metal transporter gene has also been cloned. Traditional methods of breeding have improved various cultivars that have resilience against biotic and abiotic stresses.[49]

Special breeding traits to enhance the consumption and cultivation of small millets are developing machine-harvestable cultivars, cultivars that are suitable for making value-added products, shade-tolerant genotypes, quick-growing genotypes for inter-cropping and enhancing the nutritive value of grain.[50]

1.4 Sorghum

Sorghum belongs to the family *Poaceae*, subfamily *Panicoideae*, and the genus of the tribe Andropogoneae. Sorghum is a self-pollinating C4 plant that has a high efficiency of photosynthesis. Sorghum is a stable food crop generally grown in several parts of Asia and Africa and other low-income regions, either as a fodder or feed crop for livestock. Until January 2021, the annual world production of sorghum was 61.62 million metric tonnes. The major countries producing sorghum are the USA, Nigeria, Ethiopia, Sudan, Mexico and India, followed by China, Argentina, Brazil and Niger. In Asian and African countries, sorghum is used as whole grains, decorticated grains and flour. Sorghum can be used to produce products such as thick or thin fermented porridge, boiled products like those that are made from maize and rice grits, flat breads, *etc.* Sorghum is also used in the formulation of non-alcoholic fermented beverages in Africa, Mexico and China. It is used as a substitute for gluten-free foods.[51]

1.4.1 Nutritional Composition

The nutritional composition of sorghum is similar to that of corn except for the lower content of oil. The grain has a starch content of 65% to 76%, a protein content of about 8% to 12% and a fiber content of 2%. The germ of sorghum (28%) is a rich source of oil and protein (19%). The major bran portion consists of cellulose and hemicellulose, and some amount of starch is deposited over the mesocarp tissue. The lipid of bran is more of a wax kind rather than an oil. The nutritional composition of sorghum may vary due to different sources, which include the type of hybrid, crop management techniques and soil and climatic conditions. The older varieties of sorghum differed in the size of the kernel and the relative amounts of different parts of the grain. New hybrids and better use of irrigation led to grains of larger size that have a better quantity of starch and lower content of protein. Sorghum is harder than other cereal grains, contributing to the presence of prolamin protein in higher amount, *i.e.* 3.6% to 5.1%. The lysine content of sorghum ranges from 1.06% to 3.64%. On fractionation of protein, the albumin–globulin, prolamin and glutelin content were 15%, 26% and 44%, respectively, of the total nitrogen. The amount of total sugars ranges from 0.7% to 4.2%, while reducing sugar ranges from 0.05% to 0.53%. Sorghum's fat content ranges from 2.1% to 7.6%, ash from 1.3% to 3.3% and crude fiber from 1.0% to 3.4%. The major fatty acid constituents of sorghum lipids are linoleic and oleic acid. Sorghum grain is generally eaten with the testa, thereby retaining the majority of the nutrients. Sorghum is a good source of minerals, and it is also influenced by the environmental conditions.[52]

1.4.2 Agronomic Conditions

Sorghum can be grown in drought conditions as it is a drought tolerant crop and is grown in arid and semi-arid areas of sub-Saharan Africa, the Middle East, Central America, Australia, Asia and North America. Sorghum can also survive in cold weather conditions of 12–15 °C, in waterlogged conditions (if for a shorter period) and in areas with high rainfall.[51]

1.4.3 Breeding

Sorghum is a self-pollinating crop in which the outcrossing ranges from 2% to 25%. The genome size of sorghum is comparatively smaller than maize or sugar cane. The genome of sorghum is fully sequenced, providing opportunities for plant breeders. The objective of breeding includes resistance to diseases (smut, grain mold, anthracnose, downy milder, rust, bacterial blight, *etc.*), drought resistance, resistance to acidic soil, improved quality (for use in porridge, beverages, bread and snacks), high grain and fodder yield potential, resistance to insect pests (shoot fly, midge and stalk borer) and wide adaptation. Generally, breeders can develop two types of cultivars: (a) open pollination/pure line cultivators (in developing countries) and (b) hybrid cultivars (for industrialized countries when the seed system is well developed). Methods of open pollination are very different from hybrid cultivar development. For open pollination, recurrent selection schemes are used, while for hybrid development, pure lines are used.[53]

Generally, the two types of sorghum cultivated are the open-pollinated varieties/landrace and hybrids. Hybrids of the sorghums are developed by crossing parents with resulting progeny that exhibit an increase in size, yield and vigor over the parents. For hybrids, a sterility system is required. The system involves three types of lines: (1) lines (restorer or R-) are used for restoring the fertility of a sterile line and these are the male parent of a hybrid; (2) maintainer (B-) lines do not restore the fertility of a sterile line but are used to develop and increase seed of the A-line; and (3) A-lines are sterile that do not produce any viable pollen and are the female parent of the hybrid (Chisi and Peterson 2019).

1.5 Health Benefits of Millets

Millets can be suitable alternatives to the most common cereal grains. Millets are gluten-free grains that are the best potential option for people who suffer from celiac disease or face irritation after consuming gluten-rich foods.[7] Several epidemiological studies have stated that whole grain cereals are useful in protecting our body from age-related diseases, *i.e.* diabetes, cancers, cardiovascular diseases and other problems. Millets are rich in many nutritional compounds (vitamins, essential fatty acids, fibers, resistant starch, antioxidants and minerals), allowing them to be accepted as functional foods, and they are useful in maintaining human health. Millets possess various health benefits such as prevention of cancer and cardiovascular diseases, lowering of blood pressure, delaying gastric emptying, supplying gastrointestinal bulk, reducing incidences of tumors, preventing several heart diseases and reducing cholesterol.[55] Millets are a food that is alkaline in nature, and these types of diets are often recommended for achieving optimal health conditions when combined with digestive enzymes. They are helpful in the maintenance of a healthy range of pH in the body and prevent crucial illness.[1]

Proso millet possesses high protein content and finger millet is a rich source of dietary fiber. Dietary fiber is known to improve gut health and helps in digesting carbohydrates slowly, which prevents the occurrence of type 2 diabetes. Millets, being rich in iron and folic acids, protect children and pregnant women from anemia. Osteoporosis can be prevented in the

Table 1.2 Health benefits of millets.

Millet	Health benefits	Ref.
Pearl millet	Anti-inflammatory, anticarcinogenic, antihypertensive properties, reduces risk of heart diseases, prevents atherosclerosis, reduces inflammatory bowel disease, potential for treating stomach ulcers, helpful in respiratory problems like asthma, reduces occurrence of migraine attacks, helps in weight loss, good for diabetic patients, suitable for people suffering from celiac disease, reduces cholesterol, prevents gall stones, anti-allergic properties	92 and 93
Finger millet	Reduces fractures related to osteoporosis, reduces risk of diabetes, improves gut health, reduces blood glucose levels by 30%, nephron-protecting characteristics, anti-cataractogenic properties, improves hemoglobin levels, anti-microbial, protection from mucosal ulceration	56 and 93
Foxtail millet	Low glycemic index, prevention from type 2 diabetes, reduces cardiovascular diseases	56
Proso millet	Elevates HDL, regulates cholesterol metabolism, reduces effects of type 2 diabetes, reduces obesity, protects from liver injury	56
Little millet	Hypoglycemic, hypolipidemic, management of diabetes and cardiovascular diseases, reduces constipation	56
Barnyard millet	Suitable for lactating and pregnant women, prevents obesity and type 2 diabetes, anti-carcinogenic, anti-inflammatory, antimicrobial and antioxidant properties	56
Kodo millet	Reduces postprandial hyperglycemia, reduces cataract diseases, provides antioxidant activity in health, aging and metabolic syndrome, reduces risk of colon and breast cancer, decreases risk of cardiovascular diseases, body weight reduction, cures arthritis, antidiabetic effects, improves neurological disorders, immunostimulant, immunomodulatory	40 and 94

elderly through the consumption of finger millet, as it is a rich source of calcium. Millets possess several phytochemicals along with their associated antioxidant activity, which helps in the prevention of several chronic lifestyle disorders, thereby consumption of millets offers many health benefits and diversifies the diet.[56] The health benefits of millets are given in Table 1.2.

1.5.1 Millets Against Diabetes Mellitus

Diabetes is identified by hyperglycemia, *i.e.* high blood sugar levels, and is one of the most common endocrine disorders that is caused by insufficient insulin hormone production. Millets possess a significant amount of magnesium, which is useful in the regulation of insulin and the glucose receptors in the body, thus preventing the occurrence of diabetes. Food developed from finger millet tends to have a low glycemic index as it is rich in amylase.[57]

1.5.2 Millets Against Cardiovascular Disease

Millets are good sources of magnesium, and consumption of minor millets can help lower blood pressure and reduce the threat of heart attacks caused by atherosclerosis. Potassium in millets can also help to lower blood pressure levels, thus reducing the chances of cardiovascular problems. Millets are rich in fibers and they play a role in lowering cholesterol levels. They remove low density lipoprotein (LDL) from the system and improve the effects of good cholesterol, high-density lipoprotein (HDL). Phytochemicals and lignins present in the millets are strong antioxidants, which are also helpful in the prevention of heart related issues.[58]

1.5.3 Millets with Anti-cancer Properties

Millets are rich in antioxidants, which include phenolic compounds, and might have anticancer properties. The phenolic compounds have specific flavonoids that are needed for the prevention of tumor growth. Tannins and polyphenols of sorghum have anti-carcinogenic and anti-mutagenic properties, and these tend to take actions against human melanoma cells and have some positive melanoma properties. The phytochemicals and antioxidants of the millets help reduce the risk of breast and stomach cancer. Millet fiber is suitable for the prevention of breast cancer in women, as it reduces the risk of occurrence by 50% when 30 grams of fiber is consumed daily.[59]

1.5.4 Millets Against Alimentary Tract Disorder and Malnutrition

The presence of high amounts of fiber in millets is helpful in healing several disorders such as excess gas, cramping, constipation and bloating. Millet fiber improves digestion through promoting regular bowel movements and nourishes the beneficial bacteria in the gut, along with lowering various gastrointestinal problems such as colon cancer and gastric ulcers. Finger millet contains a good combination of soluble and insoluble fiber that resists breaking down during the digestive process. It helps in healing gastrointestinal tract disorders, colon cancer and heart disease. Soluble fibers help to smooth and calm a swollen digestive tract. Fibers and polyphenols of millets help reduce inflammation in the stomach and prevent or treat peptic ulcers.[60]

1.5.5 Millets with Antioxidant and Anti-aging Properties

Phytochemicals generally act as antioxidants which protect the cells from oxidative damage, resulting in maintenance of their structural integrity and physiological stability. Millets are reported to have several bioactive compounds that provide a wide range of health benefits. Finger millet has a strong radical-scavenging molecule that is comparable to other grains. Incorporating finger millet in the diet can enhance the activity of the antioxidant enzymes, *i.e.* glutathione reductase, glutathione peroxidase and catalase, which have a protective character. These are also beneficial in delaying the ageing process by decreasing the rigidity of the elasticity in the skin, tendons, fetal tissue and periodontal ligaments.[61]

1.6 Value-added Products of Millets

Millets have been used in our diets since ancient times. Recently, millets have been explored for the development of gluten-free products. Products based on millets can be categorized as RTE and RTC foods. RTE foods are Ready to Eat foods, *i.e.* snacks (popcorn and extruded snacks), instant food (flakes), and bakery foods (breads and biscuits). RTC foods are Ready to Cook foods, *i.e.* instant foods (vermicelli, khichdi, noodles, pasta, porridge), mixes (idly, upma, dosa), beverages (malt drink mix, weaning foods) and multi grain flour. Millets can be considered as storehouses of nutrition as they are ahead of rice and wheat in their nutritional content. Also, the fiber content of millets is greater than that of rice and wheat.[62] Millets can be consumed after dehusking and can be cooked like rice. They can be ground into flour and further formed into *chapatti*. Also, they can be consumed as a thin porridge. Millet grains can be consumed in parched form, and the product, being similar to popcorn, is also suitable for malt preparation.[21]

Several traditional products are prepared from millets. Appalu, a product made from a combination of pearl millet and Bengal gram flour, is made by making dough, along with the addition of spices, which is shaped and deep-fried. A drink is prepared from roasting finger millet, followed by the addition of inoculum known as *madua*. Another fermented drink, *koozh,* is prepared from a combination of finger millet and rice flour. It is a two-stage fermented drink that is cooked, leading to thick porridge.[63]

Ready-to-eat (RTE) snacks: several RTE snacks are formulated from millets, *i.e.* puffed and toasted snacks from barnyard millet,[64] extruded snacks from barnyard millet[65] (Suri *et al.* 2022), breakfast cereal from pearl millet,[66] chakli from millet and legume blends,[67] cookies[68] and RTE snack mix from minor millets.[69]

Fermented millet products: fermented millet products include kodo ko jaanr, *i.e.* finger millet fermented beverage;[70] koozh, *i.e.* fermented millet beverage;[71] fermented millet-based probiotic dairy products[72] and fermented millet porridge and drinks.[73]

Bakery products: bakery products developed from millets include barnyard millet-based muffins,[74] bread from composite-millet flours,[75] little millet bread,[76] composite millet breads,[77] foxtail millet bread[78] and bread from pearl millet and sorghum millet.[79]

Ready-to-cook products: RTC millet products include finger millet incorporated noodles,[80,81] foxtail millet incorporated noodles[82] and pearl millet based upma.[83]

1.7 Future Scope of Millets

Millets, also known as "nutri-cereals", have a huge corner in the palates of many consumers. They are rich in many nutrients, thereby they are consumed by many consumers, and the potential of millet demand is increasing day by day. Millets have even attracted farmers as they require fewer investments in water, pesticides, fertilizers, *etc.*, and they generate good income from the market access. Also, the Minimum Support Price (MSP) of millets is increasing, which positively affects the farmers' cropping pattern decisions towards the farming of millet.[84]

Millets are crops that are grown in challenging environmental conditions, *i.e.* char, barind tracts, hilly areas and saline. Millet's adoption must be increased by resolving several issues. The major problems faced by farmers are poor storage facilities, inadequate research and development, and limited market opportunities. To encourage the cultivation of millet, sustainable production systems must be maintained, as they will be necessary for the corporate sectors, civil society groups and government to work together. Millets should be promoted as an alternative to other cereal crops, and supply chains must be developed for connecting the farmers with consumers and markets. Cultivation of millets helps in the attainment of sustainable development goals (SDGs) especially SDG 1, SDG 2, SDG 3 and SDG 13, which are related to no poverty, zero hunger, good health and well-being and climate action, respectively.[10]

1.8 Conclusion

Millets are staple food crops that are consumed mainly by the people of Asia and Africa. Millets are a rich source of nutrients as they are high in carbohydrate content, minerals, vitamins and phytochemicals. They are nutritious, making them useful in dietary consumption and maintaining nutritional balance in food. A combination of millets with other cereal crops will compensate for the deficiency of lysine that is possessed by other cereals. The agro-climatic conditions for growing millet grains are wide, as they can be grown in drought, saline, dry, *etc.,* areas. Breeding can be carried out to improve the existing varieties. The nutritional characteristics of millet make them suitable to provide several health benefits, *i.e.* prevention from cardiovascular diseases, type 2 diabetes, *etc.* Thereby, millets have high potential to be added to the daily diets of the population to improve their living conditions.

References

1. E. S. Sarita and E. Singh, Potential of millets: nutrients composition and health benefits, *J. Sci. Innovative Res.*, 2016, **5**(2), 46–50.
2. L. Yousaf, D. Hou, H. Liaqat and Q. Shen, Millet: A review of its nutritional and functional changes during processing, *Food Res. Int.*, 2021, **142**, 110197.

3. S. Nithiyanantham, P. Kalaiselvi, M. F. Mahomoodally, G. Zengin, A. Abirami and G. Srinivasan, Nutritional and functional roles of millets—A review, *J. Food Biochem.*, 2019, **43**(7), e12859.
4. N. Sachdev, S. Goomer, L. R. Singh, V. M. Pathak, D. Aggarwal and R. K. Chowhan, Current status of millet seed proteins and its applications: a comprehensive review, *Appl. Food Res.*, 2023, **3**(1), 100288.
5. K. Ambati and K. V. Sucharitha, Millets-review on nutritional profiles and health benefits, *Int. J. Recent Sci. Res.*, 2019, **10**(7), 33943–33948.
6. P. Kaur, S. S. Purewal, K. S. Sandhu, M. Kaur and R. K. Salar, Millets: A cereal grain with potent antioxidants and health benefits, *J. Food Meas. Charact.*, 2019, **13**, 793–806.
7. H. Kaur and S. Sharma, An overview of Barnyard millet (Echinochloa frumentacea), *J. Pharmacogn. Phytochem.*, 2020, **9**(4), 819–822.
8. I. Amadou, M. E. Gounga and G. W. Le, Millets, nutritional composition, some health benefits and processing, *Emirates J. Food Agric.*, 2013, **25**(7), 501–508.
9. H. Macauley, & T. Ramadjita, Cereal crops: Rice, maize, millet, sorghum, wheat, 2015.
10. S. A. Kheya, S. K. Talukder, P. Datta, S. Yeasmin, M. H. Rashid, A. K. Hasan, *et al.*, *Millets: The future crops for the tropics-Status, challenges and future prospects*, Heliyon, 2023.
11. A. D. Sukumaran Sreekala, P. Anbukkani, A. Singh, B. Dayakar Rao and G. K. Jha, Millet production and consumption in India: Where do we stand and where do we go?, *Natl. Acad. Sci. Lett.*, 2023, **46**(1), 65–70.
12. M. Singh and U. Nara, Genetic insights in pearl millet breeding in the genomic era: challenges and prospects, *Plant Biotechnol. Rep.*, 2023, **17**(1), 15–37.
13. M. Chaudhary and D. Singh, Biofortification of Pearl Millet for Nutritional Enrichment, *Biofortif. Food Crops*, 2016, 399–408.
14. S. Singh, D. Yadav, M. Beckmann, A. Naveen, P. I. Gangashetty, L. A. Mur and R. S. Yadav, Variation in protein and amino acids in global collection of pearl millet (Pennisetum glaucum) germplasm, *J. Food Compos. Anal.*, 2024, **134**, 106557.
15. K. Adéoti, S. H. Kouhoundé, P. A. Noumavo, F. Baba-Moussa and F. Toukourou, Nutritional value and physicochemical composition of pearl millet (*Pennisetum glaucum*) produced in Benin, *J. Microbiol., Biotechnol. Food Sci.*, 2017, 7, 92–96.
16. F. P. Suma and A. Urooj, Impact of household processing methods on the nutritional characteristics of pearl millet (Pennisetum typhoideum): A review, *J. Food Process. Technol.*, 2017, **4**(1), 28–32.
17. O. Halilou, Y. Assefa, H. Falalou, H. Abdou, B. F. Achirou, S. M. Karami and S. K. Jagadish, Agronomic performance of pearl millet genotypes under variable phosphorus, water, and environmental regimes, *Agrosyst., Geosci. Environ.*, 2020, **3**(1), e20131.
18. R. K. Varshney, C. Shi, M. Thudi, C. Mariac, J. Wallace, P. Qi and X. Xu, Pearl millet genome sequence provides a resource to improve agronomic traits in arid environments, *Nat. Biotechnol.*, 2017, **35**(10), 969–976.

19. N. Anuradha, C. T. Satyavathi, C. Bharadwaj, M. Sankar and L. Pathy, Association of agronomic traits and micronutrients in pearl millet, *Int. J. Chem. Stud.*, 2018, **6**(1), 181–184.
20. D. D. Serba, R. Perumal, T. T. Tesso and D. Min, Status of global pearl millet breeding programs and the way forward, *Crop Sci.*, 2017, **57**(6), 2891–2905.
21. S. Yadav, H. C. Singh and D. R. Singh, Nutritional quality of millets and their value added products with the potential health benefits: A review, *Int. J. Curr. Microbiol. Appl. Sci.*, 2021, **10**(10), 163–175.
22. R. K. Singh, M. Muthamilarasan and M. Prasad, *Foxtail millet: an introduction*, in *The foxtail millet genome*, 2017, pp. 1–9.
23. O. Panaud, Foxtail millet, in *Cereals and millets*, Berlin, Heidelberg, Springer Berlin Heidelberg, 2006, pp. 325–332.
24. N. Sharma and K. Niranjan, Foxtail millet: Properties, processing, health benefits, and uses, *Food Rev. Int.*, 2018, **34**(4), 329–363.
25. L. He, B. Zhang, X. Wang, H. Li and Y. Han, Foxtail millet: nutritional and eating quality, and prospects for genetic improvement, *Front. Agric. Sci. Eng.*, 2015, **2**(2), 124–133.
26. R. Rajasekaran, N. Francis, V. Mani and J. Ganesan, Proso millet (Panicum miliaceum L.), in *Neglected and underutilized crops*, Academic Press, 2023, pp. 247–278.
27. S. Das, R. Khound, M. Santra and D. K. Santra, Beyond bird feed: Proso millet for human health and environment, *Agriculture*, 2019, **9**(3), 64.
28. J. O. Narciso and L. Nyström, The genetic diversity and nutritional quality of proso millet (Panicum miliaceum) and its Philippine ecotype, the ancient grain "kabog millet": a review, *J. Agric. Food Res.*, 2023, **11**, 100499.
29. C. Habiyaremye, J. B. Matanguihan, J. D'Alpoim Guedes, G. M. Ganjyal, M. R. Whiteman, K. K. Kidwell and K. M. Murphy, Proso millet (Panicum miliaceum L.) and its potential for cultivation in the Pacific Northwest, US: a review, *Front. Plant Sci.*, 2017, **7**, 1961.
30. A. Seetharam, Genetic improvement of small millets In India during Pre and Post Crop Coördinated Project era, Indian Institute of Millets Research (IIMR), 2013.
31. S. Shobana, K. Krishnaswamy, V. Sudha, N. G. Malleshi, R. M. Anjana, L. Palaniappan and V. Mohan, Finger millet (Ragi, Eleusine coracana L.): a review of its nutritional properties, processing, and plausible health benefits, *Adv. Food Nutr. Res.*, 2013, **69**, 1–39.
32. M. Nirmala, M. S. Rao and G. Muralikrishna, Carbohydrates and their degrading enzymes from native and malted finger millet (Ragi, Eleusine coracana, Indaf-15), *Food Chem.*, 2000, **69**(2), 175–180.
33. P. Singh and R. S. Raghuvanshi, Finger millet for food and nutritional security, *Afr. J. Food Sci.*, 2012, **6**(4), 77–84.
34. A. Dass, S. Sudhishri and N. K. Lenka, Integrated nutrient management to improve finger millet productivity and soil conditions in hilly region of Eastern India, *J. Crop Improv.*, 2013, **27**(5), 528–546.

35. S. S. Krishna, Y. N. Reddy and R. R. Kumar, Assessment of traits for grain yield under drought in finger millet, *Plant Physiol. Rep.*, 2021, **26**(1), 84–94.
36. N. Mirza and S. S. Marla, Finger millet (Eleusine coracana L. Gartn.) breeding, in *Advances in Plant Breeding Strategies: Cereals: Volume 5*, 2019, pp. 83–132.
37. V. G. Renganathan, C. Vanniarajan, A. Karthikeyan and J. Ramalingam, Barnyard millet for food and nutritional security: Current status and future research direction, *Front. Genet.*, 2020, **11**, 500.
38. S. Sood, R. K. Khulbe, A. K. Gupta, P. K. Agrawal, H. D. Upadhyaya and J. C. Bhatt, Barnyard millet–a potential food and feed crop of future, *Plant Breed.*, 2015, **134**(2), 135–147.
39. H. D. Upadhyaya, M. Vetriventhan, S. L. Dwivedi, S. K. Pattanashetti and S. K. Singh, Proso, barnyard, little, and kodo millets, in *Genetic and genomic resources for grain cereals improvement*, Academic Press, 2016, pp. 321–343.
40. D. S. Bunkar, S. K. Goyal, K. K. Meena and V. Kamalvanshi, Nutritional, functional role of kodo millet and its processing: a review, *Int. J. Curr. Microbiol. Appl. Sci.*, 2021, **10**(01), 1972–1985.
41. S. S. Deshpande, D. Mohapatra, M. K. Tripathi and R. H. Sadvatha, Kodo millet-nutritional value and utilization in Indian foods, *J. Grain Process. Storage*, 2015, **2**(2), 16–23.
42. S. Dey, A. Saxena, Y. Kumar, T. Maity and A. Tarafdar, Understanding the antinutritional factors and bioactive compounds of kodo millet (Paspalum scrobiculatum) and little millet (Panicum sumatrense), *J. Food Qual.*, 2022, **2022**(1), 1578448.
43. T. E. Nagaraja, C. Nandini, S. Bhat and S. G. Parveen, Artificial hybridization techniques in small millets—a review, *Front. Plant Sci.*, 2023, **14**, 1112117.
44. K. N. Ganapathy, Improvement in finger millet: status and future prospects, in *Millets and sorghum: Biology and genetic improvement*, 2017, pp. 87–111.
45. P. Kathare, Biochemical and molecular characterization of water stress responsive genes in little millet, barnyard millet and rice, 2019.
46. U. Bhavani, M. M. Reddy, B. Padmaja and D. Saidanaik, Effect of nitrogen levels and varieties on economics and yield of little millet (Panicum sumatrense) under rainfed conditions, *Indian J. Agron.*, 2021, **66**(4), 509–511.
47. V. P. Sushmita, J. Lakshmi and K. Lakshmi, Nutrient Composition of Little Millet Varieties, *Andhra Agric. J.*, 2020, **67**(1), 73–76.
48. N. Sarala, M. H. Kumar, L. Madhavilatha, B. Vajantha and T. Hemalatha, Effect of spacing and fertilizer levels on growth and yield of little millet, *Andhra Pradesh Journal of Agricultural Sciences*, 2020, **6**(2), 71–74.
49. A. Sao, V. Pali, & H. E. Patil, Genetic improvement for yield, quality, biotic, and abiotic stresses in little millet (Panicum sumatrense Roth. ex Roem. and Schult), in *Genetic improvement of small millets*, Springer Nature Singapore, Singapore, 2024, pp. 571–599.

50. M. Vetriventhan, V. C. Azevedo, H. D. Upadhyaya, A. Nirmalakumari, J. Kane-Potaka and S. Anitha, *et al.*, Genetic and genomic resources, and breeding for accelerating improvement of small millets: current status and future interventions, *Nucleus*, 2020, **63**, 217–239.
51. A. Khoddami, V. Messina, K. Vadabalija Venkata, A. Farahnaky, C. L. Blanchard and T. H. Roberts, Sorghum in foods: Functionality and potential in innovative products, *Crit. Rev. Food Sci. Nutr.*, 2023, **63**(9), 1170–1186.
52. C. R. Abah, C. N. Ishiwu, J. E. Obiegbuna and A. A. Oladejo, Sorghum grains: nutritional composition, functional properties and its food applications, *Eur. J. Nutr. Food Saf.*, 2020, **12**(5), 101–111.
53. Prabhakar, R. Madhusudhana and C. Aruna, Sorghum Breeding, in *Fundamentals of Field Crop Breeding*, Springer Nature Singapore, Singapore, 2022, pp. 367–447.
54. M. Chisi and G. Peterson, Breeding and agronomy, in *Sorghum and millets*, AACC International Press, 2019, pp. 23–50.
55. A. S. Saleh, Q. Zhang, J. Chen and Q. Shen, Millet grains: nutritional quality, processing, and potential health benefits, *Compr. Rev. Food Sci. Food Saf.*, 2013, **12**(3), 281–295.
56. R. Venkateswarlu, V. M. Malathi, J. Jacob, C. V. Ratnavathi and M. Kumar, Health benefits of millets, *Indian Farming*, 2023, **73**(1), 91–93.
57. P. Lakshmi Kumari and S. Sumathi, Effect of consumption of finger millet on hyperglycemia in non-insulin dependent diabetes mellitus (NIDDM) subjects, *Plant Foods Hum. Nutr.*, 2002, **57**, 205–213.
58. S. H. Lee, I. M. Chung, Y. S. Cha and Y. Park, Millet consumption decreased serum concentration of triglyceride and C-reactive protein but not oxidative status in hyperlipidemic rats, *Nutr. Res.*, 2010, **30**(4), 290–296.
59. S. Rotela, S. Borkar and A. Borah, Health benefits of millets and their significance as functional food: A review, *J. Pharm. Innovation*, 2021, **10**, 158–162.
60. T. Thompson, The nutritional quality of gluten-free foods, in *Gluten-free food science and technology*, 2009, pp. 42–51.
61. F. Shahidi, P. K. Janitha and P. D. Wanasundara, Phenolic antioxidants, *Crit. Rev. Food Sci. Nutr.*, 1992, **32**(1), 67–103.
62. K. Lokesh, C. R. Dudhagara, A. B. Mahera and S. Kumar, Millets: The future smart food, *Pharm. Innovation J.*, 2022, **11**(4), 75–84.
63. R. Sharma, S. Sharma, B. N. Dar and B. Singh, Millets as potential nutricereals: a review of nutrient composition, phytochemical profile and techno-functionality, *Int. J. Food Sci. Technol.*, 2021, **56**(8), 3703–3718.
64. R. V. Jaybhaye and P. P. Srivastav, Development of barnyard millet ready-to-eat snack food: Part II, 2015.
65. S. Suri, A. Dutta, N. C. Shahi, A. Singh, R. S. Raghuvanshi and C. S. Chopra, Formulation, process optimization, and quality evaluation of barnyard millet (Echinochloa frumentacea)-based ready-to-eat (RTE) extruded snack, *J. Food Process. Preserv.*, 2022, **46**(11), e16936.

66. R. Kumari, K. Singh, R. Singh, N. Bhatia and M. S. Nain, Development of healthy ready-to-eat (RTE) breakfast cereal from popped pearl millet, *Indian J. Agric. Sci.*, 2019, **89**(5), 877–881.
67. S. Deshpande and K. Jha, Development of Millet and Soybean Based Ready-to-Eat Snack Food Chakli, *J. Agric. Eng.*, 2014, **51**(3), 19–23.
68. P. Virginia, P. Ajit, K. Pratibha and T. Ajano, Utilization of coarse grains and millet in preparation of ready to eat snack (Cookies), *J. Pharmacogn. Phytochem.*, 2018, **7**(3), 3539–3541.
69. P. M. Pradeep, U. Dharmaraj, B. V. Sathyendra Rao, A. Senthil, N. S. Vijayalakshmi, N. G. Malleshi and V. Singh, Formulation and nutritional evaluation of multigrain ready-to-eat snack mix from minor cereals, *J. Food Sci. Technol.*, 2014, **51**, 3812–3820.
70. S. Thapa and J. P. Tamang, Product characterization of kodo ko jaanr: fermented finger millet beverage of the Himalayas, *Food Microbiol.*, 2004, **21**(5), 617–622.
71. S. Ilango and U. Antony, Assessment of the microbiological quality of koozh, a fermented millet beverage, *Afr. J. Microbiol. Res.*, 2014, **8**(3), 308–312.
72. M. Samtiya, P. C. Badgujar, G. A. Chandratre, R. E. Aluko, A. Kumar, B. Bhushan and T. Dhewa, Effect of selective fermentation on nutritional parameters and techno-functional characteristics of fermented millet-based probiotic dairy product, *Food Chem.:X*, 2024, **22**, 101483.
73. V. Lei and M. Jakobsen, Microbiological characterization and probiotic potential of koko and koko sour water, African spontaneously fermented millet porridge and drink, *J. Appl. Microbiol.*, 2004, **96**(2), 384–397.
74. D. Goswami, R. K. Gupta, D. Mridula, M. Sharma and S. K. Tyagi, Barnyard millet based muffins: Physical, textural and sensory properties, *LWT-Food Sci. Technol.*, 2015, **64**(1), 374–380.
75. K. P. Singh, A. Mishra and H. N. Mishra, Fuzzy analysis of sensory attributes of bread prepared from millet-based composite flours, *LWT-Food Sci. Technol.*, 2012, **48**(2), 276–282.
76. M. Mannuramath, N. Yenagi and V. Orsat, Quality evaluation of little millet (Panicum miliare) incorporated functional bread, *J. Food Sci. Technol.*, 2015, **52**, 8357–8363.
77. A. Chhavi and S. Sarita, Evaluation of composite millet breads for sensory and nutritional qualities and glycemic response, *Malays. J. Nutr.*, 2012, **18**(1), 89.
78. S. Sarabhai, T. Tamilselvan and P. Prabhasankar, Role of enzymes for improvement in gluten-free foxtail millet bread: It's effect on quality, textural, rheological and pasting properties, *LWT*, 2021, **137**, 110365.
79. D. Mounika and G. Sireesha, Development of Multi Millet Bread with Pearl Millet and Sorghum Millet, *Wesleyan J. Res.*, 2021, **80**(14), 90–100.
80. K. Shukla and S. Srivastava, Evaluation of finger millet incorporated noodles for nutritive value and glycemic index, *J. Food Sci. Technol.*, 2014, **51**, 527–534.

81. B. D. M. P. B. Dissanayake and H. S. Jayawardena, Development of a method for manufacturing noodles from finger millet, *Procedia Food Sci.*, 2016, **6**, 293–297.
82. M. Meherunnahar, T. Ahmed, R. S. Chowdhury, M. A. S. Miah, K. Sridhar and B. S. Inbaraj, *et al.*, Development of novel foxtail millet-based nutri-rich instant noodles: Chemical and quality characteristics, *Foods*, 2023, **12**(4), 819.
83. S. Balasubramanian, D. N. Yadav, J. Kaur and T. Anand, Development and shelf-life evaluation of pearl millet based upma dry mix, *J. Food Sci. Technol.*, 2014, **51**, 1110–1117.
84. D. R. Pravallika, B. D. Rao, D. S. Chary and N. Devi, Market Strategies for Promotion of Millets: A Critical Analysis on Assessment of Market Potential of Ready to Eat (RTE) and Ready to Cook (RTC) Millet Based Products in Hyderabad, *Asian J. Agric. Ext., Econ. Sociol.*, 2020, **38**(12), 147–155.
85. A. A. Kulthe, S. S. Thorat and S. B. Lande, Characterization of pearl millet cultivars for proximate composition, minerals and anti-nutritional contents, *Adv. Life Sci.*, 2016, **5**(11), 4672–4675.
86. T. Yang, S. Ma, J. Liu, B. Sun and X. Wang, Influences of four processing methods on main nutritional components of foxtail millet: A review, *Grain Oil Sci. Technol.*, 2022, **5**(3), 156–165.
87. R. Devisetti, S. N. Yadahally and S. Bhattacharya, Nutrients and antinutrients in foxtail and proso millet milled fractions: Evaluation of their flour functionality, *LWT-Food Sci. Technol.*, 2014, **59**(2), 889–895.
88. P. Nazni and R. S. Devi, Effect of processing on the characteristics changes in barnyard and foxtail millet, 2016.
89. V. P. Sushmita, J. Lakshmi and K. Lakshmi, Nutrient Composition of Little Millet Varieties, *Andhra Agric. J.*, 2020, **67**(1), 73–76.
90. P. Salunke, U. Chavan, P. M. Kotecha and S. B. Lande, Studies on nutritional quality of barnyard millet cookies, *Int. J. Chem. Sci.*, 2019, **7**(4), 651–657.
91. T. Tamilselvan and A. Kushwaha, Effect of traditional processing methods on the nutritional composition of sorghum (Sorghum bicolour L. Moench) flour, *Eur. J. Nutr. Food Saf.*, 2020, **12**(7), 69–77.
92. A. M. Dias-Martins, K. L. F. Pessanha, S. Pacheco, J. A. S. Rodrigues and C. W. P. Carvalho, Potential use of pearl millet (Pennisetum glaucum (L.) R. Br.) in Brazil: Food security, processing, health benefits and nutritional products, *Food Res. Int.*, 2018, **109**, 175–186.
93. B. Dayakar Rao, K. Bhaskarachary, G. D. Arlene Christina, G. Sudha Devi, A. T. Vilas and A. Tonapi, Nutritional and health benefits of millets, ICAR_Indian Institute of Millets Research (IIMR) Rajendranagar, Hyderabad, 2017, vol. 2.
94. S. Dey, A. Saxena, Y. Kumar, T. Maity and A. Tarafdar, Understanding the antinutritional factors and bioactive compounds of kodo millet (Paspalum scrobiculatum) and little millet (Panicum sumatrense), *J. Food Qual.*, 2022, **2022**(1), 1578448.

CHAPTER 2

Millet Protein: A Comprehensive Overview

T. TAMILSELVAN,[a,b] J. BHARATH[a] AND PICHAN PRABHASANKAR*[a,b]

[a] Flour Milling Baking and Confectionery Technology Department, CSIR-Central Food Technological Research Institute, Mysore – 570020, India;
[b] Academy of Scientific & Innovative Research (AcSIR), Ghaziabad – 201002, India
*Email: psankar@cftri.res.in

2.1 Millets – An Overview

The first green revolution in 1970 primarily aimed to improve crop productivity and food availability to meet the mounting requirements of the growing global population. However, the next agricultural revolution is likely to focus on addressing two critical challenges: malnutrition and sustainability in the context of mitigating climate change. Climate change has now become a pressing reality, posing significant threats to agricultural productivity and the nutritional quality of food. As a result, the need for sustainable, resilient crops has intensified.

In this context, millets, often referred to as "nutri-cereals", have gained considerable attention from researchers and policymakers alike. Despite being underutilized in the past few decades, these small-grain cereals are now widely recognized for their ability to contribute to sustainable agriculture and nutritional security. The term "millet" includes a diverse collection of small-seeded cereal species, with kernel sizes ranging from one-tenth to one-quarter that of wheat. Among the 11 types of millets grown globally, their economic importance varies, but their role in guaranteeing

Food Chemistry, Function and Analysis No. 48
Millet Proteins: Nutritional, Structural and Functional Attributes
Edited by Shweta Suri, Anuradha Dutta and Tarun Belwal

Published by the Royal Society of Chemistry, www.rsc.org

nutritional security is universally acknowledged, predominantly in regions facing scarcity of food.

Millets are exceptionally resilient crops, capable of thriving with minimal agricultural inputs and withstanding harsh environmental conditions such as heat and drought. Botanically, millets belong to two primary subfamilies: the *Chloridoideae* (tribe *Eragrostideae*) and the *Panicoideae*. These crops are traditionally cultivated in warm, tropical climates across the world, where they serve as staple foods and are utilized in the preparation of various types of beverages. Given their adaptability and nutritional profile, millets play a critical part in assuring food security and promoting sustainability.

2.2 The History of Millets

Millets are one of the oldest cultivated crops, with their origins tracking back over 6500 years to northern China, where proso millet and foxtail millet were first domesticated about 6000 BC.[1,2] These hardy small-seeded grasses spread rapidly across Eurasia, reaching Central Asia, Eastern Europe, and eventually, the Iberian Peninsula during the Bronze Age.[3] As drought-resistant crops, millets thrived in arid and semi-arid regions, becoming staple foods for many ancient civilizations. In South Asia, millets from Africa significantly influenced early agriculture, particularly in western India, while in India itself, they hold a deep historical significance, with over 28 species referenced in ancient texts like the Vedas and Puranas.[4] Evidence of millet cultivation also appears in the Caucasus during the Middle Bronze Age and in medieval Europe, where it was widely used in Italian cuisine.[5] Their resilience, short duration, and ability to cultivate in adverse soil conditions made millets an essential crop for densely populated areas facing resource constraints. Today, millets are recognized not only for their historical importance but also for their potential to address modern issues relating to food security, as they offer a sustainable alternative in regions with limited agricultural inputs (Table 2.1).

2.3 Climate Resilience of Millets

Climate change is now a reality, presenting tremendous challenges, particularly in decreasing crop productivity. Nations are moving toward sustainable and resilient cropping systems. Millets, as climate-resilient crops, have gained attention for their adaptability to diverse ecological conditions and low input requirements. Belonging to the subfamily *Panicoideae*, these C4 grasses, including foxtail millet (*Setaria italica*), thrive in arid and semi-arid regions with minimal irrigation, making them well-suited for the challenges posed by climate change. Unlike major cereals, millets require fewer synthetic fertilizers, are less vulnerable to environmental stresses, and are more efficient in utilizing water and nitrogen. For instance, foxtail millet can produce biomass using just 257 grams of water per gram of dry weight, while maize and wheat need significantly more, making it a sustainable

Table 2.1 Millet common names, production and uses globally.[6]

	Scientific name	Common names	Major production areas
Sorghum	*Sorghum bicolor*	Great Millet, Jowar, Cholam, Jola, Jonna, Durra, Egyptian Millet	USA, Nigeria, Sudan, Mexico, Ethiopia, India, Argentina, China, Niger, Australia
Pearl Millet	*Pennisetum glaucum*	Bajra, Cattail, Bulrush, Candlestick	India, Western & Central Africa, Eastern & Southern Africa
Finger Millet	*Eleusine coracana*	Ragi, African, Bird's Foot	India, Ethiopia, Nepal, Uganda, Malawi, Burundi, Sri Lanka, Rwanda
Foxtail Millet	*Setaria italica*	Italian, German, Hungarian, Siberian, Kangani, Navane	China, Myanmar, India, Eastern Europe
Proso Millet	*Panicum miliaceum*	Common, Hog, Broom, Samai, Russian, Panivaragu	Russia, USA, Ukraine, South Korea, Kazakhstan, France, Poland, Belarus, India, Iran
Little Millet	*Panicum sumatrense*	Blue Panic, Heen Meneri	India
Kodo Millet	*Paspalum scrobiculatum*	Varagu, Ditch, Naraka, Water Couch, Indian Paspalum, Creeping Paspalum	India
Barnyard Millet	*Echinochloa frumentacea*	Japanese, Sanwa, Sawan	India, Japan, China, Malaysia
Browntop Millet	*Urochloa ramosa (L.)*	American Millet, Choti Kangni, Pala Pul, Korale, Andu Korralu, Kulasaama, Kadu-baragu	India, USA
Teff	*Eragrostis tef*	Abyssinian Lovegrass	Ethiopia, Eritrea, Australia
Fonio	*Digitaria exilis*	Fundi, Hungry Rice, Acha	West Africa, Sudan, Ethiopia, Nigeria, Niger, Togo, Senegal, Mali

option in water-scarce areas. Additionally, their C4 photosynthetic pathway enhances their photosynthetic efficiency in warm climates, allowing them to grow under higher temperatures with reduced photorespiration.

Millets also offer significant nutritional benefits, as they are high in dietary fibers, resistant starches, and vital bioactive compounds. This, combined with their ecological resilience, positions them as important crops for addressing food security in the face of changing climate conditions. While crops like foxtail millet have been extensively studied for their stress biology and bioenergy potential, further research into other millet species could unlock additional climate-resilient traits. With their short growing cycle, deep root systems, and ability to withstand drought, salinity, and heat, millets are poised to play a crucial role in sustainable agriculture, particularly in regions facing increasing environmental stressors.

2.4 Millets as Nutri-cereals

Millets generally consist of 50–88% carbohydrates, 10–15% protein, and 1–6% lipid content, with variations depending on the type, species, and variety. Notably, millets have a high amylose content, around 34%, which is higher than that of most cereals. This higher amylose content is associated with the lower digestibility of millet starch, as amylose has a more compact structure compared to amylopectin, resulting in a lower surface area for enzymatic action. Millets contain about 24–30% resistant starch, 22–28% slowly digestible starch, and 11–15% rapidly digestible starch.[7]

Millets are a superior source of dietary fiber, with barnyard millet having the highest fiber content at around 10.6%. The dietary fiber in millets includes cellulose, hemicellulose, arabinoxylans, lignin, and β-glucan. One of the key benefits of consuming millets is their low glycemic index (GI), which is largely attributed to their high fiber content. The insoluble dietary fibers contribute antioxidant properties due to the presence of polyphenolic compounds. These fibers also slow down gut transit time, helping to prevent rapid spikes in blood glucose levels.[8] Additionally, millet fibers act as prebiotics for gut microbiota, promoting the generation of short-chain fatty acids sustaining gut homeostasis, controlling the gut–brain axis, and modulating the immune system.[9]

Millets are a rich source of bioactive compounds such as phenolics, flavonoids, and phytosterols linked to antioxidant and anti-inflammatory activities. However, one challenge associated with millets is the anti-nutritional compounds such as phytates, alkaloids, protease inhibitors, and tannins, that can interfere with nutrient absorption (Table 2.2).

2.5 Millets as a Protein Source

Millets are recognized for their moderate protein content, which ranges from 7.7% to 11.8% depending on the species.[11] Millet proteins are classified into three primary segments, namely: albumin and globulin, prolamin, and glutelin. While albumin and globulin contribute 8.5–16.26% of the total protein in most small millets, prolamins account for 15–30% and glutelins 45–55%. Foxtail millet is an exception, with a higher prolamin content (60%) compared to glutelin (15.23%). In pearl millet, prolamins range from 33–49.5%, glutelins 30–45%, and globulins combined with albumins make up 18–26%.[12,13] This varied protein composition plays a significant role in millet's nutritional profile, especially its value in vegetarian diets due to the high methionine content, an essential amino acid often lacking in other cereals.

Table 2.2 The nutritional value of millets.[10]

Nutrients	Sorghum	Pearl millet	Finger millet	Kodo millet	Proso millet	Foxtail millet	Little millet	Barnyard millet
Carbohydrates (g)	67.7	61.8	66.8	66.2	70.4	60.1	65.5	65.5
Protein (g)	9.9	10.9	7.2	8.9	12.5	12.3	10.1	6.2
Fat (g)	1.73	5.43	1.92	2.55	1.1	4.3	3.89	2.2
Energy (kcal)	334	347	320	331	341	331	346	307
Dietary fiber (g)	10.2	11.5	11.2	6.4	—	—	7.7	—
Calcium (mg)	27.6	27.4	364	15.3	14	31	16.1	20
Phosphorous (mg)	274	289	210	101	206	188	130	280
Magnesium (mg)	133	124	146	122	153	81	91	82
Zinc (mg)	1.9	2.7	2.5	1.6	1.4	2.4	1.8	3
Iron (mg)	3.9	6.4	4.6	2.3	0.8	2.8	1.2	5
Thiamine (mg)	0.35	0.25	0.37	0.29	0.41	0.59	0.26	0.33
Riboflavin (mg)	0.14	0.2	0.17	0.2	0.28	0.11	0.05	0.1
Niacin (mg)	2.1	0.9	1.3	1.5	4.5	3.2	1.3	4.2
Folic acid (μg)	39.4	36.1	34.7	39.5	—	15	36.2	—

Millets contain essential amino acids, particularly sulfur-containing ones like methionine and cysteine, which contribute to their nutritional significance. Finger millet is notable for its sulfur-rich amino acid content, enhancing its dietary value. However, similar to other cereals, millet proteins contain limited lysine, an important essential amino acid, though this varies depending on the species and cultivation factors. For example, pearl millet has a relatively higher lysine content (up to 6.5 g per 100 g protein), while proso and foxtail millets have lower lysine levels but are rich in leucine.[14] In terms of overall amino acid profiles, millets offer better protein quality compared to many other cereals, projecting them as a valuable source of nutrition, predominantly in areas where protein intake is inadequate.

The protein content in millets can vary significantly due to agro-geographical factors and cultivation practices. Proso and little millet, for example, contain between 10–15% protein, a level comparable to or higher than many other cereal grains. This variability is influenced by factors such as soil nitrogen levels and environmental conditions. The relatively high protein content of millets, coupled with their diverse amino acid profiles, makes them ideal for developing nutritionally fortified food products, especially for malnourished populations. Additionally, combining millet proteins with lysine-rich vegetable or animal proteins creates a balanced amino acid profile with high biological value, further enhancing their role in combating nutritional deficiencies (Table 2.3).

Table 2.3 Amino acid profile of different millets.

Amino acids (per 100 g of protein)	Proso millet[15]	Foxtail millet[16]	Finger millet[17]	Pearl millet[18]
Essential amino acids				
Isoleucine	4.1	4.59	4.3	5.1
Phenylalanine	5.5	6.27	6	7.6
Threonine	3	3.68	4.3	3.3
Valine	5.4	5.81	6.3	4.2
Leucine	12.2	13.6	10.8	14.1
Lysine	1.5	1.59	2.2	0.5
Methionine	2.2	3.06	2.9	1
Histidine	2.1	2.11	2.3	1.7
Tryptophan	0.8	NA	NA	1.2
Non-essential amino acids				
Alanine	10.9	9.3	6.1	8.1
Arginine	3.2	3	3.4	0.9
Serine	6.3	4.56	5.3	5.4
Tyrosine	4	2.44	8.6	2.7
Aspartic acid	6.2	7.71	5.7	6.2
Cysteine	NA	0.45	NA	0.8
Glutamic acid	21.3	22	23.2	22.8
Glycine	2.1	2.91	3.3	0.7
Proline	7.3	5.54	9.9	8.2
Protein efficiency ratio (PER) (1)	1.1	0.8	2	1.6

2.6 The Need for Alternative Proteins

In the modern era of healthy lifestyles, the need for a protein-rich diet substantially increased the demand and search for alternative sources, which must be cost effective and sustainable options. Compared to traditional plant and animal-based protein sources, proteins in millets are a superior source of essential amino acids that can fit the need of the hour. In comparison to animal proteins, plant proteins are highly hydrophobic, cheap, widely prevalent, have minimal pathogen contamination and, irrespective of dietary limitations, the majority of the global population can consume them.[19] The factors influencing the global demand for alternative protein sources, especially millets, are summarized in Figure 2.1. The challenges associated with millet proteins are lower stability, poor solubility and faster degradation, which need to be overcome for market acceptability. The utilization of millet protein or its different fractions is very limited in food industries due to minimal research on its physicochemical, structural, and functional properties.

2.6.1 Millet Protein Distribution and Composition

In millets, protein is the major macromolecule present next to starch. Various factors such as variety, species, cultivating regions, agricultural practices followed, distribution in seed, growing temperatures, soil quality, seed maturity stage and water content affect the protein quality, amount, and composition.[21] In comparison with major cereals such as rice and wheat, the

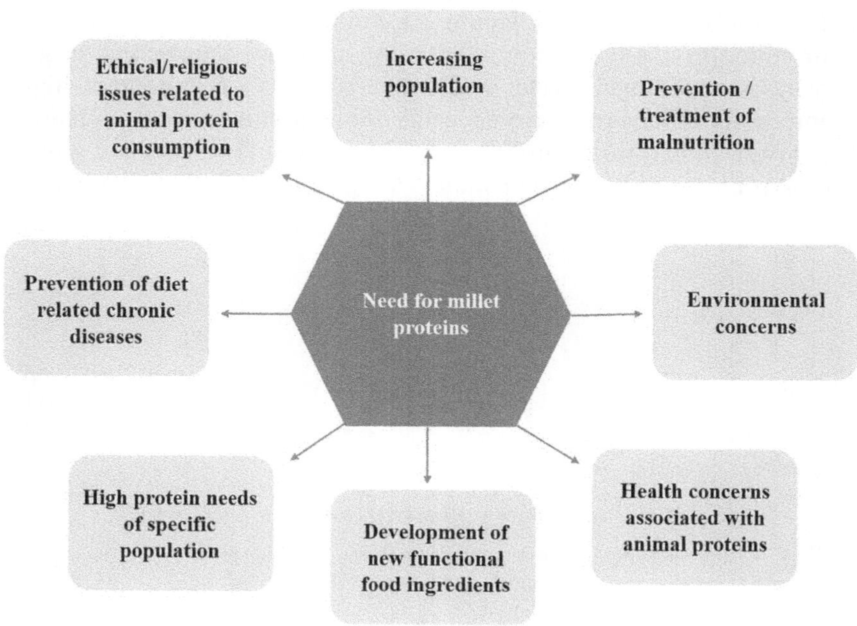

Figure 2.1 Major factors driving the millet protein demand.[20]

proteins in millets are incomplete and lack sufficient amounts of essential amino acids such as lysine, threonine, and tryptophan. Millet proteins are majorly distributed in the endosperm and embryo. They are non-allergenic and gluten-free in nature. In the cytoplasmic proteinaceous matrix space, millet starch granules (amyloplast) coincide with protein bodies in endospermic proteins. Also, enzymes present in bran and germ fractions influence the final protein concentration.[22] In wheat, glutenin is present as a major fraction (28.92% to 49.72%), whereas rice contains higher glutenin content, *i.e.*, brown rice (75–81%), milled rice (79–83%), and rice bran (22–45%).[23] Millets, especially proso millet which contains prolamin as a predominant fraction, can be utilized as an alternative protein source for gluten-related disorders. Therefore, millet is safe for consumption by persons with various gluten-related disorders such as celiac disease, non-celiac gluten sensitivity, dermatitis herpetiformis, gluten ataxia and wheat allergy.[24] Celiac patients were fed with a sorghum-based food product for 5 days and it was found that patients did not experience any gastrointestinal and non-gastrointestinal symptoms, and no variation was observed in the anti-transglutaminase antibody levels. The results suggest that sorghum and millet proteins can be safely consumed by celiac patients.

Based on the difference in solubility, the four major fractions of millet proteins are albumin (water-soluble), globulin (salt-soluble), prolamins (alcohol-soluble) and glutelins (alkali and alcohol soluble). These fractions function as metabolic proteins, storage proteins and protective proteins, and their composition and content varies among millet crops. Out of these fractions, glutelins accounted for more than 50 g per 100 g of total millet proteins present.[25] The average protein content of millets and their fractions based on solubility is given in Figure 2.2. Among the protein fractions present in millet bran, albumins and globulins have lower molecular weights, are easily extracted using water and saline solutions, and contain higher concentrations of essential amino acids than prolamins and glutelins.[26] Except foxtail millet, minor millets vary in prolamin (15–30%) and glutelin (45–55%) content. Foxtail millet contains a higher content of prolamin (60%)

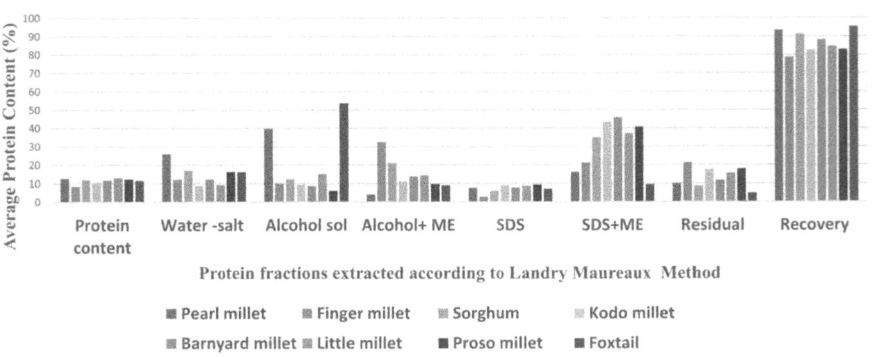

Figure 2.2 Protein content in different millets and their fractions based on Landry & Moureaux method. Reproduced from ref. 21 with permission from Elsevier, Copyright 2023.

and a lower glutelin content (15.23%). A higher content of albumin and globulin fractions are observed in proso millet (2%) and foxtail millet (1.8%), and the lowest in kodo millet (0.90%).[27]

2.7 Thermal Processing of Millets and Its Effects

Millets are not staple foods in many parts of the world, but their nutritional properties are leveraged through value addition. Like other cereals, the processing of millets aims to improve digestibility, reduce anti-nutritional factors, and enhance sensory properties. In millets, a major focus is on reducing anti-nutritional compounds and mitigating undesirable bitter flavors. Common pre-treatment methods include soaking, germination, and fermentation. These are often followed by thermal treatments such as popping, extrusion, roasting, and hydrothermal processes like parboiling.

Thermal treatments have a profound impact on the protein structure of millets. At temperatures exceeding the protein denaturation point, the three-dimensional structure of proteins unfolds, exposing hydrophobic regions. This promotes hydrophobic interactions, sulfur linkages, and hydrogen bonding,[28,29] which improve functional properties like gelation and emulsification. However, these changes can also decrease protein digestibility. The reduction in digestibility is attributed to the formation of complexes with anti-nutritional compounds, which limit the access of proteolytic enzymes.[30]

Sharma *et al.*,[30] and Bora *et al.*[31] found that thermal treatments, such as dry and wet heating, parboiling, and popping, reduce the digestibility of millet proteins due to protein aggregation and interactions with other millet components like phenolic compounds. These processes result in structural changes, such as the formation of hydrophobic protein aggregates, which hinder enzyme access to protein bodies, thereby reducing digestibility. For example, Gulati *et al.*[32] observed that both dry and wet heat treatments of proso millet significantly decreased protein digestibility, likely because of increased surface hydrophobicity and protein aggregation. Similarly, Muyonga *et al.*[33] found that popping reduced digestibility in amaranth grains as a result of amino acid breakdown and the formation of intramolecular disulfide bridges during high-temperature processing.

On the positive side, thermal treatments can enhance the functional properties of millet proteins. The exposure of hydrophobic regions improves gelation and emulsification properties. High-temperature processes like autoclaving can eliminate anti-nutritional compounds, thereby improving digestibility. In food matrices such as dough, denatured proteins interact with gelatinized starch, enhancing the rheological properties.[34]

2.7.1 Microwave Treatment

Microwave (MW) treatment notably impacted the functional properties of finger millet protein (FMP). An increase in emulsifying and water absorption capacities was observed with MW and Hot-air Assisted Radio Frequency

(HARF) treatments compared to untreated FMP, while the oil absorption capacity decreased. However, foaming properties and protein purity declined post-treatment. These effects are likely due to changes in the protein's structural attributes, such as reduced crystallinity and alterations in secondary structures, specifically, an increment in β-turns and a reduction observed in α-helices and β-sheets. These structural modifications were confirmed through Fourier-transform infrared spectroscopy (FTIR) and X-ray diffraction (XRD) analyses.[35] Treatment of foxtail millet flour results in an increase in water absorption capacity and a decrease in oil absorption capacity. This is attributed to the heat induced protein denaturation, which results in unfolding of proteins, thus leading to the exposure of polar side chains[36] (Table 2.4).

2.8 Effect of Non-thermal Processing Methods

2.8.1 High-pressure Processing (HPP)

This is an innovative technique that applies high pressures of 400–600 MPa to eliminate microorganisms without significantly compromising the nutritional quality of foods. HPP impacts the secondary and tertiary structures of proteins by modifying hydrogen bonds and other non-covalent interactions, leaving covalent bonds intact. This structural alteration enhances functionalities such as solubility, water and oil holding capacity, emulsification, foaming, and gelation.

According to Zhang *et al.*,[45] HPP treatments between 100–500 MPa affect the secondary structure of proteins, reducing hydrophobicity and increasing thermal stability due to enhanced structural order. The reduction in hydrophobicity makes HPP-processed proteins ideal candidates for the delivery of water-soluble and hydrophilic bioactive compounds. Additionally, Zhang *et al.*[46] found that HPP facilitated enzymatic hydrolysis of proteins and isolated antioxidants, showing the highest *in vitro* antioxidant activity at 300 MPa. This highlights the potential of using low-cost millet proteins as functional ingredients in food formulations.

2.8.2 Irradiation

Irradiation has been found to improve the protein digestibility and nutritional quality of various millet species. Sujatha *et al.*[47] reported that treating foxtail millet, pearl millet, and sorghum with 1 to 2.5 KGy enhanced *in vitro* protein digestibility, likely due to the protein's hydrolysis into amino acids and the cleavage of disulfide linkages, which unfolds protein structures and increases their exposure to proteolytic enzymes. Similarly, Mohamed *et al.*[48] observed that irradiating proso millet at doses up to 2 KGy led to an increase in crude protein content and free amino acids, further boosting its nutritional profile.

Reddy *et al.*[49] found that gamma irradiation of finger millet proteins at doses up to 15 KGy significantly enhanced protein content, possibly due to

Table 2.4 Effects of thermal processing on millets.

Protein source	Processing method and conditions	Effects	Ref.
Foxtail millet	Hydrothermal treatment at 40 °C for 6 hours	Protein ranged from 9.41% to 12.72%; removal of endogenous proteins by 88.59–92.27%; increased water absorption and swelling power. Endogenous proteins negatively affected the rheological behavior.	37
Proso millet and little millet	Microwave treatment at 600 W for 6 minutes and 30 minutes	Protein increased to 10.32 g per 100 g in treated samples (compared to 4.48 g per 100 g in raw samples); enhanced water and oil absorption, improved functional properties; FTIR confirmed protein presence with branched granule formation.	38
Sorghum grains	Microwave treatment at 600 W for 6 minutes	Protein concentration rose by 0.60%; thermal and pasting behavior enhanced; glycemic and hydrolysis indexes decreased by 3.15% and 5.74%, respectively.	39
Pearl millet	Decortication using a modern dehuller	Crude protein dropped from 12.36% to 11.50% due to removal of bran; decreased tannin and phytic acid levels; nitrogen-free extract content in grains increased.	40
Pearl millet	Fermentation with pure *Lactobacillus plantarum* at 37 °C for 96 hours	Protein increased from 8.7% to 20.54% with starter culture fermentation, and to 20.21% with natural fermentation; lipid and carbohydrate levels dropped; anti-nutrient levels were reduced.	41
Foxtail millet	Fermentation with *L. paracasei* in solid-state at 37 °C for 48 hours	Crude protein content increased by 20.51%; crystalline structure and thermal behavior altered; total carbohydrate content decreased, while tannin and soluble sugars remained stable.	42
Foxtail millet	Germination with soaking period of 13.18–46.82 hours at temperatures of 19.55–46.55 °C	Protein concentration increased to 13.75 g from 10.60 g; optimal germination at 15.84 minutes of soaking and 25 °C; higher phenolic and flavonoid content; fat content reduced from 3.86 g per 100 g to 2.78 g per 100 g.	43
Kodo millet	Germination at soaking time of 13.81 hours, temperature of 38.75 °C, and germination time of 35.82 hours	Protein content increased from 6.7% to 7.9%; improved mineral and GABA content; reduced phytates and tannins levels.	44

peptide bond cleavage that disrupts protein conformation. However, ElShazali *et al.*[50] noted that irradiation alone had no significant effect on reducing anti-nutritional factors in millets. When combined with cooking, however, irradiation substantially improved digestibility by reducing these factors. Gowthamraj *et al.*[51] also highlighted that treating finger millet with doses between 2.5 KGy and 10 KGy effectively reduced anti-nutritional factors, suggesting a synergistic role of irradiation in enhancing protein bioavailability when combined with additional processing.

2.8.3 Ultrasonic Treatment

Ultrasonication is an efficient method to enhance the functional and structural properties of millet proteins by using high-frequency sound waves, typically above 20 kHz. This process involves the formation of zones of high pressure and temperature, which are followed by rapid cooling, leading to structural modifications in the proteins. These mechanical waves, particularly through cavitation, break down non-covalent bonds, partially degrading the secondary and tertiary structures without significantly affecting the primary structure.[52] As a green and sustainable technique, ultrasonication offers advantages over conventional thermal treatments, including greater energy efficiency, lower production costs, and enhanced functional properties like solubility, emulsification, and thermal stability.[53]

Studies, such as those by Nazari *et al.*,[54] have shown that ultrasonication at different intensities of 18.4, 29.58, and 73.95 W cm^{-2} for durations ranging from 5 to 20 minutes significantly improved the emulsion activity index, solubility, and foaming properties of millet proteins. This enhancement is attributed to reductions in molecular weight, increased surface charge, and the exposure of hydrophilic and hydrophobic groups due to partial unfolding of the protein structure. Peng *et al.*[55] observed that treatment with 450 W ultrasonication altered the protein structure of foxtail millet bran, increasing random coils and β-turns while decreasing α-helix and β-sheet structures, resulting in a flexible protein structure. This change exposes more hydrophobic groups, improving solubility and stability and enhancing antioxidant capacity by releasing active peptides. Ultrasound treatment has also been reported to reduce the average particle size, thereby exposing polar and non-polar amino groups, improving solubility, and enhancing water–protein interactions.[30] Through these structural modifications, ultrasonication has shown potential to enhance the functional and nutritional profile of millet proteins, making them suitable as functional food ingredients.

2.8.4 Pulsed Electric Field (PEF)

PEF treatment has notable impacts on the structural and functional properties of millet proteins, particularly by modifying secondary and tertiary structures. PEF interrupts hydrogen bonds and other interactions that stabilize protein secondary structures, leading to substantial structural changes. Studies have shown that PEF treatment increases β-sheet content while decreasing α-helix levels, as confirmed by FTIR analysis through shifts in specific subpeak areas related to β-sheets and α-helices.[56,57] These structural changes create a more ordered arrangement in the protein, as random coils and β-turns transition towards β-sheets under moderate-intensity PEF (MIPEF). Additionally, MIPEF promotes the formation of soluble aggregates, resulting in larger protein particles.[58]

PEF treatment also enhances the hydrophobicity and solubility of millet proteins, particularly in alcohol solutions, by exposing more hydrophobic groups like tyrosine and tryptophan on the protein surface. This increased surface hydrophobicity initially raises fluorescence intensity and supports the formation of stable hydrophobic clusters, although the effect can diminish at higher field strengths due to altered electrostatic interactions.[59] Functionally, PEF enhances millet protein's solubility, water- and oil-holding capacities, and emulsifying and foaming properties. For instance, PEF-treated proteins show increased solubility with rising pulse frequency and voltage.[60] Emulsification and foaming capacities also improve with PEF treatment, achieving maximum values at optimized field strengths and treatment durations.[61] These enhancements are driven by structural modifications that increase the protein's surface activity, making PEF a promising technique for improving the functionality of millet proteins in food formulations.

2.8.5 Cold Plasma

Cold plasma treatment, a novel, chemical-free technique, has demonstrated notable modifications to the structural and functional properties of millet proteins. It involves ionizing gases at ambient temperatures, producing reactive oxidants like ozone and molecular oxygen, which oxidize amino acids, leading to modifications such as dimerization, nitration, sulfoxidation, and hydroxylation.[62,63] Cold plasma causes secondary structure alterations, reducing α-helix content while increasing β-turns and β-sheets, which influence hydration and protein–protein interactions. Studies on plasma-treated millet have also shown decreased protein content and splitting of covalent bonds within protein fragments, likely due to the generation of free radicals.[64]

Functionally, cold plasma treatment enhances the emulsifying and foaming capacities of millet proteins, with improved stability in foaming properties observed at higher plasma frequencies. The process also significantly increases protein solubility, likely due to reduced particle size and enhanced electrostatic repulsion between protein molecules, as cold plasma promotes structural unfolding that exposes hydrophilic sites.[65] Increased surface hydrophobicity aids in emulsification, making cold plasma-treated millet proteins more effective in food applications such as foams and emulsions. Additionally, cold plasma treatment decreases the thermal stability of millet proteins, as observed with other proteins, facilitating their functional integration into various food products.[66]

2.8.6 Milling and Decortication

Decortication and milling are primary processing methods used to remove the outer layers of millet grains, enhancing their palatability and consumer acceptance. However, this process affects the nutritional composition of millets. Research indicates that protein recovery slightly decreases in finger millet[67] and proso millet[68] after decortication. This reduction is likely due to protein denaturation and polyphenolic complex interactions, which reduce glutelin fractionation while increasing the extractability of globulin and prolamin fractions. Additionally, milling tends to lower the levels of protein, fat, and fiber, with barnyard millet exhibiting a 14% moisture reduction post-processing.[69]

In pearl millet, abrasive decortication, especially with extended pearling, decreases total polypeptides and iron levels.[70] Comparatively, brown millet showed an increase in protein and phenolic content following milling, as bran removal reduced anti-nutritional compounds such as phytates and tannins, thus enhancing protein digestibility.[71] Similarly, dehulling pearl millet increased protein content by up to 17.5%, alongside improved digestibility and functionality, although excessive dehulling led to a decline. Decorticated foxtail millet showed a rise in protein content from 11 to 13.4 g per 100 g in milled grains and 11.6 g per 100 g in polished grains. Proso millet followed a similar trend, with protein rising from 13.6 to approximately 14.1–14.18 g per 100 g in milled and polished forms, likely due to bran layer removal.[71]

The removal of anti-nutritional compounds like polyphenols in the hull during decortication significantly improves protein bioavailability and mineral absorption. For example, decortication of finger millet decreased phytates and polyphenols by 39.8% and 74.7%, respectively.[72] Likewise, decortication in sorghum resulted in moisture and protein reduction due to the removal of the bran layer, which reduced protein content and other nutrient concentrations. Variability in decortication outcomes, including nutrient retention and protein loss, can be attributed to the use of different decortication techniques, millet varieties, and processing conditions, with the extent of hull removal playing a significant role.[73]

2.8.7 Fermentation

Fermentation, a traditional technique used to enhance the nutritional profile of grains, including millets, involves the transformation of food substrates *via* microorganisms. The fermentation process improves protein content and mineral extractability in millet grains, likely due to the breakdown of complex structures such as starch and proteins, which increases bio-accessibility and *in vitro* protein digestibility (IVPD).[74] Proteolytic enzyme synthesis during fermentation can lead to slight degradation of protein complexes, reducing hydrophilic properties and altering protein structure.

In a study reported in ref. 75, finger millet flour was treated with different treatments such as addition of lemon juice (T1), sodium chloride (T2), combination of T1 and T2 (T3), and fermentation of slurry for 36 hours at 37 °C using yogurt followed by T3 (T4). Due to these treatments, the protein content significantly reduced from 5.76 g per 100 g in the control to 3.57 g per 100 g, 3.29 g per 100 g, 2.99 g per 100 g, and 3.77 g per 100 g, respectively. This might be due to due to leaking of soluble proteins and an increase in protein solubility through amino acid charge alteration, which facilitated the breakdown of complex proteins into simpler peptides. The study concluded that fermentation time, and treatments resulted in drastic changes in the microstructural properties, and protein quality characteristics of the finger millet flour.[75]

Fermentation with specific microbial strains has been shown to improve the functional properties of proteins in millet flours. For instance, fermentation using pure cultures with pearl millet flour improved protein efficiency and digestibility, likely due to microbial enzyme activity that degrades anti-nutritional factors and converts complex, insoluble proteins into more digestible forms.[76] Additionally, lactic acid fermentation with *Lactobacillus* species reduces tannin and phytic acid levels, enhancing the protein bioavailability in finger millet by disrupting protein–tannin complexes through microbial proteolytic enzymes.[77]

In little millet, fermentation by strains like *Lactobacillus acidophilus* and *Saccharomyces boulardii* increased protein content from 9.86% to 10.95%. This increase is ascribed to the carbohydrate breakdown by microorganisms, which biosynthesizes amino acids necessary for microbial growth and development.[78] Similarly, fermentation of foxtail millet with *Lactobacillus paracasei* led to protein hydrolysis, producing hydrophobic amino acids that enhance digestibility.[44]

2.8.8 Germination

Germination, a process involving the soaking of grains under controlled time and temperature conditions, significantly enhances the nutritional profile of millet grains. Research demonstrates that germination (40 h at 25 °C) improves protein functionality and digestibility in foxtail millet by increasing the amino acid profile.[44] For instance, germination increased the protein content in foxtail millet due to an improved amino acid profile.[43]

In pearl millet, germination elevated protein levels from 11.4% to 13.2% in one variety and 14.4% to 16.3% in another.[79] Other studies showed the protein content of pearl millet increased from 14% to 26% with germination,[76] while proso millet exhibited a similar increase at 96 h of sprouting followed by germination. A combination of processes such as soaking and germination in pearl millet mobilize stored nitrogen, further enhancing protein functionality.

Although a reduction in protein was observed in foxtail millet, from 11.2 to 8.9 g per 100 g, possibly due to leaching during soaking, germination in finger millet and pearl millet has been shown to increase acidity and protein hydrolysis, resulting in simpler protein compounds.[80] Germinated pearl millet grains show a 50% increase in protein content attributed to amino acid synthesis during the process. Similarly, germination of foxtail millet led to an increase in protein and phenolic content by 29.72% and 77.42%, respectively, while reducing anti-nutritional factors such as tannins and phytic acid by 45.07% and 39.02%, likely due to enzymatic biosynthesis of stable, free molecules.[81]

Table 2.5 The effect of non-thermal treatment on millet protein.

Processing conditions	Millet type	Effects	Ref.
Ultrasonication Amplitude: 5 & 10%; Time: 5, 10, & 20 min	Foxtail Millet	- Ultrasound unfolded and partially denatured protein structures, enhancing functionality. - Improved heat denaturation, emulsifying and foaming capacity, zeta potential, and surface hydrophobicity. - Treatment duration had less impact than amplitude.	83
Ultrasonication Power: 400 W; Amplitude: 60, 80, & 100%; Time: 5, 12.5, or 20 min	Proso Millet Bran	- Protein digestibility remained stable post-treatment. - Cooking led to hydrophobic protein clumps, decreasing digestibility. - Ultrasound unfolded protein structures, enhancing accessibility.	68
Ultrasonication Intensities: 18.4, 29.58, & 73.95 W cm^{-2}; Time: 5, 12.5, & 20 min	Proso Millet	- Increased protein solubility and reduced foaming capacity. - Optimal conditions (73.95 W cm^{-2} for 12.5 min) achieved improved functional properties. - Ultrasound modified molecular structure.	54
Pulsed Electric Field Pulse width: 75 μs; Pulse number: 50; Interval: 100 ms; Field strength: 1, 2, & 3 kV cm^{-1}	Foxtail Millet	- Enhanced protein solubility, with reduced α-helix and β-turn structures. - Weaker hydrophobicity observed at 3 kV cm^{-1}. - Moderate field intensity expanded tertiary protein structure.	56
High-Pressure Processing (HPP) Pressure: 400 MPa; Temperature: 60 °C; Time: 60 min	Foxtail Millet	- Protein content remained stable. - Combined HPP and germination improved functional properties and reduced antinutritional factors. - Protein–starch interactions possibly formed covalent or ionic bonds.	84
HPP Pressure: 300 MPa & 600 MPa; Room temperature; Time: 10 min	Foxtail Millet	- No significant change in amino acid and nitrogen content. - Storage studies showed a reduction in nitrogen content, possibly from oxidative breakdown.	85

Table 2.5 (*Continued*)

Processing conditions	Millet type	Effects	Ref.
HPP Pressure: 350 MPa; Temperature: 20 °C; Time: 10 min	Finger Millet	- Non-enzymatic browning and cross-linking decreased protein digestibility. - Enhanced nutritional and sensory properties without altering physicochemical attributes. - Protein networks formed with intra- and intermolecular S–S linkages.	86
HPP Pressure: 100, 300, & 500 MPa; Temperature: 25 °C; Time: 10 min	Millet	- Increased polarity in gliadin's polar groups and reduced hydrophobicity. - Increased protein particle size and decreased surface charge. - Improved thermal stability and solubility as pressure increased.	45
HPP Pressure: 200, 400, & 600 MPa	Amaranth	- Proteins showed high susceptibility at 400 MPa, leading to formation of insoluble aggregates.	87
HPP Pressure: 300 MPa; Temperature: 20 °C; Time: 15 min	Sorghum	- Increased protein digestibility in raw sorghum flour.	88

Additionally, germination modifies protein structure and morphology and enhances the availability of hydrolytic enzymes.[16] For example, germinating foxtail millet at room temperature boosted protein and phenolic content by 29.72% and 77.42%, respectively, with improved 2,2-diphenyl-1-picrylhydrazyl (DPPH) radical scavenging activity.[82] Optimal protein, antioxidant activity (90.5%), total phenolic content (TPC at 45.67 mg Gallic Acid Equivalent (GAE) per 100 g), total flavonoids (TFC at 30.52–43.96 mg Rutin Equivalent (RU) per g), and dietary fiber (27.42 g) were achieved with a 15.84-minute soaking time and germination temperature of 25 °C[43] (Table 2.5).

2.9 Conclusion

Millet proteins hold significant potential for addressing global nutritional and food security challenges, especially in regions prone to climate variability and in regions with limited agricultural resources. While millets generally have lower total protein content compared to major cereals like wheat or maize, their balanced amino acid composition, particularly the presence of essential amino acids such as methionine and cysteine, makes them a valuable dietary component. As the demand for sustainable and alternative protein sources increases, millets are emerging as a promising solution. Their adaptability to arid environments, minimal input requirements, and resilience to pests and diseases underscore their potential in sustainable agriculture. Additionally, their protein content, when integrated into diverse food systems, can contribute to reducing protein-energy malnutrition, especially in under-resourced regions. Processing methods play a pivotal role in enhancing the nutritional quality and functional properties of millet proteins. Techniques such as fermentation, germination, and non-thermal processing methods can significantly improve protein digestibility, reduce antinutritional factors, and enhance bioavailability. However, each method impacts protein structure and composition differently, demanding careful optimization to balance nutritional enhancement with functional and sensory qualities. In conclusion, millet proteins represent a nutritionally rich, environmentally sustainable, and culturally significant food source with vast untapped potential. Future research should focus on improving protein extraction and processing techniques, protein functionality, exploring novel applications in food product development, and integrating millet-based proteins into global food systems.

References

1. L. Martin, E. Messager, G. Bedianashvili, N. Rusishvili, E. Lebedeva, C. Longford, R. Hovsepyan, L. Bitadze, M. Chkadua, N. Vanishvili, F. Mort, K. Kakhiani, M. Abramishvili, G. Gogochuri, B. Murvanidze, G. Giunashvili, V. Licheli, A. Salavert, G. André and E. Herrscher, *Sci. Rep.*, 2021, **11**, 20494.
2. X. Liu, M. Jones, Z. Zhao, G. Liu and T. C. O'Connell, *Am. J. Phys. Anthropol.*, 2012, **149**, 283–290.
3. N. Miller, R. Spengler and M. Frachetti, *Holocene*, 2016, **26**, 1566–1575.
4. S. Weber, *Anim. Nutr.*, 1998, **39**, 267–274.
5. G. Ganzarolli, M. Alexander, A. Arnau and O. Craig,*J. Archaeol. Sci.*, 2018, **96**, 124–130.
6. V. Bhat; D. B. Rao and V. A. Tonapi, ICAR-Indian Institute of Millets Research, Hyderabad, 2018.
7. B. Sharma and H. S. Gujral, *Food Chem.*, 2019, **299**, 125115.
8. M. Müller, E. Canfora and E. Blaak, *Nutrients*, 2018, **103**, 275.
9. M. Akhtar, Y. Chen, Z. Ma, X. Zhang, D. Shi, J. A. Khan and H. Liu, *Anim. Nutr.*, 2022, **8**, 350–360.
10. T. Longvah, I. Anantan, K. Bhaskarachary, K. Venkaiah and T. Longvah, National Institute of Nutrition, Indian Council of Medical Research, Hyderabad, 2017.
11. J. H. Hulse. *Polyphenols in cereals and legumes: proceedings of a Symposium*. IDRC, Ottawa, ON, CA, 1980.
12. M. Chandna and N. K. Matta, *Phytochemistry*, 1990, **291**, 3395–3399.
13. K. Parameswaran and B. Thayumanavan, *Plant Foods Hum. Nutr.*, 1995, **48**, 119–126.

14. S. Bean., L. Zhu, B. Smith, J. D. Wilson, B. Ioerger and M. Tilley, Starch and protein chemistry and functional properties, in *Sorghum and Millets*, Elsevier, Duxford, United Kingdom, 2019.
15. A. Bagdi, G. Balázs, J. Schmidt, M. Szatmári, R. Schoenlechner, E. Berghofer and S. J. A. A. Tömösközia, *Acta Aliment.*, 2011, **401**, 128–141.
16. M. T. Kamara, Z. Huiming, Z. Kexue, I. Amadou and F. Tarawalie, *Am. J. Food Technol.*, 2009, **4**, 255–267.
17. U. M. Nakarani, D. Singh, K. P. Suthar, N. Karmakar, P. Faldu and E. H. Patil, *Food Chem.*, 2021, **341**, 128271.
18. B. M. Romdhane, L. Riahi, M. Borji, J. Riahi, F. Chibani, A. Ghorbel and N. Zoghlami, *Cereal Chem.*, 2023, **1006**, 1263–1272.
19. F. Jhan, A. Gani, N. Noor and A. Shah, *ACS Food Sci. Technol.*, 2021, **8**, 1418–1427.
20. N. Sachdev, S. Goomer and L. R. Singh, *J. Sci. Food Agric.*, 2021, **101**(3), 831–842.
21. N. Sachdev, D. Goomer, L. R. Singh, V. M. D. Pathak, D. Aggarwal and D. R. K. Chowhan, *Appl. Food Res.*, 2023, **31**, 100288.
22. K. S. Poutanen, A. O. Kårlund, G.-C. Gallego, D. P. Johansson, N. M. Scheers, I. M. Marklinder, A. K. Eriksen, P. C. Silventoinen, E. Nordlund, N. Sozer, K. J. Hanhineva, M. Kolehmainen and R. Landberg, *Nutr. Rev.*, 2022, **806**, 1648–1663.
23. J. L. Balindong, L. Liu, R. M. Ward, B. J. Barkla and D. L. E. Waters, *J. Cereal Sci.*, 2016, **72**, 124–130.
24. C. Ciacci, L. Maiuri, N. Caporaso, C. Bucci, D. L. Giudice, R. D. Massardo, P. Pontieri, D. N. Fonzo, S. R. Bean, B. Ioerger and M. Londei, *Clin. Nutr.*, 2007, **266**, 799–805.
25. Y. Zhang, X. Jing, Z. Chen and X. Wang, *LWT*, 2022, **164**, 113654.
26. P. R. Shewry and N. G. Halford, *J. Exp. Bot.*, 2002, **533**, 947–958.
27. K. R. Hegde, M. R. Bhavana, M. Buvaneswaran, V. R. Sinija, A. Rawson and V. Hema, *Food Humanity*, 2025, **4**, 100504.
28. J. L. Mession, M. Chihi, N. Sok and R. Saurel, *Food Hydrocolloids*, 2015, **46**, 233–243.
29. J. Zink, T. Wyrobnik, T. Prinz and M. Schmid, *Int. J. Mol. Sci.*, 2016, **179**, 1376.
30. S. Sharma, S. Joshi, A. Gupta, R. Sharma, S. K. Sandhu and B. Singh, *J. Food Meas. Charact.*, 2023, **176**, 6236–6252.
31. P. Bora, S. Ragaee and M. Marcone, *Food Chem.*, 2019, **274**, 718–725.
32. P. Gulati, A. Li, D. Holding, D. Santra, Y. Zhang and D. J. Rose, *J. Agric. Food Chem.*, 2017, **659**, 1952–1959.
33. J. H. Muyonga, B. Andabati and G. Ssepuuya, *Food Sci. Nutr.*, 2014, **21**, 9–16.
34. A. Malik, K. Khamrui and W. Prasad, *Future Foods*, 2021, **3**, 100027.
35. M. V. Rao, C. K. Sunil and N. Venkatachalapathy, *J. Cereal Sci.*, 2023, **114**, 103774.
36. M. V. Rao, K. Akhil, C. Sunil, N. Venkatachalapathy and R. Jaganmohan, *Int. J. Chem. Stud.*, 2021, **91**, 2762–2767.
37. F. Zhang and Q. Shen, *J. Cereal Sci.*, 2021, **100**, 103255.

38. S. R. Kumar, M. B. Sadiq and A. K. Anal, *LWT*, 2020, **128**, 109465.
39. R. Li, L. Dai, H. Peng, P. Jiang, N. Liu, D. Zhang and Z. Li, *J. Food Process Eng.*, 2021, **4410**, e13804.
40. E. Babiker, B. Abdelseed, H. Hassan and O. Adiamo, *World J. Sci., Technol. Sustainable Dev.*, 2018, **153**, 278–286.
41. O. E. Chinenye, O. A. Ayodeji and A. J. Baba, *Am. J. Biosci. Bioeng.*, 2017, **51**, 12–16.
42. I. Amadou, M. E. Gounga, Y. H. Shi and G. W. Le, *Food Bioprod. Process.*, 2014, **921**, 38–45.
43. S. Sharma, D. C. Saxena and C. S. Riar, *Cogent Food Agric.*, 2015, **11**, 1081728.
44. S. Sharma, N. Sharma, S. Handa and S. Pathania, *Food Chem.*, 2017, **214**, 162–168.
45. Y. Zhang, X. Zhang, Z. Zhang, Z. Chen, X. Jing and X. Wang, *LWT*, 2022, **154**, 112755.
46. Y. Zhang, X. Jing, Z. Chen and X. Wang, *LWT*, 2022, **164**, 113654.
47. M. Sujatha, T. Hymavathi, K. U. Devi, T. P. Roberts and D. P. Kumar, *Int. J. Curr. Microbiol. Appl. Sci.*, 2017, **611**, 1014–1020.
48. E. A. Mohamed, I. Mohamed, A. Ahmed, A. Yagoub and E. E. Babiker, *Int. J. Food Sci. Technol.*, 2010, **455**, 906–912.
49. C. K. Reddy and K. Viswanath, *J. Food Sci. Technol.*, 2019, **56**, 2651–2659.
50. A. M. ElShazali, A. A. Nahid, H. A. Salma and E. B. Elfadil, *Int. Food Res. J.*, 2011, **184**, 1401.
51. G. Gowthamraj, C. Jubeena and N. Sangeetha, *Radiat. Phys. Chem.*, 2021, **183**, 109403.
52. L. Mirmoghtadaie, S. Aliabadi and S. M. Hosseini, *Food Chem.*, 2016, **199**, 619–627.
53. S. M. T. Gharibzahedi and B. Smith, *Trends Food Sci. Technol.*, 2020, **98**, 107–116.
54. B. Nazari, M. A. Mohammadifar, S.-S. Aliabadi, E. Feizollahi and L. Mirmoghtadaie, *Ultrason. Sonochem.*, 2018, **41**, 382–388.
55. Z. Peng, F. Wang, L. Yu, B. Jiang, J. Cao, Z. Sun and J. Cheng, *Ultrason. Sonochem.*, 2024, **110**, 107044.
56. Y. Zhang, X. Jing, Z. Chen and X. Wang, *Cereal Chem.*, 2023, **1002**, 360–370.
57. K. D. Mattice and A. G. Marangoni, *Food Chem.*, 2021, **338**, 128010.
58. F. G. Galindo, P. Dejmek, K. Lundgren, A. G. Rasmusson, A. Vicente and T. Moritz, *Planta*, 2009, **230**, 469–479.
59. S. G. Giteru, I. Oey and M. A. Ali, *Trends Food Sci. Technol.*, 2018, **72**, 91–113.
60. L. Zhang, L. J. Wang, W. Jiang and J. Y. Qian, *LWT*, 2017, **84**, 73–81.
61. J. Zhang, T. Tang, Z. Jiang, Y. Liu and A. Jiang, *Food Chem.*, 2019, **293**, 455–462.
62. S. Jaddu, R. C. Pradhan and M. Dwivedi, *Innovative Food Sci. Emerging Technol.*, 2022, **77**, 102957.
63. E. Takai, T. Kitamura, J. Kuwabara, S. Ikawa, S. Yoshizawa, K. Shiraki and K. Kitano, *J. Phys. D: Appl. Phys.*, 2014, **4728**, 285403.

64. R. Lokeswari, P. S. Sharanyakanth, S. Jaspin and R. Mahendran, *IEEE Trans. Plasma Sci.*, 2021, **495**, 1745–1751.
65. M. N. Mollakhalili, M. Yousefi, A. Nematollahi and N. Khorshidian, *Eur. Food Res. Technol.*, 2021, **247**, 1579–1594.
66. H. Ji, X. Tang, L. Li, S. Peng, C. Gao and Y. Chen, *Food Hydrocolloids*, 2020, **109**, 106124.
67. U. Dharmaraj and N. G. Malleshi, *LWT*, 2011, **447**, 1636–1642.
68. N. Mustač, D. Novotni, M. Habuš, S. Drakula, L. Nanjara, B. Voučko and D. Ćurić, *J. Cereal Sci.*, 2020, **91**, 102864.
69. U. C. Lohani, J. P. Pandey and N. C. Shahi, *Food Bioprocess Technol.*, 2012, **5**, 1113–1119.
70. A. Tiwari, S. K. Jha, R. K. Pal, S. Sethi and L. Krishan, *J. Food Process. Preserv.*, 2014, **383**, 1215–1223.
71. R. Devisetti, S. N. Yadahally and S. Bhattacharya, *LWT*, 2014, **592**, 889–895.
72. T. Rathore, R. Singh, D. B. Kamble, A. Upadhyay and S. Thangalakshmi, *Pharm. Innovation J.*, 2019, **84**, 283–291.
73. F. Hama, C. Icard-Vernière, J. P. Guyot, C. Picq, B. Diawara and M.-C. Rivier, *J. Cereal Sci.*, 2011, **543**, 425–433.
74. J. A. Adebiyi, A. O. Obadina, F. M.-A. Bafubiandi, O. A. Adebo and E. Kayitesi, *J. Cereal Sci.*, 2016, **70**, 132–139.
75. S. Narayanasamy, *J. Food Sci. Technol.*, 2021, **588**, 3223–3234.
76. A. S. Saleh, Q. Zhang, J. Chen and Q. Shen, *Compr. Rev. Food Sci. Food Saf.*, 2013, **123**, 281–295.
77. S. Budhwar, K. Sethi and M. Chakraborty, *Food Prod., Process. Nutr.*, 2020, **2**, 1–17.
78. P. Pampangouda, K. B. Munishamanna and H. Gurumurthy, *J. Appl. Nat. Sci.*, 2015, **71**, 260–264.
79. A. B. Hassan, I. A. M. Ahmed, N. M. Osman, M. M. Eltayeb, G. A. Osman and E. E. Babiker, *Pak. J. Nutr.*, 2006, **51**, 86–89.
80. J. O. Owheruo, B. O. Ifesan and A. O. Kolawole, *Food Sci. Nutr.*, 2019, **72**, 476–482.
81. S. Sharma, D. C. Saxena and C. S. Riar, Changes in the GABA and polyphenols contents of foxtail millet on germination and their relationship with *in vitro* antioxidant activity, *Food Chem.*, 2018, **245**, 863–870.
82. P. M. Pradeep and Y. N. Sreerama, *Food Chem.*, 2015, **169**, 455–463.
83. R. Sharma and S. Sharma, *Food Chem.*, 2022, **368**, 130815.
84. N. Sharma, S. K. Goyal, T. Alam, S. Fatma, A. Chaoruangrit and K. Niranjan, *J. Cereal Sci.*, 2018, **83**, 162–170.
85. Z. Zhou, X. Shang, Z. Wang, C. Sun and L. Zhang, *J. Food Qual.*, 2022, **1**, 2456921.
86. A. Angioloni and C. Collar, *J. Cereal Sci.*, 2012, **563**, 713–719.
87. M. C. Condés, F. Speroni, A. Mauri and M. C. Añón, *Innovative Food Sci. Emerging Technol.*, 2012, **14**, 11–17.
88. I. Correia, A. Nunes, J. A. Saraiva, A. S. Barros and I. Delgadillo, *LWT*, 2011, **444**, 1245–1249.

CHAPTER 3

Novel Millet Peptides (Millet-derived Bioactive Peptides): Functional Role, Applications, and Challenges

KIRAN MISHRA,[a] V. M. MALATHI,[b] CHINMAYA BEHERA,[c] RAHNA FATHIMA,[c] R. VENKATESWARLU,[b] C. TARA SATYAVATHI[b] AND VEDA KRISHNAN*[c]

[a] Department of Biosciences and Bioengineering, Indian Institute of Technology Bombay, Mumbai, 400076, India; [b] Biochemistry Laboratory, ICAR-Indian Institute of Millets Research, Hyderabad, 500030, India; [c] Food Matrix Lab, Division of Biochemistry, ICAR-Indian Agricultural Research Institute, New Delhi, 110012, India
*Email: vedakrishnan@iari.res.in

3.1 Introduction

Millets are archaeological staples, well known as nutri-cereals and widely grown across Asia and Africa. Generally, millets are composed of protein, carbohydrates, and dietary fibers, along with significant amounts of minerals, vitamins, and bioactive compounds.[1,2] Among the components, millet protein and its peptides, which are short chains of amino acids, have been reported to exhibit specific biological activities, making them valuable for health and nutrition. The seed protein of millet contains components such as albumin, cross-linked prolamin, β-prolamin, globulin, and glutelin. Millets are notable for their high content of essential amino acids, like lysine,

Food Chemistry, Function and Analysis No. 48
Millet Proteins: Nutritional, Structural and Functional Attributes
Edited by Shweta Suri, Anuradha Dutta and Tarun Belwal

Published by the Royal Society of Chemistry, www.rsc.org

tryptophan, methionine, and cysteine, which are often limited in many staple grains. Amino acids with sulfur-containing functional groups, such as Cys and Met, are vital for numerous biological functions.

Millet-derived bioactive peptides (MBAPs) are peptides with a size of less than 10 kDa,[3] and proteolytic enzymes hydrolyze their precursor proteins to create bioactive compounds. The levels of these peptides are known to vary based on the millet variety.[4] MBAPs are known to regulate various functions *in vitro* and *in vivo*. Numerous scientific studies have highlighted the antimicrobial, anticancer, antihypertensive, antioxidant, and antidiabetic properties of MBAPs. Their bioactivities are based on their sequence alignment and contents of amino acids. Other than the size and structure, the absorption from the gastrointestinal (GI) tract and bioavailability at target tissue also have crucial roles in the ultimate bio-functionality of MBAPs.

3.2 Extraction of Millet-derived Bioactive Peptides (MBAPs)

Different methods for extracting MBAPs have been developed, each with its advantages and limitations. In addition to extraction, identifying and characterizing MBAPs is essential for understanding their specific biological roles and potential therapeutic benefits. Before extraction, some sources need disruptive pre-treatment techniques to release the protein. This involves sonication, mechanical grinding, osmotic shock, freeze–thaw, alkaline treatment, and so on. The choice of extraction method can significantly impact the yield, purity, and functionality of the peptides,[5] hence extraction is a critical step. Various approaches, ranging from conventional methods to enzymatic and novel technologies, are employed to release bioactive peptides from millet proteins. Each method comes with its advantages and limitations, influencing the total efficiency and applicability of the extraction process. MBAPs are then purified through ultrafiltration, dialysis, or continuous chromatography. Gel Filtration Chromatography (GFC) and Reverse-phase Liquid Chromatography (RPLC) are well-known methods of peptide purification, while the challenge of separating low molecular weight (M_w) peptides has been solved using immobilized metal affinity chromatography. Further, identification, amino acid sequence determination, and peptide structure are elucidated with tools like High-performance Liquid Chromatography (HPLC), Liquid Chromatography–Mass Spectrometry (LC–MS), Mass Spectrometry (MS), and Nuclear Magnetic Resonance (NMR).

3.2.1 Conventional Extraction Methods

Traditional extraction methods are widely used to solubilize millet proteins, which are then hydrolyzed into MBAPs. However, these methods often result in protein degradation, loss of bioactivity, and environmental concerns. Two key techniques include acid or alkali extraction and thermal extraction.

In acid or alkali extraction, proteins are dissolved in acidic or alkaline solutions to hydrolyze the peptide bonds, making multi-peptides more accessible for hydrolysis. This process is simple and cost-effective for large-scale production, but it carries risks such as protein denaturation, toxic by-products, and complex peptide recovery due to the use of harsh chemicals and extreme pH conditions.[6–8] Thermal extraction involves heating millet proteins to break them down into smaller peptide chains, but prolonged exposure to high temperatures can degrade MBAPs and reduce their functional properties.[9,10] While these conventional methods are effective, their limitations in preserving MBAP bioactivity promote the development of more innovative extraction techniques.

3.2.2 Enzymatic Extraction Methods

Enzymatic extraction is recognized as a refined and efficient method for isolating MBAPs from millet proteins, offering advantages over conventional methods. Utilizing proteolytic enzymes such as trypsin, pepsin, alcalase, bromelain proteinase K, thermolysin, neutrase, and papain, individually or in combination, enables the selective hydrolysis of millet proteins into smaller peptides.[11] These enzymes are chosen based on their ability to cleave specific peptide bonds within proteins, and the process occurs under controlled conditions of temperature, pH, and enzyme concentration to ensure the release of bioactive peptides.[12] This precision ensures that the functional and structural integrity of the peptides is preserved, making enzymatic extraction a preferred method, particularly for applications requiring bioactivity, such as in food, pharmaceutical, and nutraceutical industries.

Pretreated (ground and sieved) extracted protein fine powder can be subjected to enzymatic hydrolysis. During specific enzymatic cocktail-based hydrolysis, enzymes catalyze the breakdown of proteins into peptides. Trypsin specifically cleaves at lysine and arginine carboxyl sites, while pepsin acts on aromatic amino acids like phenylalanine, tyrosine, and tryptophan. Hence, the selection and optimization of the appropriate enzyme is crucial depending on the desired peptide profile.[13] The selection of the enzyme depends on the specific peptide bonds targeted for cleavage and conditions, such as temperature, pH, and enzyme concentration, to ensure efficient protein hydrolysis. For instance, alcalase is typically used under alkaline conditions, while pepsin functions optimally in acidic environments.[14] During hydrolysis, the enzyme binds to specific sites on the protein molecule, cleaving the peptide bonds and reducing the proteins into shorter peptide chains or individual amino acids. The controlled conditions ensure that the released peptides retain their structural integrity, preventing denaturation or loss of bioactivity, which is crucial for maintaining their functional properties.[11] Controlling pH and temperature is equally important, as these factors influence the enzyme's activity. Millet gliadin was enzymatically cleaved with an optimum hydrolysis time of 50 min.[15] Agrawal, Joshi, and Gupta (2016) prepared millet peptides through hydrolysis with pig trypsin.[16] Furthermore, the concentration of the enzyme must be carefully managed to avoid incomplete hydrolysis or over-hydrolysis, which can alter the bioactivity of the peptides.

As research advances, futuristic explorations are being developed to enhance the efficiency of enzymatic extraction. Immobilized enzymes, where enzymes are fixed to a solid support, offer advantages such as reusability and stability, reducing costs in industrial applications. Co-enzymatic approaches, where multiple enzymes are used in sequence or simultaneously, can also improve the yield and diversity of bioactive peptides. For example, using a combination of trypsin and papain may yield a wider range of bioactive peptides than using a single enzyme alone.[11] Additionally, the integration of enzymatic hydrolysis with other advanced extraction technologies, such as Microwave-assisted Extraction (MAE) or Ultrasound-assisted Extraction (UAE), is gaining attention. These combined approaches are discussed further in this section. Similar to enzymolysis, microbial fermentation has also been considered a relatively mild method of extraction of MBAPs.

After the enzymatic reaction is complete, the hydrolyzed peptides are separated from the enzyme and other reaction components. This is commonly done using techniques such as ultrafiltration, which filter out larger molecules while allowing smaller peptides to pass through. Chromatography methods, including ion exchange and GFC, are then employed to purify the peptides further, ensuring that the final product consists of bioactive peptides with the desired characteristics.[14] Using ultrafiltration, gel filtration, and RP-UFLC (Reverse-phase Ultra-fast Liquid Chromatography) chromatographic techniques, two new peptides, TSSSLNMAVRGGLTR and STTVGLGISMRSASVR, were isolated from finger millet (*Eleusine coracana*) trypsin hydrolysate for the first time.[12] Thus, enzymatic hydrolysis stands out as a highly effective method for extracting bioactive peptides from millet proteins due to its specificity, mild operating conditions, and the ability to preserve peptide functionality. As the field progresses, innovations in enzyme technology and process optimization will continue to improve the efficiency and scalability of this extraction method, making it increasingly viable for industrial-scale applications.

3.2.3 Novel Extraction Methods

Novel extraction methods have significantly advanced the process of obtaining bioactive peptides from millet. These techniques focus on improving efficiency, sustainability, and peptide yield while minimizing damage to bioactive components. The most promising methods include UAE, MAE, and Supercritical Fluid Extraction (SFE).

3.2.3.1 Ultrasound-assisted Extraction (UAE)

UAE is a modern technique that utilizes ultrasonic waves to disrupt the cellular structure of millet, facilitating the release of bioactive peptides.

Ultrasonic waves, typically in the range of 20–100 kHz, generate cavitation bubbles in the solvent. When these bubbles collapse, they create localized high temperatures and pressures. This phenomenon causes mechanical stress that breaks down the cell wall and enhances the release of cellular contents, including proteins and peptides.[17,18] UAE is particularly effective because it operates at low temperatures, which is crucial for preserving the bioactivity and structural integrity of peptides. The low energy consumption and reduced solvent use make it a sustainable and environmentally friendly alternative to conventional methods. Moreover, UAE significantly reduces extraction time, making it a more efficient process overall.[17] However, the effectiveness and precision of UAE depend on several critical factors, including the frequency of the ultrasonic waves, the solvent used, the power intensity, and the time required for the extraction. Each of these parameters requires careful optimization to maximize yield and maintain the quality of the extracted peptides. Despite its advantages, the high cost of ultrasonic equipment can limit its use in smaller-scale operations or in facilities with limited budgets.[18,19]

3.2.3.2 Microwave-assisted Extraction (MAE)

MAE is another innovative method that employs microwave energy to heat the solvent and millet mixture rapidly. Microwaves make the polar molecules in the mixture vibrate and release heat through ionic conduction and dipole polarization. This rapid heating disrupts the cell walls and promotes the release of peptides and other bioactive compounds from millet.[20] MAE offers significant advantages over traditional thermal extraction methods. It provides faster extraction times and higher efficiency while using less solvent, making it both cost-effective and environmentally friendly. MAE is particularly beneficial for extracting heat-sensitive peptides, as it operates under controlled temperatures, minimizing the risk of peptide degradation. However, scaling up MAE for industrial use can be challenging. Precise control of microwave intensity and uniform heating is essential to prevent overheating, which can degrade peptides and reduce their bioactivity. The use of specialized equipment and the need for careful optimization of variables such as extraction time, microwave power, and solvent type add to the complexity of the process.[21]

3.2.3.3 Supercritical Fluid Extraction (SFE)

SFE is a cutting-edge technique that uses supercritical CO_2 as a solvent to extract bioactive peptides from millet. In its supercritical state, CO_2 exhibits properties of both a gas and a liquid, allowing it to penetrate the cellular matrix of the millet effectively and dissolve proteins and peptides.[22] The use of CO_2 is particularly advantageous because it is non-toxic, non-flammable, and leaves no harmful residues, making SFE an environment-friendly and "green" technology. SFE operates at relatively low temperatures, which is critical for preserving the structural integrity and bioactivity of peptides. The technique also allows for highly selective extraction, producing purer peptide fractions compared with other methods. By adjusting parameters such as pressure and temperature, SFE can be tailored to extract specific peptides with minimal degradation.[23]

Despite its benefits, SFE is associated with high equipment and operational costs, making it less accessible for smaller-scale or low-budget operations. The process also requires specialized knowledge to optimize the various parameters involved, such as selecting the appropriate pressure, temperature, and CO_2 flow rate. These factors contribute to the overall complexity and cost of implementing SFE on a large scale.[22,24]

3.3 Characterization of Millet-derived Bioactive Peptides (MBAPs)

Once millet proteins are successfully extracted and hydrolyzed into smaller peptides, identifying and characterizing these peptides becomes essential. Identification techniques provide insights into the peptide sequences, M_w, structures, and biological functions, making it possible to pinpoint specific peptides with beneficial health effects. Modern technologies allow for highly accurate identification of bioactive peptides that can subsequently be utilized in food products, nutraceuticals, and therapeutic applications.

3.3.1 Mass Spectrometry (MS)

MS is a powerful analytical technique widely used for distinguishing and characterizing peptides. It aids in the determination of the M_w and structure of peptides by quantifying the mass-to-charge ratio (m/z) of ionized fragments.[25] MS is typically performed with other techniques, such as liquid chromatography (LC), to enhance its resolution and accuracy. Tandem mass spectrometry (MS/MS) is a more advanced version of MS, often used to sequence peptides. In MS/MS, the peptide ions generated in the first stage are further fragmented into smaller ions, and the masses of these fragments are measured in the second stage. This process generates detailed information about the peptide's amino acid sequence, enabling researchers to identify and characterize millet-derived peptides accurately.[26] MS and MS/MS offer high sensitivity and accuracy for determining peptide M_w and sequences, with MS/MS enabling precise identification of peptide structures in complex mixtures. Many scientists have resorted to directly measuring the M_w of MBAPs using LC–MS or the peptide property calculator (http://www.pepcalc.com). The most common methods for quantifying peptides are electrospray ionization (ESI) or matrix-assisted laser desorption ionization (MALDI) in conjunction with mass spectrometry (MS). However, these techniques require costly equipment and expertise, and varying ionization efficiency can lead to incomplete peptide identification.[26] MS, particularly in MS/MS with other techniques, remains the gold standard for peptide identification and provides comprehensive insights into the structure and bioactivity of millet-derived peptides.

3.3.2 Chromatography Techniques

Chromatography is often used in conjunction with MS to separate peptides before their identification.[26] It enables the separation of complex peptide mixtures based on their chemical properties, such as charge, hydrophobicity or size. Several chromatographic techniques are used in peptide identification, with LC being the most common. LC, especially HPLC, is a key method for separating and purifying peptides before mass spectrometric analysis. In LC, peptides are dissolved in a mobile phase (liquid solvent) and made to pass through a column packed with the stationary phase particles.[27] The peptides are separated based on their affinity for the stationary phase, with hydrophobic or hydrophilic characteristics playing a critical role in determining the retention time.

RP-HPLC is one of the most extensively used forms of HPLC in peptide analysis. In RP-HPLC, the stationary phase is hydrophobic (*e.g.*, silica particles coated with alkyl chains), and peptides are separated based on their affinity for the hydrophobic molecule. Hydrophilic peptides are eluted earlier, while hydrophobic peptides will interact more strongly with the stationary phase, and are thus eluted later and separated.[27] LC efficiently separates peptides from complex mixtures and is compatible with mass spectrometry for detailed identification. RP-HPLC offers high resolution to detect subtle peptide differences. However, sample preparation and method optimization can be time-consuming, and the solvent system must be carefully chosen to prevent peptide degradation.

Ion-exchange chromatography (IEC) separates peptides based on their charge. The peptides are passed through a column containing a charged resin, which attracts peptides with opposite charges. Anionic peptides bind to cationic resins, while cationic peptides bind to anionic resins. The bound peptides are eluted by gradually increasing the concentration of a counter-ion, allowing them to be collected and analysed.[28] IEC effectively separates peptides based on their charge, aiding in the identification of specific peptide classes and is compatible with techniques like mass spectrometry. However, peptides with similar charges may co-elute, complicating separation, and precise pH control is necessary for effective binding and elution.[28] Chromatographic techniques, when combined with MS, offer a powerful approach for identifying MBAPs, allowing for both separation and detailed structural analysis.

3.3.3 SDS Polyacrylamide Gel Electrophoresis (SDS-PAGE)

SDS-PAGE is a broadly used technique for characterizing bioactive peptides.[29] This approach isolates peptides and proteins based on their M_w, enabling the identification and estimation of the size of bioactive peptides obtained through enzymatic hydrolysis or extraction from natural sources such as millet.[30] The use of sodium dodecyl sulfate (SDS) detergent ensures that the peptides are denatured and have a homogeneous negative charge, allowing them to migrate simply by size through a polyacrylamide gel matrix when an electric field is introduced. As the peptides migrate through the gel, smaller molecules travel faster and further, providing a clear band pattern correlating to the M_w, which can be estimated with the help of standard M_w ladders.[31]

In studies focusing on bioactive peptides, SDS-PAGE is often employed to monitor the peptide profile post-extraction or post-enzymatic hydrolysis. The technique can either help determine the efficiency of enzymatic breakdown or be used for studying the types of fragments that are formed post-treatment. The resulting gel bands can be visualized through staining methods such as Coomassie brilliant blue or silver staining, depending on the sensitivity required, with silver staining being the most sensitive.[32] Further analysis, such as MS, can be used to complement SDS-PAGE by identifying the specific sequences of bioactive peptides within each band, providing insights into their structure and potential biological activities.[32,33]

3.3.4 Bioinformatics Tools for Peptide Identification

In addition to the physical identification methods, bioinformatics tools are increasingly being used to predict and identify MBAPs based on their amino acid sequences. These tools allow researchers to analyze peptide sequences for potential bioactivities, such as antioxidant, antihypertensive, or antimicrobial properties, without requiring extensive experimental testing. Constructed peptide sequences can be investigated against millet protein sequences that exist in the NCBI database (https://www.ncbi.nlm.nih.gov/) using the protein BLAST tool. Several peptide databases have been developed to store and analyze information on bioactive peptides. These databases contain information on known peptide sequences, their biological activities, and their structural characteristics. Researchers can use these databases to compare newly identified MBAPs to known bioactive peptides, facilitating rapid identification and functional prediction. The most commonly used peptide databases include BIOPEP and UNIPROT.[34] *In silico* tools use computational algorithms to predict the potential bioactivity of peptides based on their amino acid sequences. By analyzing the structure and properties of the peptide, these tools can predict whether it will have specific biological effects (*e.g.*, ACE inhibition, antioxidant activity).[34,35] Bioinformatics tools complement traditional identification methods by offering rapid, cost-effective ways to predict the bioactivity of millet-derived peptides, guiding further experimental work. Figure 3.1 shows an overview of the different extraction–characterization processes associated with MBAPs.

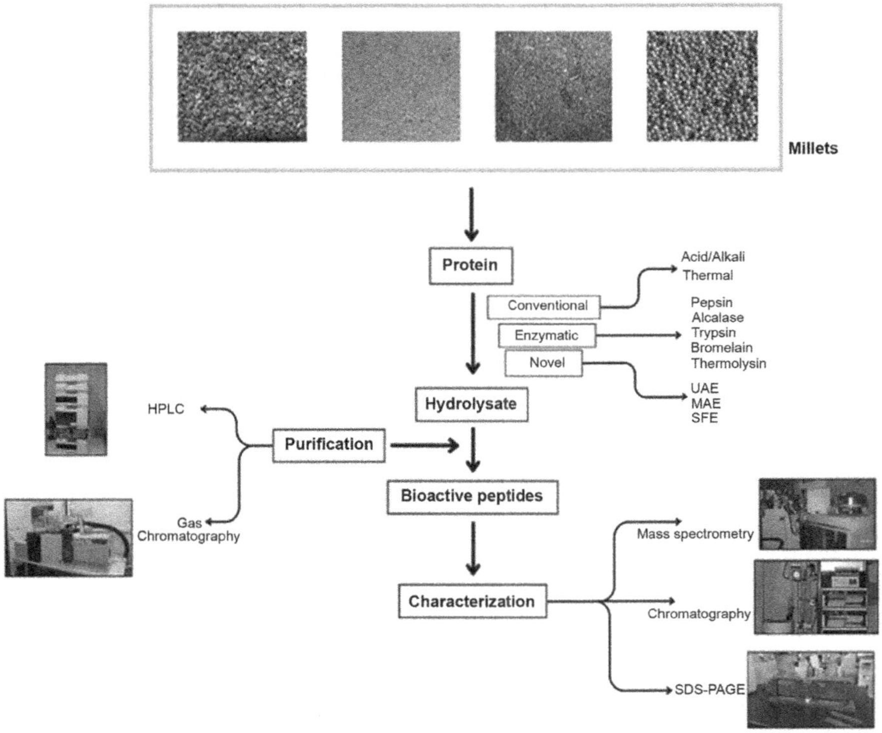

Figure 3.1 Schematic representation of the protein extraction process from various sources, followed by hydrolysis into peptides using conventional methods (acid/alkali extraction, thermal extraction), enzymatic techniques (pepsin, alcalase, trypsin, bromelain, thermolysin), and novel approaches (UAE—ultrasound-assisted extraction, MAE—microwave-assisted extraction, SFE—supercritical fluid extraction). The resulting hydrolysate is then purified using HPLC and gel filtration chromatography, followed by characterization of bioactive peptides through mass spectrometry, SDS-PAGE, and spectrometry.

3.4 Bioactivities and Health Benefits of MBAPs

Several peptides derived from millets have applications that solely aim to improve human health, such as bioactive, vital foods, and/or functional foods.[36,37] MBAPs as natural medicines or nutraceuticals have been announced to tackle metabolic syndromes by promoting antioxidant, antimicrobial, anti-inflammatory, antihypertensive, and antidiabetic activity. Figure 3.2 provides an overview of the different activities of MBAPs. Table 3.1 presents the bioactivities of MBAPs and their health benefits.

3.4.1 Antioxidant Activity

Excessive generation of free radicals and the resulting oxidative stress have been associated with the pathogenesis of various chronic ailments,

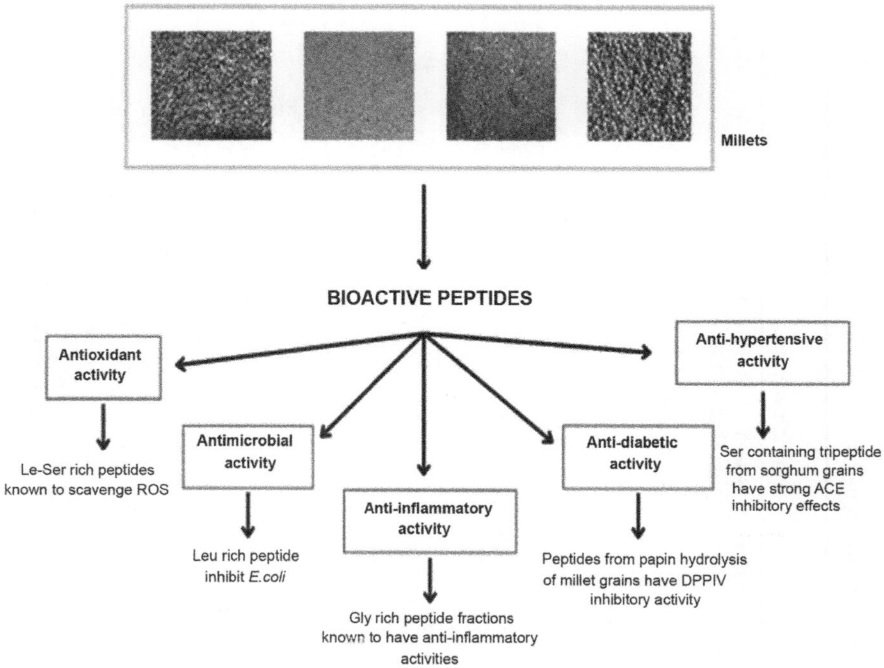

Figure 3.2 The diverse potential of bioactive peptides extracted from millets. These bioactive peptides exhibit antioxidant, antimicrobial, anti-inflammatory, antidiabetic, and antihypertensive properties. These activities hold high value for the development of novel functional foods and nutraceuticals with health-promoting benefits (ACE: Acetyl Choline Esterase).

including diabetes, hypertension, inflammation, and neurodegenerative diseases.[38,39] In addition to the endogenous system, dietary antioxidants are crucial in enhancing the redox status of the body and minimizing the risk of associated diseases. Several studies have focused on dietary or natural food-based antioxidants as they are safe compared to their synthetic counterparts. Furthermore, food components with antioxidant activity would delay oxidation of the food material, reducing the incidence of rancidity. The presence of certain amino acids, *viz.* alanine, valine, leucine, isoleucine, proline, cysteine, threonine, methionine, phenylalanine, tyrosine, tryptophan, histidine, and lysine, has been implicated in the radical scavenging activities of peptides.[45] The scavenging of reactive oxygen species (ROS) could be brought about by their ability to transfer a hydrogen atom and/or single electron.[40,41] Furthermore, the role of antioxidant peptides in regulating the endogenous antioxidant pathways, including the MAPK (mitogen-activated protein kinase), NF-κB (nuclear factor-kappa B), and so on, has also been reported.[42] The amino acid composition, N- and C-terminal residues, and hydrophobicity are among the factors that influence the antioxidant activity of the peptides.[16,22]

Table 3.1 Bioactivities of millet-derived bioactive peptides (MBAPs) and their health benefits.[a]

Millet source	Protein/enzyme hydrolysis/treatments	Activity endorsing health benefits	*In silico*/*in vitro* assays performed	Peptide sequence	Ref.
Foxtail millet	*Lactobacillus paracasei* followed by trypsin hydrolysis	Antioxidant, anti-aging	DPPH radical scavenging activity	S–G–Y–Y–M–H; L–G–T–F–Q–N; L–H–A–L–L–L	43
Pearl millet	Trypsin	Antioxidant	DPPH radical scavenging activity, ABTS, hydroxyl and metal chelation ability	S–D–R–D–L–L–G–P–N–N–Q–Y–L–P–K	16
Sorghum	Alcalase	Antioxidant	DPPH radical scavenging activity	V–A–I–T–L–T–M–K; V–S–K–S–V–L–V–K	45
Finger millet	Trypsin	Antioxidant	DPPH radical scavenging activity, ABTS, metal-chelating activity	T–S–S–S–L–N–M–A–V–R–G–G–L–T–R; S–T–T–V–G–L–G–I–S–M–R–S–A–S–V–R	12
Foxtail millet	Alcalase	Antioxidant	DPPH and ORAC radical scavenging activity	P–F–L–F; I–A–L–L–I–P–F	46
Sorghum	Nutrease	Antioxidant	DPPH radical scavenging activity, metal ion chelating activity, ORAC, ABTS.	Q–A–M–C–G–V–V; A–M–C–G–V–V–Q; S–A–S–A–L–Q–M; P–A–A–Q–A–L–T–P–L; L–P–A–A–Q–A–L–T–P; L–P–S–Y–C–T–T–P; S–A–A–I–P–P–Y–Y–C–G–L–Y–Q–L–P–S; Y–A–L–R–E–Q–T; V–A–Q–N–M–P–V–Q–S–V–V–Q; Q–P–Q–C–S–P; M–R–M–M–D–M–Q–S; G–G–G–L–Y–P–C–A–E–Y; V–A–Q–V–A–Q–N–M–P–A; A–V–A–Q–V–A–Q–N–M–P; T–P–C–A–T–S–A–A–I–P–P; F–L–Y–P–C–A–E–Y–L; V–Q–S–V–V–Q–Q–L–Q; M–D–M–Q–S–R–C–Q–A–M; M–M–D–M–Q–S–R–C–Q–A; T–P–L–A–M–A–V–A–Q–V–A–Q; T–P–L–A–M–A–V–A–Q–V–A–Q	48

Foxtail millet	Pepsin, pancreatin	Antioxidant	DPPH radical scavenging activity, reducing power, and ORAC, ABTS.	NA	49
Millet gliadin	High hydrostatic pressure, pepsin	Antioxidant	DPPH radical scavenging activity, superoxide anion scavenging	S–L–S–H–L–T–V–Q; S–L–A–H–V–T–V–Q; L–H–A–L–T–L–Q; S–G–I–L–A–P–S–P–V–L; S–H–L–T–V–Q; V–S–A–A–A–A; S–P–A–A–F	15
Sorghum	Not specified	Antiviral	Antiviral assay, tested for prophylactic effects	Molecular weight (2000 Da)	50
Foxtail millet	*Lactobacillus paracasei*	Antifungal	Well diffusion assay, anti-fungal assay	L–G–T–F–Q–N; S–G–Y–Y–M–H; L–H–A–L–L–L	43 and 48
Finger millet and proso millet	Not specified	Antibacterial	Assays for bacterial inhibition, zone of inhibition	Peptide fraction with a molecular weight of 35 k Da	53
Proso millet	Pepsin, pancreatin	Antibacterial	COX-1 and COX-2 inhibition assays, LOX inhibition assay	Globulin 7S fraction, prolamin fraction, high glycine composition	10
Foxtail millet	Alcalase	Anti-inflammatory	Nitric oxide inhibition, pro-inflammatory cytokine inhibition	Q–N–W–D–F–C–E–A–W–E–P–C–F; E–D–D–Q–M–D–P–M–A–K; P–F–L–F; I–A–L–L–I–P–F	49
Sweet sorghum	Alcalase	Antihypertensive	ACE inhibition assay & molecular docking	T–L–S	58
Foxtail millet	Digestive protease	Antihypertensive	ACE activity measurement in serum	NA	56
Proso millet	Alcalase or digestive proteases	Antihypertensive	*In vitro* ACE inhibition assay	Prolamin fraction, heat-treated prolamin, globulin 11S	55
Millet protein	Papain	Antidiabetic	DPP-IV inhibition assay, molecular docking assay	N–D–W–H–T–G–P–L–S and T–Y–P–H–Q–Q–P–P–I–L–T	61
Proso millet	Protease	Antidiabetic	α-Glucosidase and α-amylase inhibition assays	Prolamin fraction, heat-treated prolamin (100 °C), globulin 11S fraction (65 °C)	55

[a] NA: not available; DPPH: 2,2-diphenyl-1-picrylhydrazyl; ABTS: 2,2′-azino-bis(3-ethylbenzothiazoline-6-sulfonic acid); ORAC: oxygen radical absorbance capacity; ACE: angiotensin converting enzyme; COX: cyclooxygenase-2 (Cox-2); LOX: lipooxygenase.

Numerous studies have reported MBAPs with antioxidant activities in terms of *in vitro* radical scavenging ability and cell culture-based assays. Three peptides, LHALLL, SGYYMH and LGTFQN derived from foxtail millet meal fermented with *Lactobacillus paracasei* have been found to exhibit DPPH (2,2-diphenyl-1-picrylhydrazyl) and superoxide radical scavenging properties (Amadou *et al.* 2013).[43] Leu–His–Ala–Leu–Leu–Leu (678.87 Da), a Leu-rich peptide, showed the highest activity, which was explained by both its increased hydrophobicity and amino acid sequence. Agrawal *et al.*[16] also isolated the bioactive peptide (14 aa) SDRLLGPNNQYLPK from pearl millet protein hydrolysate. This peptide exhibited excellent antioxidant activity, as demonstrated by its ability to scavenge DPPH, ABTS (2,2-azino-bis-(3-ethylbenzothiazoline-6-sulfonic acid)), hydroxyl, and its metal chelation ability *in vitro*. Further, Agrawal *et al.*[45] identified and characterized seven peptides from green tender sorghum protein hydrolysate obtained by alcalase treatment for potential antioxidant capability. The peptides exhibited higher antioxidant activity than the unhydrolyzed protein. Two of those peptides, VAITLTMK and VSKSVLVK, exhibited 74.19% and 77.64% DPPH radical scavenging, respectively, demonstrating excellent antioxidant capacity. The same authors reported two antioxidant peptides, STTVGLGISMRSASVR and TSSSLNMAVRGGLTR, from finger millet.[12] In this study, the authors characterized the antioxidant activity of synthetic peptides through molecular docking studies. It was found that the serine and threonine residues of the peptide interact with free radicals, resulting in the antioxidant potential of the peptide.[12] Ji *et al.*[46] investigated the effect of pre-treatments, *viz.* ultrasonic treatment and heat, on the conformational changes of foxtail millet protein, which predominantly consists of prolamin. The treatments could bring about alterations in the secondary structure of millet protein and also enhanced the degree of hydrolysis. The sequences of the two most bioactive peptides were identified as PFLF (522.3 Da) and IALLIPF (785.5 Da). Furthermore, the role of foxtail millet prolamin-derived peptides on the inhibition of ROS damage and malondialdehyde (MDA) formation in H_2O_2-induced human keratinocyte cell culture (HaCaT) have been demonstrated by Ji *et al.*[47] Apart from the reduction in the ROS and MDA generation, three foxtail millet prolamin-derived peptide fractions (two of which are synthetic versions of the peptides from the previous study, PFLF and IALLIPF, and an alcalase hydrolysis peptide fraction of M_w below 1 kDa (MPH-A-I)) also enhanced the levels of reduced glutathione (GSH), demonstrating the antioxidant activity in the cells.[48] The peptide IALLIPF demonstrated the highest reduction in ROS and enhancement of GSH, while PFLF exhibited the highest reduction in MDA production. This study also confirmed the amino acid composition of the peptide and its hydrophobicity are important determinants of antioxidant activity. Also, the presence of Pro in the above peptides is another important determinant as its pyrimidine ring could impose conformational constraints that might enhance the availability of those amino acid residues that serve as antioxidants.[47] Other antioxidant peptides have been derived from millets, including sorghum kafirin hydrolysate.[47]

3.4.2 Antimicrobial Activity

Natural compounds with antimicrobial activity are widely preferred as functional food ingredients. MBAPs from various sources reportedly display antimicrobial capabilities. Peptides derived from millets with activities against microbes, including bacteria, viruses and fungi, have been reported. A dose-dependent anti-viral activity of a peptide fraction (M_w 2000 Da) from sorghum has been reported.[49] The peptide fraction could inhibit the replication of herpes simplex virus (EC_{50} 6.25 μM). Further, the peptide could moderately inhibit virus-induced cytopathic effects when present before the infections by the herpes simplex and bovine herpes virus, showing its prophylactic effects. An antifungal peptide (26.9 kDa) from foxtail millet with activity against *Alternaria alternate* (IC_{50} of 1.3 $\mu mol\, L^{-1}$) has been reported. The peptide also exhibited antifungal activities against other fungi, including *Botrytis cinerea, Trichoderma viride*, and *Fusarium oxysporum*.[50] Microscopic studies revealed remarkable changes in *A. alternate* hyphae in terms of retraction of cytoplasm and loss of cell wall rigidity upon treatment with the antifungal peptide. It was speculated that the antifungal activity of the peptide might be due to inference in the fungal cell wall synthesis hampering its growth and morphogenesis. It is reported that three antimicrobial peptides, *viz.* FFMp6 (Leu–Gly–Thr–Phe–Gln–Asn), FFMp4 (Ser–Gly–Tyr–Tyr–Met–His), and FFMp10 (Leu–His–Ala–Leu–Leu–Leu), derived from foxtail millet meal fermented with *L. paracasei*, show fair inhibitory activity against *E. coli*, demonstrated through the well diffusion assay.[51] While all three peptides displayed lower inhibition as compared to gentamicin, the peptide Leu–His–Ala–Leu–Leu–Leu displayed a comparatively larger zone of inhibition (13.4 ± 0.8 mm) against *E. coli* at 60 $\mu g\, mL^{-1}$, which was attributed to its higher leucine content. Crude protein extracts of finger millet (PRM-1 variety) and proso millet displayed anti-microbial activity against bacterial pathogens, including *Pseudomonas aeruginosa, Bacillus subtilis* and *E. coli*, as demonstrated by *in vitro* assays.[52] Further, the study also demonstrated a protein/peptide fraction (35 kDa) from finger millet (PRM-1) displayed antibacterial activity against *P. aeruginosa*, as shown by a zone of inhibition of 19 mm. These elucidations demonstrate the antimicrobial role of millet-derived peptides.

3.4.3 Anti-inflammatory Activity

Inflammation is an immune response to protect a host in the fight against an allergen or any other foreign invasions (pathogens or toxins or other substances).[53] However, persistent or chronic inflammation is recognized as a contributing factor associated with various conditions, including atherosclerosis, rheumatoid arthritis, obesity, and certain types of malignancies, *etc.*[46] Anti-inflammatory activity in terms of COX-1 and COX-2 inhibition have been reported from *in vitro* gastrointestinal digested globulin 7S fractions of proso millet grains heated at 65 °C.[10] The peptide fractions were found to exhibit IC_{50} values of 0.08 mg mL^{-1} and 0.12 mg mL^{-1} for COX-1 and COX-2 inhibition, respectively. The same study also reported LOX inhibitory activity ($IC_{50} = 0.14$ mg mL^{-1}) in the peptide derived from a prolamin fraction of proso millet heated at 100 °C. They reported that the peptide fractions exhibiting anti-inflammatory potential were composed of good amounts of Gly. Further, the anti-inflammatory role of bioactive peptides generated from foxtail millet subjected to *in vitro* digestion of germinated and boiled grains has been reported.[48] Two peptides with the sequences QNWDFCEAWEPCF and EDDQMDPMAK could inhibit the inflammatory markers, including nitric oxide, interleukin 6 (IL-6) and tumor necrosis factor-α (TNF- α), as demonstrated by studies in the RAW 264.7 cells.[48] Similarly, MBAPs derived from the alcalase hydrolysate prolamin fraction of foxtail millet displayed anti-inflammatory potential, as reported in a cell-based study using RAW 264.7 murine macrophages.[47] Peptide fractions, including PFLF, IALLIPF and an alcalase-treated peptide fraction of M_w below 1 kDa (MPH-A-I), could inhibit the formation of nitric oxide and the pro-inflammatory cytokines, *viz.* TNF-α, interleukin-6 and IL-1β, in a dose-dependent manner in the lipopolysaccharide stimulated RAW264.7 cells. The peptide IALLIPF at 250 μg mL^{-1} could inhibit the pro-inflammatory cytokines effectively. The anti-inflammatory activity is also determined by the amino acid composition of the peptides. The amino acids, including proline and leucine, impart anti-inflammatory capabilities.

3.4.4 Antihypertensive Activity

High blood pressure, and the consequent development of hypertension, is regarded as one of the major contributors to cardiovascular diseases. Hence, early detection and treatment of hypertension is crucial.[54] Therapeutic strategies for hypertension primarily target the angiotensin I converting enzyme (ACE), an essential enzyme of the renin–angiotensin–aldosterone system (RAAS) involved in the catalytic conversion of angiotensin I to angiotensin II.[57] The RAAS and the kallikrein–kinin system (KKS) play important roles in regulating sodium homeostasis and blood pressure.[5,55] As natural, food-based ACE inhibitors may be safe alternatives to synthetic drugs, the role of bioactive peptides as ACE inhibitors and thereby as antihypertensive agents have been explored from various sources. Sweet sorghum grain protein hydrolysis by alcalase produced a tripeptide, Thr–Leu–Ser, with high ACE inhibitory ability.[56] The ACE inhibition was attributed to the presence of the serine residue in the peptide, which interacted with the enzyme at the S1 and S2 pockets, as demonstrated through molecular docking studies. Zielińska *et al.* reported the antihypertensive potential of foxtail millet protein hydrolysates in spontaneously hypertensive rats.[54] They found that administration of the raw, extruded, and fermented protein hydrolysates generated by subjecting the corresponding foxtail millet protein to digestive protease could lower the serum ACE activity and angiotensin II levels in hypertensive rats. The hydrolysates from raw and extruded foxtail millet showed a more significant impact compared to those from fermented millet. Moreover, a decrease in the percentage of heart weight was noted in the hypertensive rats treated with the hydrolysates, indicating the antihypertensive effect of the peptides.[54] The antihypertensive potential of protein hydrolysates and peptide fractions derived from heat-treated proso millet grains in terms of ACE inhibition has been studied by Karaś *et al.*[55] The study illustrated that the peptides from the prolamin fraction, prolamin fraction treated at 100 °C and globulin 11S treated at 65 °C displayed *in vitro* ACE inhibitory effects with IC_{50} values of 0.42 mg mL^{-1}, 0.33 mg mL^{-1} and 0.38 mg mL^{-1}, respectively.

3.4.5 Antidiabetic Activity

Type 2 diabetes mellitus (T2DM) is one of the most widespread metabolic disorders across the globe. It is predicted that 693 million people will be affected by T2DM by 2045.[59] Diabetic drugs include dipeptidyl peptidase-IV (DPP-IV) inhibitors. DPP-IV is an endogenous proteolytic enzyme that degrades incretin hormones such as GLP-1 and GIP, which promote insulin secretion, pancreatic β-cell proliferation, and postprandial plasma glucose levels. Thus, protecting the incretins by inhibiting DPP-IV could ameliorate hyperglycemia associated with T2DM.[60,61] Also, inhibition of carbohydrate digestive enzymes in the gastrointestinal tract, such as α-amylase and α-glucosidase, could be exploited in the management of postprandial hyperglycemia.[62] Consequently, lots of research efforts have been made to identify natural compounds from various sources that can inhibit these enzymes. However, limited studies are available in this direction based on millet-derived peptides at present. Gu *et al.* (2021) reported two DPP-IV inhibitory peptides, NDWHTGPLS and TYPHQQPPILT, resulting from papain hydrolysate of millet protein.[61] These peptides could interact with S1 and S2 subsites of the active center of DPP-IV through H-bonds and π–π interactions, thus inhibiting the enzyme, as revealed by molecular docking studies. Further, peptide fractions from heat-treated proso millet grains, including peptides from the prolamin fraction, prolamin fraction treated at 100 °C and 65 °C treated globulin 11S, could display *in vitro* inhibition of α-glucosidase and α-amylase enzymes.[55]

3.4.6 Bioavailability of MBAPs Using *In Vitro* and *In Vivo* Models

Many peptides possessing antioxidant, antimicrobial, antihypertensive, and anti-diabetic properties have been identified from food materials through *in vitro* studies and are called MBAPs.[63] Despite their bioactivities in *in vitro* models, the evidence for intestinal absorption and bioavailability of peptides other than dipeptides and tripeptides at target tissues is lacking. PepT1 H+/peptide co-transporter is a well-characterized di- and tripeptide transport system in intestinal epithelium, providing direct evidence for the absorption of these peptides across intestinal epithelium.[64] Recently, a 35 kDa protein was isolated and purified from foxtail millet bran using cell-based screening and was found to have anti-cancer properties. This protein, foxtail millet bran protein (FMBP), was homologous to peroxidase as revealed by MS and gene sequencing analysis. The FMBP could suppress xenografted tumor growth in nude mice and colon cancer cell growth *via* induction of apoptosis.[65] However, the delivery of the protein was through a systemic route to achieve bioavailability at the target tissue. Identification of the most effective small-size fragment of FMBP might pave the way for oral formulation using emerging techniques of peptide delivery.

3.5 Techno-functional Properties of MBAPs

The techno-functional properties of MBAPs refer to specific physical and chemical characteristics that influence their behavior in food systems, significantly impacting the quality and functionality of the end products. MBAPs, derived from protein hydrolysis, have unique functionalities, such as emulsification, foaming, water-holding, and gelation capacities. These properties are crucial in developing various food products like baked goods, beverages, and functional foods, as they can improve texture, stability, and nutrient richness.[66] The bioactive nature of MBAPs adds to their appeal, offering health benefits such as antioxidant, antihypertensive, and antimicrobial effects. Understanding the techno-functional attributes of MBAPs is essential for food scientists and product developers who aim to use these compounds in innovative and health-promoting food formulations, especially given the growing demand for foods that are plant-based, gluten-free, and nutritionally enhanced.

In contrast to animal proteins, plant-based proteins are less commonly used in food products due to specific limits in their functional features, including water-holding capacity (WHC), gelation, solubility, emulsifying, and foaming properties. These techno-functional attributes are crucial for creating plant-based food products. Additionally, plant proteins contain antinutritional factors (ANFs) and are typically less digestible than animal proteins. Plant protein isolates, concentrates, and hydrolysates can benefit from innovative processing methods, such as microfluidization, high-pressure processing, pulsed electric fields, electrospraying, irradiation, ultrasound, and cold plasma, to improve their quality, functionality, and digestibility. These techniques improve the use of plant-derived proteins as emulsifiers, encapsulating agents, texturizers, foam stabilizers, and film formers.[67] Different techniques are used extensively to enhance the techno-functional properties of peptides.

3.5.1 Ultrasound-mediated Modifications and Their Effect on Techno-functional Properties

Ultrasound is a highly effective method for modifying proteins using sound waves with frequencies above 20 kHz. It is known for being simple, cost-effective, energy-saving, and environmentally friendly.[68] Ultrasound can be categorized into high-energy (20–100 kHz, 10–1000 W cm^{-2}) and low-energy (5–10 MHz, 1 W cm^{-2}) types based on frequency and intensity. High-energy ultrasound induces cavitation, creating extreme pressure (up to 1000 atm) and temperature (up to 5000 K), which significantly alters the mechanical, physical, and biochemical properties of proteins, making them suitable for food processing. On the other hand, low-energy ultrasound has non-destructive effects and is used to ensure food safety and quality. High-power ultrasound can be applied through sonication baths or ultrasonic probes, offering versatility in a range of food industry applications.[69,70] In the research conducted by Sharma *et al.* (2023), foxtail millet protein fractions, specifically concentrate, prolamin, and glutelin, were subjected to high-intensity ultrasound (HIU) using a 1 cm probe in pulse mode at amplitudes of 5% and 10% for durations of 5, 10, and 20 min at room temperature.[71] The impact of HIU was analyzed through Fourier-transform infrared (FTIR) spectroscopy, which revealed shifts in peaks related to hydrogen bonding and C–H vibrations, suggesting alterations in protein hydration and partial unfolding. The results indicated structural changes, such as an increase in β-sheet content and a decrease in α-helix content, alongside variations in folding and aggregation patterns. These modifications resulted in noteworthy improvements in functional properties, including reduced particle size, increased solubility, especially at pH 9, and enhanced surface hydrophobicity, particularly in concentrates, ultimately leading to better emulsifying properties and FC. An ultrasonic homogenizer with a titanium probe was used to treat foxtail millet bran protein hydrolysates (FMBPHs) in a different investigation.[72] The samples were sonicated for 30 minutes at a frequency of 20 kHz and at different power levels (60 W to 600 W). According to the FTIR results, ultrasound altered the secondary structure of FMBPHs without forming new chemical bonds. Shear forces and turbulence caused shifts in distinctive peaks, which indicated structural alterations. As the ultrasound power increased, the content of α-helix and β-sheet structures decreased while random coil and β-turn structures increased, indicating a transition from rigid to flexible conformations. Notably, the treatment also raised the fluorescence intensity and surface hydrophobicity, peaking at 450 W, while free sulfhydryl content increased before declining at higher power levels. Scanning electron microscopy (SEM) demonstrated that ultrasound transformed compact aggregates into thinner, more porous structures, enhancing functional properties, including improved dispersion and stability. Finally, a study conducted by Nazari *et al.* investigated the effects of high-power ultrasound on proso millet protein concentrate,[73] using intensities of 18.4, 29.58, and 73.95 W cm^{-2} for various durations. This study found that ultrasound significantly enhanced solubility, particularly at 73.95 W cm^{-2} for 12.5 minutes, although longer treatments at high intensity led to decreased solubility due to aggregate formation. Additionally, increased FC and improved emulsifying properties were observed, attributed to structural modifications that enhanced protein adsorption at oil–water interfaces.

Overall, these studies collectively highlight the effectiveness of ultrasound treatment in improving the techno-functional properties of millet proteins.

3.5.2 Microwave Mediated Modifications and Their Effect on Techno-functional Properties

Microwave treatment (MWT) is a non-ionizing radiation technique that generates rapid, volumetric heat through intermolecular dipole molecules. This results in ionic conduction or dipolar rotation, causing rapid heating. MWT is effective for inactivating enzymes and modifying functional properties, such as reducing peak viscosity, making it useful in enhancing the storage stability of food products.[74] The analysis led by Rao *et al.* (2023) illustrated the effects of MWT and Hot Air Radiofrequency (HARF) treatments in foxtail millet flour and its protein isolate.[75] The study found that higher power levels and longer heating times led to significant reductions in peak viscosity up to 75.50% due to starch degradation and amylose-lipid complexation. From this treatment, it was concluded that MWT significantly enhanced storage stability by reducing free fatty acid (FFA) levels by 59.7%. However, microwave treated flour showed a decrease in protein purity due to denaturation, while structural analyses (FTIR and XRD) indicated reduced crystallinity and increased β-turns in proteins. Techno-functional properties, including improved emulsion stability (ES) of 31.00% and emulsion activity index (EAI) of 42.67%, were enhanced, although foaming stability (FS) and foaming capacity (FC) declined due to protein aggregation. Oil absorption capacity (OAC) and WAC were significantly increased by 5.50% and 3.90%, respectively, attributed to greater protein side chain exposure.

3.5.3 Hot Air Radio Frequency (HARF) Mediated Modifications and Their Effect on Techno-functional Properties

Radiofrequency (RF), including hot air radiofrequency (HARF) heating, uses electromagnetic waves to create heat by polarizing molecules. It provides uniform heating and is particularly effective for enzyme inactivation. RF techniques also influence the pasting properties of food products by altering their molecular structure, and improving functionality and shelf life.[76] In research by Rao *et al.* (2023), the effects of HARF treatment on foxtail millet (FM) flour were examined.[75] HARF treatment was applied using a pilot-scale RF system at 40.68 MHz with a hot-air supply. The grains, conditioned at various moisture levels (12–24%), underwent one or two passes through the system. Significant variations in surface mean temperatures and heating uniformity index (HUI) were observed, with optimal values found at moderate moisture levels of 15% and 18%. HARF-treated samples exhibited significant reductions in peak viscosity, with the lowest peak viscosity observed in samples treated at lower moisture levels. These reductions were linked to the depolymerization of starch and the retention of simple sugars.[77] HARF treatment also improved the storage stability of FM flour, reducing FFA levels by 46.27%. While this was less effective than MWT, HARF-treated samples exhibited higher protein purity (75.50 ± 2.50%) compared to microwave treated samples, though still lower than untreated flour due to protein denaturation. Structural changes observed through FTIR and XRD analyses showed reductions in secondary structure content, with formation of β-turns and decreased crystallinity, though the effects were less pronounced than in microwave treated samples. Techno-functional properties were notably improved, with enhanced ES of 22.67% and EAI of 46.33%, while FC and FS declined due to protein aggregation. HARF-treated samples also exhibited better WAC of 5.23% and OAC of 4.18% compared to microwave treated samples, likely due to prolonged heating exposure that enhanced protein interaction and hydration.

3.5.4 High Hydrostatic Pressure Mediated Modifications and Their Effect on Techno-functional Properties

High hydrostatic pressure (HHP) is emerging as a promising non-thermal technology in the food production industry, particularly for enhancing the functional properties of proteins during processing, transportation, marketing, and storage. HHP provides several benefits, such as ease of use, quick processing durations, and the retention of nutritional quality.[78] HHP primarily affects hydrogen bonds, noncovalent bonds and ionic bonds in the three-dimensional structure of the macromolecules, while covalent bonds remain largely unaffected.[79] This selective influence can lead to the alteration of protein conformation, impacting their functional characteristics like solubility, gelation, and emulsification. In an experiment conducted by Zhang *et al.*, millet gliadin was extracted by mixing defatted millet flour with distilled water, followed by ethanol precipitation and centrifugation.[15] The extracted gliadin was then treated with high hydrostatic pressure (HHP) at 100, 300, and 500 MPa for 10 minutes at 25 °C. The study examined how HHP affects millet gliadin's properties. It was found that solubility peaked at 70 $mL\,dL^{-1}$ ethanol but decreased at higher concentrations due to the nonpolar amino acid composition. HHP treatment led to increased particle size and aggregation, attributed to structural changes from α-helices to β-sheets and reduced zeta potential. Thermal stability improved, with the denaturation temperature rising from 73 °C to 96 °C and enthalpy change increasing, indicating a more ordered protein structure. SEM confirmed that HHP treatment resulted in larger, aggregated particles, highlighting its impact on millet gliadin's structural and functional characteristics.

3.5.5 Cold Plasma-induced Changes and Their Effect on Techno-functional Properties

Cold plasma (CP) treatment is a promising alternative technique that is utilized in the food industry.[76] CP is a non-thermal technology in which gases are ionized at ambient temperature to produce plasma containing reactive species such as ROS and reactive nitrogen species (RNS).[77] These molecules interfere with biomolecules and induce physical and chemical changes. CP has been widely used for microbial disinfection, detoxification, germination enhancement, and, more recently, for enhancing the techno-functional properties of proteins in millets. These alterations render CP-treated sprouted foxtail millet protein concentrate (SFMPC) acceptable for application in food industries, including stabilizers, emulsifiers, and fortifiers in plant-derived products.[80] SFMPC was treated with dielectric barrier discharge (DBD) plasma at different voltages in different exposure times and the proteins were examined for structural and functional changes using FTIR, SEM, and Intrinsic Fluorescence (ITF) Spectroscopy. The changes in secondary structure, such as loss of α-helices, β-turns and a formation of random coils and β-sheets, indicated unfolding of protein from exposure to CP treatment. A longer duration of CP exposure caused more unfolding. Thus, emulsifying properties are impacted by the exposure of hydrophobic regions, which would disrupt the oil–water adsorption interface, to CP treatment. DLS is used to elucidate the particle size and zeta (ζ) potential properties of untreated and treated SFMPCs. Regardless of voltage and exposure duration, CP treatment led to a consistent decrease in the particle size.[77] The free carbonyl group (C=O) reveals that proteins have undergone oxidation at this site, and these proteins are exposed to free radicals and reactive species like oxides of nitrogen. Proteins in SFMPC also undergo oxidation after exposure to CP. This alters protein solubility, which shows a noticeable linear increase at different voltages. CP treatment also enhances properties like WAC and OAC. Likely the EAI and stability of SFMPC after CP treatment, particularly at higher voltages and longer exposure time, is enhanced.[80] By inducing disulfide bond formation or dissociation of protein molecules, CP may promote protein aggregation, according to protein profiling of SFMPC using SDS-PAGE. Despite the fact that CP significantly enhances the characteristics and capabilities of millet proteins, it faces difficulties in adhering to national laws for human consumption.[80]

3.5.6 Influence of γ-irradiation on Techno-functional Properties

The enhanced nutritional benefits of millets have garnered significant interest in the market.[81] Poor storage conditions of millets lead to post-harvest losses, particularly during rainy seasons when sprouting and fungal development cause spoiling.[82] Irradiation provides a solution to the issues associated with the breaking of primary and secondary chemical interactions that sustain the native structure of biopolymers, such as starch, proteins, and enzymes, caused by free radical generation. Meanwhile, it alters the techno-functional properties such as WAC and OAC exhibiting significant reduction across millet varieties.[82] However, the FS of the air–liquid interface diminished as radiation increased, attributed to the degradation of long protein chains. Gowthamraj, Jubeena, and Sangeetha (2021) attribute the increase in FC to the enhanced protein solubility resulting from the irradiation treatment of millet flours.[83] The light transmittance of millet flours also showed variations over time.[82] The light transmittance decreased significantly over time, regardless of the treatments applied and was significantly high in irradiated samples of millet flours. Therefore, infant formulas may benefit from the greater light transmittance values in the irradiation samples for quick food product dispersibility.[82] Following irradiation treatment, IR spectroscopic investigations show a significant switch in functional groups of millet biopolymers.[82] This indicates that gamma irradiation affects the biopolymers in the flours, potentially altering their functionality in food systems. SEM images depict a clear alteration in grain morphology due to irradiation, with grains becoming more compact and showing more distinct surface granules. This structural compactness can contribute to changes in water absorption and other functional properties.[82] Adding to this microbial analysis after irradiation shows a reduction in the microbial counts, thus preventing the post-harvest losses that result from inadequate storage infrastructure.[82]

3.6 Challenges in the Stability, Bioavailability, Bitterness, Delivery, and Commercialization of MBAPs

The stability of bioactive peptides from millet during processing, storage, and in the gastrointestinal environment is a critical concern. Peptides may degrade or lose their bioactivity due to environmental factors like temperature, pH, or enzymatic activity. Ensuring the stability of these peptides from production to consumption is essential for maintaining their health benefits. High hygroscopy of MBAPs also reduces their stability as it accelerates microbial and chemical degradation.[84] These challenges span across their stability, bioavailability, sensory characteristics (such as bitterness), effective delivery systems, and eventual commercialization. These difficulties must be overcome before MBAPs can completely achieve their market potential. The use of encapsulation techniques, such as microencapsulation or nanoencapsulation, can be used to protect peptides during processing and storage. Optimizing processing methods, like freeze-drying or spray-drying, could also be a logical step to retain peptide integrity. The low bioavailability of MBAPs is also a major limitation to address. Bioavailability refers to the extent and rate at which peptides are absorbed and utilized in the body. Many bioactive peptides face poor absorption in the gastrointestinal tract, or they may be broken down into inactive forms before reaching the target site. Developing peptide derivatives or employing delivery systems that enhance their absorption could be a possible strategy to improve bioavailability.

The use of bio-enhancers or modifying peptide sequences can also improve their resistance to digestive enzymes.

3.6.1 Sensory Challenges Like Bitterness

MBAPs have limitations in palatability due to bitter taste, which can negatively impact consumer acceptance when used in supplements or functional foods. Bitterness can limit their direct application in food products unless effective taste-masking strategies are employed. Heiniö *et al.* reported that the small molecular weight peptides are responsible for the bitterness of rye grains.[85] While these small peptides often exhibit enhanced biological activity, their use in the food industry presents a trade-off between flavor and nutritional benefits. Employing taste-masking agents, such as sweeteners or flavor additives, to counteract bitterness could be employed. The possibility of a capsule rather than liquid formulation could also limit the issue of bitterness. Enzymatic modifications or peptide hydrolysis to reduce bitterness while retaining bioactivity have also been thoroughly tried and tested. Encapsulation techniques that balance the masking of off-taste without having a significant effect on solubility are needed for oral delivery of bitter peptide mixtures.

3.6.2 Challenges in Commercialization

Commercializing MBAPs involves scaling up production while ensuring consistency, cost-effectiveness, and compliance with regulatory standards. Peptide extraction and purification processes can be complex and expensive, which may hinder large-scale production. Additionally, regulatory approvals for novel nutraceutical ingredients can be time-consuming and challenging, especially in different markets. Improving cost-efficient extraction and purification methods, such as enzyme-based extraction, could make peptide production more scalable. Rigorous clinical trials should be conducted to validate health claims and meet regulatory requirements in various regions. Peptide-based products can be developed and commercialized through collaborations between food and pharmaceutical industries. Another significant challenge in producing bioactive peptides lies in the relatively low protein content found in millet compared with other food sources like meat, legumes, and milk.[5] Since bioactive peptides are derived from proteins, the limited protein availability in millets may reduce the yield and concentration of these functional peptides. This constraint necessitates innovative approaches in processing techniques to enhance protein extraction and optimize peptide production while ensuring the preservation of the nutritional and bioactive properties unique to millets.

Ensuring that MBAPs are delivered effectively to target tissues or organs is another challenge. Conventional oral delivery may not always be effective, as peptides can be degraded in the stomach or intestines. The development of advanced delivery systems, such as liposomes, nanoparticles, or hydrogels, to protect peptides can ensure effective targeted delivery. Formulation into functional foods or beverages that enhance peptide absorption and bioactivity is another suitable strategy that has been adopted widely. While the bioactivity of these peptides is largely dependent on their size and structure, their absorption from the gastrointestinal tract and their bioavailability in target tissues are also critical factors. This challenge is primarily due to their molecular size, lipophilicity, ionic charge, and mode of administration.[86] The oral route is the most favored method for delivering bioactive peptides due to its convenience and relatively low production cost. However, before MBAPs can reach their target site and produce physiological effects, they must overcome significant challenges such as gastrointestinal breakdown, difficulty permeating membranes in cells, instability, and a short plasma half-life due to pH changes.[87,88] MBAPs must be resistant to gastrointestinal barriers and high processing temperatures to fully exploit their health benefits. Enhancers for absorption are commonly administered along with MBAPs to improve the efficacy of their permeability through epithelial cell membranes by relaxing tight junctions, altering plasma membrane fluidity, and reducing mucus viscosity.[89] Various carrier systems like nanoparticle emulsions, liposomes, and microspheres have been found to impact the absorption and enhance the oral bioavailability of peptides.[90] Recently, a permeation-enhancing molecule SNAC (sodium *N*-(8-[2-hydroxybenzoyl]-amino)caprylate) was used for oral delivery of a GLP-1 receptor agonist (GLP-1RA) of 4.1 KDa size.[91] This is a significant advancement, being the first FDA-approved treatment wherein a peptide was proven to be absorbed through the GI tract, achieving a consistent therapeutic effect associated with the systemic delivery route. The physiological effect of bioactive peptides is generally assessed as the ability to inhibit or activate enzymes in key metabolic pathways using *in vitro* enzyme assays. For many bioactive peptides, absorption and bioavailability estimation are still challenges, hence the need for novel interventions. Currently, the most effective preliminary method for studying the delivery and absorption of bioactive peptides at the intestinal level involves the use of the Caco-2 model (derived from human colon carcinoma cells), which simulates mature human enterocytes, making it a useful *in vitro* tool for assessing drug and nutrient permeability, absorption kinetics, and metabolism at the intestinal level.[92]

Apart from bioactive peptides derived from grain storage proteins, proteins themselves are also utilized for encapsulation and oral delivery. A stabilized emulsion delivery system used in supplements, food, cosmetics or pharmaceuticals is created through the "Pickering" mechanism, which involves creating a stable blend of oil, water, and proteins. Proteins stick to the droplet surface during lipid homogenization, creating a useful hindrance.[93] Because millet proteins are highly hydrophobic and poorly soluble in their natural state, emulsion-based delivery techniques require physical, chemical, or enzymatic modifications. The hydrophobic core is an efficient colloidal delivery device for encapsulating chemicals due to its surface-active qualities and adsorptive qualities.[94] Wheat is frequently used for encapsulating lipophilic compounds in nanoparticles and films, followed by barley, maize and sorghum proteins, due to their hydrophobic nature and high cysteine levels, which encourage cross-linking.[95]

The extractability of protein from millets is largely influenced by factors such as the cultivar, seed pre-treatment, protein content, isolation technique used, composition of amino acid, type and enzyme concentrations, and the composition of the solvent system. Additionally, working conditions like the flour-to-solvent ratio, extraction time, pH and the physical cell disruption technique play a significant role. Even slight alterations in these parameters can result in significant changes in protein structure, subsequently impacting the functional properties of the isolated proteins. Strong collaboration among biopolymer manufacturers, protein producers, and millet processors is crucial for successful marketing of peptides. Millets are an affordable, nutrient-rich crop that is easy to grow and widely produced, offering a promising opportunity for effective utilization and value addition.[96]

MBAPs are often classified under novel food ingredients or biologically active substances, requiring rigorous safety evaluations by regulatory agencies like the FDA or EFSA. This involves proving their safety for consumption, environmental impact, and long-term effects on health, which is time-consuming and costly. The complexity of natural products further complicates this process, as variations in microbial strains, fermentation methods, and processing can lead to inconsistent results, making regulatory approval even harder to attain.

3.7 Conclusion

In summary, while MBAPs have great potential in nutraceuticals and supplements, overcoming challenges in stability, bioavailability, bitterness, delivery, and commercialization will be key to their success in the market. Continuous innovation in processing technologies and formulation strategies, along with regulatory compliance, will help pave the way for their wider adoption. While MBAPs show promise in areas such as antimicrobial properties, immune modulation, and health benefits, further investigation is needed to fully establish their efficacy. Robust clinical trials and in-depth studies are necessary to understand their mechanisms of action, therapeutic doses, interactions with the human body, and potential side effects. Without solid scientific evidence, stakeholders may hesitate to invest in their commercialization or risk market launch without enough supporting data.

Abbreviations

MBAPs	Millet-derived bioactive peptides
RPLC	Reverse-phase liquid chromatography
M_w	Molecular weight
MS	Mass spectrometry
LC–MS	Liquid chromatography–mass spectrometry

HPLC	High-performance liquid chromatography
NMR	Nuclear magnetic resonance
UAE	Ultrasound-assisted extraction
MAE	Microwave-assisted extraction
RP-UFLC	Reverse-phase ultra-fast liquid chromatography
SFE	Supercritical fluid extraction
m/z	Mass to charge ratio
MS/MS	Tandem mass spectrometry
ESI	Electrospray ionization
MALDI	Matrix-assisted laser desorption ionization
LC	Liquid chromatography
RP-HPLC	Reverse-phase HPLC
IEC	Ion-exchange chromatography
SDS	Sodium dodecyl sulfate
SDS-PAGE	SDS polyacrylamide gel electrophoresis
ROS	Reactive oxygen species
MDA	Malondialdehyde
GSH	Glutathione
Gly	Glycine
TNF-α	Tumor necrosis factor-α
IL-6	Interleukin-6
ACE	Angiotensin I converting enzyme
RAAS	Renin–angiotensin–aldosterone System
KKS	Kallikrein–kinnin system
DPP-IV	Dipeptidyl peptidase-IV
GLP-1	Glucagon-like peptide 1
GIP	Glucose-dependent insulinotropic polypeptide
MAPK	Mitogen associated protein kinase
NF-κB	Nuclear factor-kappa B
WHC	Water-holding capacity
ANFs	Antinutritional factors
HIU	High-intensity ultrasound
FTIR	Fourier-transform infrared spectroscopy
FC	Foaming capacity
FMBPHs	Foxtail millet bran protein hydrolysates
SEM	Scanning electron microscopy
MWT	Microwave treatment
HARF	Hot air radio frequency
FFA	Free fatty acid
XRD	X-ray diffraction
ES	Emulsion stability
EAI	Emulsion activity index
FS	Foaming stability
WAC	Water absorption capacity
OAC	Oil absorption capacity
RF	Radiofrequency

FM	Foxtail millet
HUI	Heating uniformity index
HHP	High hydrostatic pressure
CP	Cold plasma
RNS	Reactive nitrogen species
SFMPC	Sprouted foxtail millet protein concentrate
DBD	Dielectric barrier discharge
ITF	Intrinsic fluorescence
DLS	Dynamic light scattering
PTDs	Protein transduction domains
CPPs	Cell-penetrating peptides
SNAC	Sodium *N*-(8-[2-hydroxybenzoyl] amino) caprylate
GLP-1RA	Glucagon like peptide-1 receptor agonist
COX-1	Cyclooxygenase-1
COX-2	Cyclooxygenase-2
LOX	Lipooxygenase

References

1. V. Krishnan, Millets: superfood of the century–challenges and opportunities, *Front. Nutr.*, 2024, **11**, 1462567.
2. K. Srilekha, T. Kamalaja, K. U. Maheswari and R. N. Rani, Evaluation of physical, functional and nutritional quality parameters of kodo millet flour, *J. Pharmacogn. Phytochem.*, 2019, **8**(4), 192–195.
3. N. Farrokhi, J. P. Whitelegge and J. A. Brusslan, Plant peptides and peptidomics, *Plant Biotechnol. J.*, 2008, **6**(2), 105–134.
4. T. K. Virupaksha, G. Ramachandra and D. Nagaraju, Seed proteins of finger millet and their amino acid composition, *J. Sci. Food Agric.*, 1975, **26**(8), 1237–1246.
5. A. Majid and P. C. G. Priyadarshini, Millet derived bioactive peptides: A review on their functional properties and health benefits, *Crit. Rev. Food Sci. Nutr.*, 2020, **60**(19), 3342–3351.
6. N. D. Patil, A. Bains, K. Sridhar, M. Bhaswant, S. Kaur, M. Tripathi, D. Lanterbecq, P. Chawla and M. Sharma, Extraction, Modification, Biofunctionality, and Food Applications of Chickpea (Cicer arietinum) Protein: An Up-to-Date Review, *Foods*, 2024, **13**(9), 1398.
7. J. Tang, D. Yao, S. Xia, L. Z. Cheong and M. Tu, Recent progress in plant-based proteins: From extraction and modification methods to applications in the food industry, *Food Chemistry: X*, 2024, 101540.
8. M. N. Perović and M. G. Antov, The influence of enzymatic pre-treatment of chickpea on properties of protein nanoparticles prepared by heat treatment, *LWT*, 2022, **16Hou3**, 113545.
9. S. Suri, K. S. Balasaheb, D. K. Yadav, S. Malakar, P. Choudhary, A. Mohapatra and P. Dhurve, Overview of Millet proteins: Quality characteristics, effect of thermal/non-thermal processing and applications, *Food Biosci.*, 2023, 103434.

10. M. Karaś, A. Jakubczyk, U. Szymanowska, K. Jęderka, S. Lewicki and U. Złotek, Different temperature treatments of millet grains affect the biological activity of protein hydrolyzates and peptide fractions, *Nutrients*, 2019, **11**(3), 550.
11. D. E. Cruz-Casas, C. N. Aguilar, J. A. Ascacio-Valdés, R. Rodríguez-Herrera, M. L. Chávez-González and A. C. Flores-Gallegos, Enzymatic hydrolysis and microbial fermentation: The most favorable biotechnological methods for the release of bioactive peptides, *Food Chem.:Mol. Sci.*, 2021, **3**, 100047.
12. H. Agrawal, R. Joshi and M. Gupta, Purification, identification and characterization of two novel antioxidant peptides from finger millet (*Eleusine coracana*) protein hydrolysate, *Food Res. Int.*, 2019, **120**, 697–707.
13. (a) M. Akbarian, A. Khani, S. Eghbalpour and V. N. Uversky, Bioactive peptides: Synthesis, sources, applications, and proposed mechanisms of action, *Int. J. Mol. Sci.*, 2022, **23**(3), 1445. (b) Atta-ur-Rahman and M. Iqbal Choudhary, *Applications of NMR Spectroscopy*, Betham Science, 2016, vol. 5.
14. T. A. Aderinola and K. G. Duodu, Production, health-promoting properties and characterization of bioactive peptides from cereal and legume grains, *BioFactors*, 2022, **48**(5), 972–992.
15. Y. Zhang, X. Zhang, Z. Zhang, Z. Chen, X. Jing and X. Wang, Effect of high hydrostatic pressure treatment on the structure and physicochemical properties of millet gliadin, *LWT*, 2022, **154**, 112755.
16. H. Agrawal, R. Joshi and M. Gupta, Isolation, purification and characterization of antioxidative peptide of pearl millet (*Pennisetum glaucum*) protein hydrolysate, *Food Chem.*, 2016, **204**, 365–372.
17. N. Mehta, P. Kumar, A. K. Verma, P. Umaraw, S. K. Khatkar, A. B. Khatkar, D. Pathak, U. Kaka and A. Q. Sazili, Ultrasound-assisted extraction and the encapsulation of bioactive components for food applications, *Foods*, 2022, **11**(19), 2973.
18. K. Kumar, S. Srivastav and V. S. Sharanagat, Ultrasound assisted extraction (UAE) of bioactive compounds from fruit and vegetable processing by-products: A review, *Ultrason. Sonochem.*, 2021, **70**, 105325.
19. L. Shen, S. Pang, M. Zhong, Y. Sun, A. Qayum, Y. Liu, A. Rashid, B. Xu, Q. Liang, H. Ma and X. Ren, A comprehensive review of ultrasonic assisted extraction (UAE) for bioactive components: Principles, advantages, equipment, and combined technologies, *Ultrason. Sonochem.*, 2023, 106646.
20. R. G. Araujo, R. M. Rodríguez-Jasso, H. A. Ruíz, M. Govea-Salas, M. Pintado and C. N. Aguilar, Recovery of bioactive components from avocado peels using microwave-assisted extraction, *Food Bioprod. Process.*, 2021, **127**, 152–161.
21. P. S. Mali and P. Kumar, Optimization of microwave assisted extraction of bioactive compounds from black bean waste and evaluation of its antioxidant and antidiabetic potential *in vitro*, *Food Chem. Adv.*, 2023, **3**, 100543.

22. P. A. Uwineza and A. Waśkiewicz, Recent advances in supercritical fluid extraction of natural bioactive compounds from natural plant materials, *Molecules*, 2020, **25**(17), 3847.
23. F. Herzyk, D. Piłakowska-Pietras and M. Korzeniowska, Supercritical Extraction Techniques for Obtaining Biologically Active Substances from a Variety of Plant By products, *Foods*, 2024, **13**(11), 1713.
24. Q. W. Zhang, L. G. Lin and W. C. Ye, Techniques for extraction and isolation of natural products: A comprehensive review, *Chin. Med.*, 2018, **13**, 1–26.
25. A. L. Rockwood, M. M. Kushnir and N. J. Clarke, Mass Spectrometry, in *Elsevier eBooks*, 2018, pp. 33–65.
26. A. N. Neagu, M. Jayathirtha, E. Baxter, M. Donnelly, B. A. Petre and C. C. Darie, Applications of tandem mass spectrometry (MS/MS) in protein analysis for biomedical research, *Molecules*, 2022, **27**(8), 2411.
27. R. E. Aluko, Antihypertensive peptides from food proteins, *Annu. Rev. Food Sci. Technol.*, 2018, **6**(1), 235–262.
28. M. Mori, Ion-exclusion chromatography, in *Ion-Exchange Chromatography and Related Techniques*, Elsevier, 2024, pp. 163–179.
29. A. G. Abril, M. Pazos, T. G. Villa, P. Calo-Mata, J. Barros-Velázquez and M. Carrera, Proteomics Characterization of Food-Derived Bioactive Peptides with Anti-Allergic and Anti-Inflammatory Properties, *Nutrients*, 2022, **14**(20), 4400.
30. A. S. Pavlova, E. S. Dyudeeva, M. S. Kupryushkin, N. V. Amirkhanov, D. V. Pyshnyi and I. A. Pyshnaya, SDS-PAGE procedure: Application for characterization of new entirely uncharged nucleic acids analogs, *Electrophoresis*, 2017, **39**(4), 670–674.
31. A. Takemori, D. S. Butcher, V. M. Harman, P. Brownridge, K. Shima, D. Higo, J. Ishizaki, H. Hasegawa, J. Suzuki, M. Yamashita, J. A. Loo, R. R. O. Loo, R. J. Beynon, L. C. Anderson and N. Takemori, PEPPI-MS: Polyacrylamide-Gel-Based Prefractionation for Analysis of Intact Proteoforms and Protein Complexes by Mass Spectrometry, *J. Proteome Res.*, 2020, **19**(9), 3779–3791.
32. J. A. Welsh, L. M. Jenkins, J. Kepley, G. C. Lyons, D. M. Moore, T. Traynor, J. A. Berzofsky and J. C. Jones, High Sensitivity Protein Gel Electrophoresis Label Compatible with Mass-Spectrometry, *Biosensors*, 2020, **10**(11), 160.
33. E. O. Mensah, L. Nadtochii, P. Adadi and D. Agyei, Chia derived bioactive peptides: Extraction, characterization, pharmacological activities and potential food applications, *Food Biosci.*, 2024, 103975.
34. J. Bechaux, V. Ferraro, T. Sayd, C. Chambon, J. F. Le Page, Y. Drillet, P. Gatellier and V. Santé-Lhoutellier, Workflow towards the generation of bioactive hydrolysates from porcine products by combining *in silico* and *in vitro* approaches, *Food Res. Int.*, 2020, **132**, 109123.
35. N. A. Pearman, E. Ronander, A. M. Smith and G. A. Morris, The identification and characterisation of novel bioactive peptides derived from porcine liver, *Curr. Res. Food Sci.*, 2020, **3**, 314–321.

36. J. Jacob, V. Krishnan, C. Antony, M. Bhavyasri, C. Aruna, K. Mishra, T. Nepolean, C. T. Satyavathi and K. B. Visarada, The nutrition and therapeutic potential of millets: an updated narrative review, *Front. Nutr.*, 2024, **11**, 1346869.
37. V. Krishnan, P. Verma, S. Saha, B. Singh, T. Vinutha, R. R. Kumar, A. Kulshreshta, S. P. Singh, T. Sathyavathi, A. Sachdev and S. Praveen, Polyphenol-enriched extract from pearl millet (*Pennisetum glaucum*) inhibits key enzymes involved in post prandial hyper glycemia (α-amylase, α-glucosidase) and regulates hepatic glucose uptake, *Biocatal. Agric. Biotechnol.*, 2022, **43**, 102411.
38. C. Wen, J. Zhang, H. Zhang, Y. Duan and H. Ma, Plant protein-derived antioxidant peptides: isolation, identification, mechanism of action and application in food systems: a review, *Trends Food Sci. Technol.*, 2020, **105**, 308–322.
39. X. Gong, Q. An, L. Le, F. Geng, L. Jiang, J. Yan, D. Xiang, L. Peng, L. Zou, G. Zhao and Y. Wan, Prospects of cereal protein-derived bioactive peptides: Sources, bioactivities diversity, and production, *Crit. Rev. Food Sci. Nutr.*, 2022, **62**(11), 2855–2871.
40. H. Górska-Warsewicz, W. Laskowski, O. Kulykovets, A. Kudlińska-Chylak, M. Czeczotko and K. Rejman, Food products as sources of protein and amino acids—The case of Poland, *Nutrients*, 2018, **10**(12), 1977.
41. R. Esfandi, M. E. Walters and A. Tsopmo, Antioxidant properties and potential mechanisms of hydrolyzed proteins and peptides from cereals, *Heliyon*, 20192019, **5**(4), e01538.
42. R. Lv, Y. Dong, Z. Bao, S. Zhang, S. Lin and N. Sun, Advances in the activity evaluation and cellular regulation pathways of food-derived antioxidant peptides, *Trends Food Sci. Technol.*, 2022, **122**, 171–186.
43. I. Amadou, M. E. Gounga and G. W. Le, Millets, nutritional composition, some health benefits and processing, *Emirates J. Food Agric.*, 2013, **25**(7), 501–508.
44. H. Fan, H. Liu, Y. Zhang, S. Zhang, T. Liu and D. Wang, Review on plant-derived bioactive peptides: biological activities, mechanism of action and utilizations in food development, *J. Future Foods*, 2022, **2**(2), 143–159.
45. H. Agrawal, R. Joshi and M. Gupta, Isolation and characterisation of enzymatic hydrolysed peptides with antioxidant activities from green tender sorghum, *LWT*, 2017, **84**, 608–616.
46. Z. Ji, R. Feng and J. Mao, Separation and identification of antioxidant peptides from foxtail millet (*Setaria italica*) prolamins enzymatic hydrolysate, *Cereal Chem.*, 2019, **96**(6), 981–993.
47. Z. Ji, J. Mao, S. Chen and J. Mao, Antioxidant and anti-inflammatory activity of peptides from foxtail millet (*Setaria italica*) prolamins in HaCaT cells and RAW264. 7 murine macrophages, *Food Biosci.*, 2020, **36**, 100636.
48. S. Xu, Y. Shen, G. Chen, S. Bean and Y. Li, Antioxidant Characteristics and Identification of Peptides from Sorghum Kafirin Hydrolysates, *J. Food Sci.*, 20192019, **84**(8), 2065–2076.

49. S. Hu, J. Yuan, J. Gao, Y. Wu, X. Meng, P. Tong and H. Chen, Antioxidant and anti-inflammatory potential of peptides derived from *in vitro* gastrointestinal digestion of germinated and heat-treated foxtail millet (*Setaria italica*) proteins, *J. Agric. Food Chem.*, 2020, **68**(35), 9415–9426.
50. I. Camargo-Filho, D. A. G. Cortez, T. Ueda-Nakamura, C. V. Nakamura and B. P. Dias-Filho, Antiviral activity and mode of action of a peptide isolated from *Sorghum bicolor*, *Phytomedicine*, 2008, **15**, 202–208.
51. W. Xu, L. Wei, W. Qu, Z. Liang, J. Wang, X. Peng, Y. Zhang and K. Huang, A novel antifungal peptide from foxtail millet seeds, *J. Sci. Food Agric.*, 2011, **91**, 1630–1637.
52. I. Amadou, M. E. Gounga and G. W. Le, Millets, nutritional composition, some health benefits and processing, *Emirates J. Food Agric.*, 2013, **25**(7), 501–508.
53. A. N. J. A. L. I. Bisht, M. A. N. I. S. H. A. Thapliyal and A. J. E. E. T. Singh, Screening and isolation of antibacterial proteins/peptides from seeds of millets, *Int. J. Curr. Pharm. Res.*, 2016, **8**(3), 96–99.
54. E. Zielińska, B. Baraniak and M. Karaś, Antioxidant and anti-inflammatory activities of hydrolysates and peptide fractions obtained by enzymatic hydrolysis of selected heat-treated edible insects, *Nutrients*, 2017, **9**(9), 970.
55. M. Karaś, A. Jakubczyk, U. Szymanowska, K. Jęderka, S. Lewicki and U. Złotek, Different temperature treatments of millet grains affect the biological activity of protein hydrolyzates and peptide fractions, *Nutrients*, 2019, **11**(3), 550.
56. J. Chen, W. Duan, X. Ren, C. Wang, Z. Pan, X. Diao and Q. Shen, Effect of foxtail millet protein hydrolysates on lowering blood pressure in spontaneously hypertensive rats, *Eur. J. Nutr.*, 2017, **56**, 2129–2138.
57. M. M. Ibrahim, RAS inhibition in hypertension, *J. Hum. Hypertens.*, 2006, **20**(2), 101–108.
58. Q. Wu, J. Du, J. Jia and C. Kuang, Production of ACE inhibitory peptides from sweet sorghum grain protein using alcalase: Hydrolysis kinetic, purification and molecular docking study, *Food Chem.*, 2016, **199**, 140–149.
59. N. H. Cho, J. E. Shaw, S. Karuranga, Y. Huang, J. D. da Rocha Fernandes, A. W. Ohlrogge and B. I. D. F. Malanda, IDF Diabetes Atlas: Global estimates of diabetes prevalence for 2017 and projections for 2045, *Diabetes Res. Clin. Pract.*, 2018, **138**, 271–281.
60. A. B. Nongonierma and R. J. FitzGerald, An *In Silico* Model to Predict the Potential of Dietary Proteins as Sources of Dipeptidyl Peptidase IV (DPP-IV) Inhibitory Peptides, *Food Chem.*, 2014, **165**, 489–498.
61. H. Gu, J. Gao, Q. Shen, D. Gao, Q. Wang, M. Tangyu and X. Mao, Dipeptidyl peptidase-IV inhibitory activity of millet protein peptides and the related mechanisms revealed by molecular docking, *LWT*, 2021, **138**, 110587.
62. C. N. Kunyanga, J. K. Imungi, M. W. Okoth, H. K. Biesalski and V. Vadivel, Total phenolic content, antioxidant and antidiabetic properties of

methanolic extract of raw and traditionally processed Kenyan indigenous food ingredients, *LWT-Food Sci. Technol.*, 2012, **45**(2), 269–276.
63. S. Li, L. Liu, G. He and J. Wu, Molecular targets and mechanisms of bioactive peptides against metabolic syndromes, *Food Funct.*, 2018, **9**(1), 42–52.
64. Y. J. Fei, Y. Kanai, S. Nussberger, V. Ganapathy, F. H. Leibach, M. F. Romero and M. A. Hediger, Expression cloning of a mammalian proton-coupled oligopeptide transporter, *Nature*, 1994, **368**(6471), 563–566.
65. S. Shan, Z. Li, I. P. Newton, C. Zhao, Z. Li and M. A. Guo, A novel protein extracted from foxtail millet bran displays anti-carcinogenic effects in human colon cancer cells, *Toxicol. Lett.*, 2014, **227**(2), 129–138.
66. R. Sharma, S. Sharma, B. N. Dar and B. Singh, Millets as potential nutri-cereals: A review of nutrient composition, phytochemical profile and techno-functionality, *Int. J. Food Sci. Technol.*, 2021, **56**(8), 3703–3718.
67. A. Dhiman, K. Thakur, V. Parmar, S. Sharma, R. Sharma, G. Kaur, B. Singh and R. Suhag, New insights into tailoring physicochemical and techno-functional properties of plant proteins using conventional and emerging technologies, *J. Food Meas. Charact.*, 2023, **17**(4), 3845–3873.
68. D. Knorr, A. Froehling, H. Jaeger, K. Reineke, O. Schlueter and K. Schoessler, Emerging Technologies in Food Processing, *Annu. Rev. Food Sci. Technol.*, 2011, **2**(1), 203–235.
69. C. Arzeni, K. Martínez, P. Zema, A. Arias, O. E. Pérez and A. M. R. Pilosof, Comparative study of high intensity ultrasound effects on food proteins functionality, *J. Food Eng.*, 2012, **108**(3), 463–472.
70. O. A. Higuera-Barraza, C. L. Del Toro-Sanchez, S. Ruiz-Cruz and E. Márquez-Ríos, Effects of high-energy ultrasound on the functional properties of proteins, *Ultrason. Sonochem.*, 2016, **31**, 558–562.
71. N. Sharma, J. K. Sahu, A. Choudhary, M. Meenu and V. Bansal, High intensity ultrasound (HIU)-induced functionalization of foxtail millet protein and its fractions, *Food Hydrocolloids*, 2023, **134**, 108083.
72. Z. Peng, F. Wang, L. Yu, B. Jiang, J. Cao, Z. Sun and J. Cheng, Effect of ultrasound on the characterization and peptidomics of foxtail millet bran protein hydrolysates, *Ultrason. Sonochem.*, 2024, **110**, 107044.
73. B. Nazari, M. A. Mohammadifar, S. Shojaee-Aliabadi, E. Feizollahi and L. Mirmoghtadaie, Effect of ultrasound treatments on functional properties and structure of millet protein concentrate, *Ultrason. Sonochem.*, 2018, **41**, 382–388.
74. D. N. Yadav, T. Anand, J. Kaur and A. K. Singh, Improved Storage Stability of Pearl Millet Flour Through Microwave Treatment, *Agric. Res.*, 2012, **1**(4), 399–404.
75. M. V. Rao, C. K. Sunil and N. Venkatachalapathy, Effect of microwave and hot air radiofrequency treatments on physicochemical and functional properties of foxtail millet flour and its protein isolate, *J. Cereal Sci.*, 2023, **114**, 103774.
76. B. Ling, S. Ouyang and S. Wang, Radio-frequency treatment for stabilization of wheat germ: Storage stability and physicochemical properties, *Innovative Food Sci. Emerging Technol.*, 2019, **52**, 158–165.

77. H. Zhu, L. Yang, X. Fang, Y. Wang, D. Li and L. Wang, Effects of intermittent radio frequency drying on structure and gelatinization properties of native potato flour, *Food Res. Int.*, 2021, **139**, 109807.
78. L. Liao, T. Liu, M. Zhao, H. Zhao and C. Cui, Aggregation behavior of wheat gluten during carboxylic acid deamidation upon hydrothermal treatment, *J. Cereal Sci.*, 2011, **54**(1), 129–136.
79. C. Panchapakesan, N. Sozer, H. Dogan, Q. Huang and J. L. Kokini, Effect of different fractions of zein on the mechanical and phase properties of zein films at nano-scale, *J. Cereal Sci.*, 2012, **55**(2), 174–182.
80. V. Monica, R. Anbarasan and R. Mahendran, Cold Plasma-Induced Changes in the Structural and Techno-functional Properties of Sprouted Foxtail Millet Protein Concentrate, *Food Bioprocess Technol.*, 2024, 1–18.
81. A. S. M. Saleh, Q. Zhang, J. Chen and Q. Shen, Millet Grains: Nutritional Quality, Processing, and Potential Health Benefits, *Compr. Rev. Food Sci. Food Saf.*, 2013, **12**(3), 281–295.
82. H. M. Wani, P. Sharma, K. Gul, J. P. Bishnoi, I. A. Wani, S. L. Kothari and A. A. Wani, Influence of α-irradiation and genotype on the structural and techno-functional properties and microbial quality of Millet flour, *Appl. Food Res.*, 2022, **2**(2), 100181.
83. G. Gowthamraj, C. Jubeena and N. Sangeetha, The effect of γ-irradiation on the physicochemical, functional, proximate, and anti-nutrient characteristics of finger millet (CO14 & CO15) flours, *Radiat. Phys. Chem.*, 2021, **183**, 109403.
84. K. Sarabandi, P. Gharehbeglou and S. M. Jafari, Spray-drying encapsulation of protein hydrolysates and bioactive peptides: Opportunities and challenges, *Drying Technol.*, 2020, **38**(5–6), 577–595.
85. R.-L. Heiniö, E. Nordlund, K. Poutanen and J. Buchert, Use of enzymes to elucidate the factors contributing to bitterness in rye flavour, *Food Res. Int.*, 2012, **45**(1), 31–38.
86. T. Shantha Kumar, K. Soppimath and S. Nachaegari, Novel Delivery Technologies for Protein and Peptide Therapeutics, *Curr. Pharm. Biotechnol.*, 2006, **7**(4), 261–276.
87. M. Segura-Campos, L. Chel-Guerrero, D. Betancur-Ancona and V. M. Hernandez-Escalante, Bioavailability of Bioactive Peptides, *Food Rev. Int.*, 2011, **27**(3), 213–226.
88. A. Muheem, F. Shakeel, M. A. Jahangir, M. Anwar, N. Mallick, G. K. Jain, M. H. Warsi and F. J. Ahmad, A review on the strategies for oral delivery of proteins and peptides and their clinical perspectives, *Saudi Pharm. J.*, 2016, **24**(4), 413–428.
89. B. J. Aungst, Absorption Enhancers: Applications and Advances, *AAPS J.*, 2012, **14**(1), 10–18.
90. H. Malhaire, J.-C. Gimel, E. Roger, J.-P. Benoît and F. Lagarce, How to design the surface of peptide-loaded nanoparticles for efficient oral bioavailability?, *Adv. Drug Delivery Rev.*, 2016, **106**, 320–336.

91. V. R. Aroda, L. Blonde and R. E. Pratley, A new era for oral peptides: SNAC and the development of oral semaglutide for the treatment of type 2 diabetes, *Rev. Endocr. Metab. Disord.*, 2022, **23**(5), 979–994.
92. F. Tonolo, M. Sandre, S. Ferro, A. Folda, V. Scalcon, G. Scutari, E. Feller, O. Marin, A. Bindoli and M. P. Rigobello, Milk-derived bioactive peptides protect against oxidative stress in a Caco-2 cell model, *Food Funct.*, 2018, **9**(2), 1245–1253.
93. L. Zou, A. Xie, Y. Zhu and D. J. McClements, Cereal proteins in nanotechnology: formulation of encapsulation and delivery systems, *Curr. Opin. Food Sci.*, 2019, **25**, 28–34.
94. L. Wang, P. Gulati, D. Santra, D. Rose and Y. Zhang, Nanoparticles prepared by proso millet protein as novel curcumin delivery system, *Food Chem.*, 2018, **240**, 1039–1046.
95. J. A. Tapia-Hernández, C. L. Del-Toro-Sánchez, F. J. Cinco-Moroyoqui, J. E. Juárez-Onofre, S. Ruiz-Cruz, E. Carvajal-Millan, G. A. López-Ahumada, D. D. Castro-Enriquez, C. G. Barreras-Urbina and F. Rodríguez-Felix, Prolamins from cereal by-products: Classification, extraction, characterization and its applications in micro- and nanofabrication, *Trends Food Sci. Technol.*, 2019, **90**, 111–132.
96. N. Sachdev, S. Goomer, L. R. Singh, V. M. Pathak, D. Aggarwal and Dr. R. K. Chowhan, Current status of millet seed proteins and its applications: A comprehensive review, *Appl. Food Res.*, 2023, **3**(1), 100288.

CHAPTER 4

Nutritional Value, Digestibility and Effect of Millet Protein on the Gut Microbiome

MEENA KUMARI PALANI KUMAR[a] AND MUTHUKUMAR SERVA PEDDHA*[b]

[a] Postdoctoral Fellow, Department of Internal Medicine, University of Iowa, USA, Email: meena-palanikumar@uiowa.edu; [b] Chief Scientist, Department of Biochemistry, CSIR-Central Food Technological Research Institute, India
*Email: muthukumar@cftri.res.in

4.1 Introduction

Millets, a diverse group of ancient grains, have long been recognized for their resilience in arid climates and their ability to thrive in regions where other staple crops, like rice and wheat, struggle to grow. While traditionally valued for their role in food security, millets are now gaining global recognition for their impressive nutritional profile and potential health benefits.[1] Millets are an excellent source of essential nutrients, including protein, carbohydrates, healthy fats, vitamins, minerals, and bioactive compounds like polyphenols and phytates. These bioactive compounds exhibit a wide range of biological effects, including antioxidant, anti-inflammatory, and anti-hyperglycemic properties, as well as positive impacts on cholesterol levels and blood pressure.[2] Moreover, they have the ability to modulate gut microbiota, contributing to overall health.[3] Millet grains, however, also contain antinutrients like phytates and tannins, which can interfere with mineral absorption, though processing methods such as soaking,

Food Chemistry, Function and Analysis No. 48
Millet Proteins: Nutritional, Structural and Functional Attributes
Edited by Shweta Suri, Anuradha Dutta and Tarun Belwal

Published by the Royal Society of Chemistry, www.rsc.org

fermentation, and milling can reduce their presence and enhance nutrient bioavailability.[4] What sets millet apart from other grains like rice and wheat is its relatively high content of certain nutrients, particularly protein and fiber, as well as its rich profile of micronutrients. Millet protein is of particular interest because it is a complete protein, containing all nine essential amino acids, with a notably high methionine content. While it is somewhat limited in lysine and threonine, millet still provides a well-rounded source of protein, which is crucial for body maintenance, muscle repair, and immune function.[5] Beyond its basic nutritional properties, millet has a unique effect on the gut microbiome. The human gut microbiome, consisting of trillions of microorganisms, plays a crucial role in digestion, nutrient absorption, immune system regulation, and even mental health. Diet is one of the most influential factors shaping the composition and diversity of gut microbes, and protein intake is particularly important in this regard.[6] Millet protein has been shown to promote the growth of beneficial bacteria such as *Bifidobacterium* and *Bacteroides*, both of which are vital for maintaining a healthy gut environment. These bacteria help to ferment undigested protein, producing short-chain fatty acids (SCFAs) as byproducts.[3] SCFAs are critical for gut health, as they serve as an energy source for colon cells and have anti-inflammatory effects. Incorporating millet into the diet not only provides essential nutrients but may also offer prebiotic benefits through its impact on gut bacteria. This, in turn, can support overall digestive health, improve immune function, and even contribute to better metabolic health.[7] Further research into the specific mechanisms of how millet protein interacts with the gut microbiome will undoubtedly provide deeper insights, but current evidence suggests that millet holds considerable promise as a functional food in promoting gut health and general well-being. This chapter explores the intricate relationship between millet's nutritional properties and its potential to influence gut microbiome composition, emphasizing its role as a valuable dietary component in supporting holistic health.

4.2 The Nutritional Value of Millets

Millets are increasingly recognized for their significant nutritional value, offering a wide array of health benefits due to their unique composition. They contain 60%–70% dietary carbohydrates, 6%–19% protein, 1.5%–5% fat, 12%–20% dietary fiber, 2%–4% minerals, and various phytochemicals. The nutritional values will change depending on the growth climate and the species being used.[2] Table 4.1 shows the nutrient composition of millets. This nutritional profile positions millets as an energy-rich food, while also providing essential nutrients such as protein and minerals. They offer a considerable amount of nutrition, especially when processed with techniques that preserve their digestibility.[8] Millets, often termed "miracle grains" or "nutri-cereals," are known for their high antioxidant content, abundant fiber, and rich mineral composition. Sorghum, one of the major types of millet, offers about 10.4% protein, 1.3% fat, and 73% carbohydrates, making

Table 4.1 Nutrient composition of millets (per 100 g edible portion). Data from ref. 72.

Millet	Protein[a] (g)	Fat (g)	Ash (g)	Crude fiber (g)	Carbohydrate (g)	Energy (kcal)	Ca (mg)	Fe (mg)	Thiamin (mg)	Riboflavin (mg)	Niacin (mg)
Maize	9.2	4.6	1.2	2.8	73.0	358	26	2.7	0.38	0.20	3.6
Sorghum	10.4	3.1	1.6	2.0	70.7	329	25	5.4	0.38	0.15	4.3
Pearl millet	11.8	4.8	2.2	2.3	67	363	42	11.0	0.38	0.21	2.8
Finger millet	7.7	1.5	2.6	3.6	72.6	336	350	3.9	0.42	0.19	1.1
Foxtail millet	11.2	4.0	3.3	6.7	63.2	351	31	2.8	0.59	0.11	3.2
Common millet	12.5	3.5	3.1	5.2	63.8	364	8	2.9	0.41	0.28	4.5
Little millet	9.7	5.2	5.4	7.6	60.9	329	17	9.3	0.30	0.09	3.2
Barnyard millet	11.0	3.9	4.5	13.6	55.0	300	22	18.6	0.33	0.10	4.2
Kodo millet	9.8	3.6	3.3	5.2	66.6	353	35	1.7	0.15	0.09	2.0

[a] All values except protein are expressed on a dry weight basis.

it a robust energy source with 349 kcal per 100 g. Sorghum also contains significant quantities of iron, calcium, and dietary fiber, which are critical for preventing nutrient deficiencies, particularly in developing countries.[9] Other millet varieties, such as finger millet, stand out for their high calcium content (344 mg per 100 g), making them an essential food for bone health.[10] Pearl millet, rich in resistant starch, dietary fibers, and antioxidants, is also a potent source of iron, aiding in the fight against anemia.[11] Millets also shine as a source of vital vitamins, including niacin (B3), thiamine, riboflavin, and folic acid, which are essential for energy metabolism and overall health. Foxtail millet is particularly noteworthy for its high zinc and iron content, contributing to immune function and cognitive health.[12] Additionally, millets like kodo and little millet are rich in polyunsaturated fatty acids and dietary fibers, promoting cardiovascular health and aiding digestion. Their polyphenol content further enhances their antioxidant activity, which is associated with anti-aging effects and the prevention of oxidative stress-related diseases.[13] The seed coat of millet contains pectin, cellulose, and hemicelluloses, which are resistant to breakdown by digestive enzymes, making them a valuable source of dietary fiber. This dietary fiber plays a significant role in human health by influencing various metabolic and digestive processes, including glucose absorption and cholesterol levels. The high fiber content of millet seed coats offers health benefits and can be utilized as an excellent source of dietary fiber. Additionally, the by-products derived from millet seed coats are often used in the preparation of composite flour for making biscuits, providing both nutritional and functional value in food products.[14] Phenolic compounds are naturally present in millets and other plant sources and play a vital role in scavenging free radicals and neutralizing reactive oxygen species (ROS), which helps prevent and treat diseases associated with oxidative stress. Polyphenols, including phenolic acids, flavonoids, stilbenes, tannins, and lignans, act as free radical scavengers, activate antioxidant enzymes, and inhibit oxidases, making them effective natural antioxidants and therapeutic agents. Millets, known for their rich content of bioactive compounds such as phenolics, flavonoids, vitamins, and minerals, offer significant health-promoting properties. These bioactive compounds help protect against various diseases, especially cardiovascular diseases and some types of cancer. The major phenolic compounds present in millets include gallic acid, caffeic acid, ferulic acid, syringic acid, and several others, which contribute to the free radical-scavenging, antimicrobial, and anticancer activities of millets, enhancing their nutritional and therapeutic value.[15] Millets present a powerful combination of nutrients and bioactive compounds, making them ideal for diverse dietary needs. Their role in traditional diets, especially in developing countries, ensures nutritional security and provides a versatile base for various food products. With increasing awareness of their health benefits, millets are gradually being integrated into global food systems, promoting both nutrition and sustainable agriculture.[16]

4.3 Beneficial Effects of Millets

Research has consistently shown that plant-based diets play a protective role against various degenerative diseases, including cancer, cardiovascular conditions, diabetes, metabolic syndrome, and Parkinson's disease.[17,18] Additionally, robust epidemiological evidence indicates that whole-grain cereals help safeguard against age-related illnesses such as diabetes, cardiovascular diseases, and certain types of cancer.[19] While the health benefits of whole grains were historically attributed to their vitamins, minerals, essential fatty acids, and fiber content, recent studies suggest that other bioactive compounds also contribute to their positive effects. These include resistant starch, oligosaccharides, lipids, antioxidants (*e.g.*, phenolic acids, avenanthramides, flavonoids), lignans, phytosterols, and even antinutritional factors like phytic acid and tannins.[20] Millets, too, should be recognized as functional foods and nutraceuticals due to their provision of dietary fibers, proteins, energy, minerals, vitamins, and antioxidants essential for human health. They offer several potential health benefits, such as cancer and cardiovascular disease prevention, tumor reduction, lowering blood pressure and cholesterol, reducing fat absorption rates, delaying gastric emptying, and enhancing gastrointestinal health.[21] Notably, the U.S. Department of Agriculture's nutritional guidelines highlight the importance of grains and grain-based products in a balanced diet, placing them at the foundation of the food guide pyramid to promote optimal health.[22]

4.3.1 Cardio-protective Effects

Cardiovascular diseases (CVDs) are a significant global health concern, with their prevalence rising due to factors such as obesity, smoking, unhealthy diets, and physical inactivity. Research has increasingly highlighted the potential of millets as functional foods in preventing CVD and related metabolic disorders. Millets are rich in essential nutrients like magnesium, potassium, fiber, and bioactive compounds that contribute to heart health. Magnesium helps lower blood pressure and reduces the risk of strokes, particularly in conditions like atherosclerosis, while potassium acts as a vasodilator, maintaining healthy blood pressure levels and mitigating cardiovascular risks. The plant lignans in millets, metabolized by gut microflora into animal lignans, offer protection against heart disease and certain cancers. Additionally, the high fiber content in millets aids in lowering cholesterol by eliminating low-density lipoprotein (LDL) cholesterol and enhancing the effects of high-density lipoprotein (HDL) cholesterol. Empirical studies further underscore the cardioprotective effects of millets. Diets incorporating barnyard millet starch have been shown to reduce blood glucose, serum cholesterol, and triglyceride levels significantly compared to rice and other minor millets. Similarly, feeding proso millet protein to genetically obese, type 2 diabetic mice under high-fat diets improved plasma adiponectin levels and HDL cholesterol, while finger millet and proso millet reduced serum triglycerides in hyperlipidemic rats more effectively than white rice and sorghum.[23] Grain sorghum lipid extract (GSL) has been shown to lower non-HDL cholesterol by inhibiting cholesterol absorption and reducing liver cholesterol ester concentration, attributed to its plant sterols and policosanol content, which also inhibit endogenous cholesterol synthesis.[24] Phenolic extracts from various millets, including kodo, finger, proso, foxtail, little, and pearl millet, have shown significant inhibition of lipid peroxidation *in vitro*, with kodo millet exhibiting superior antioxidant effects comparable to synthetic antioxidants. Epidemiological evidence reinforces these findings, showing that regular consumption of whole grains, including millets, reduces the risk of CVD. Long-term studies have revealed that incorporating whole grains into the diet can significantly lower the risk of coronary heart disease and total mortality. For instance, a study involving 27 000 men found that consuming 40 g of whole grains daily reduced the risk of coronary heart disease by 20%. Among millets, pearl millet is particularly notable for its lignins and phytonutrients, which act as powerful antioxidants to protect heart health. Finger millet and proso millet also stand out for their ability to reduce serum triglycerides and improve cardiovascular health, while barnyard millet effectively lowers blood glucose, cholesterol, and triglycerides. These findings collectively highlight the potential of millets as a vital component of a heart-healthy diet, emphasizing their role in reducing cardiovascular risks and promoting overall health.[25]

4.3.2 Anti-diabetic Effects

Millet varieties have been extensively studied for their antidiabetic properties, primarily due to their ability to inhibit key digestive enzymes such as α-amylase and α-glucosidase. For instance, finger millet-based diets have demonstrated reduced starch digestibility and absorption, leading to lower glycemic effects.[26] The bioactive compounds in millets modulate glucose-induced oxidative stress and inhibit starch-digesting enzymes, showcasing their potential in diabetes management. Protein concentrates from Korean foxtail millet and proso millet have improved glycemic responses in type 2 diabetic mice, significantly decreasing insulin levels while increasing plasma adiponectin and HDL cholesterol.[23,27] Proso millet protein concentrates also exhibit protective effects against D-galactosamine-induced liver injury in rats. Similarly, aqueous extracts of foxtail millet seeds have demonstrated anti-hyperglycemic and anti-lipidemic activities in streptozotocin-induced diabetic rats, lowering triglycerides, LDL, and VLDL (very low-density lipoprotein) cholesterol while increasing HDL cholesterol. Sorghum, another millet variety, contains slow digestible starch (SDS), which prolongs carbohydrate digestion and absorption, making it favorable for managing diabetes and hyperlipidemia.[28] Studies have shown that sorghum-based foods possess a low glycemic index (GI), reducing postprandial blood glucose and glycosylated hemoglobin levels. Variations in cooking and processing methods, such as chapati preparation from white or yellow jowar flour, influence the GI and offer dietary advantages for diabetes management.[29] Pearl millet is known for enhancing insulin sensitivity and lowering triglyceride levels. Its high fiber content slows digestion and glucose release, aiding in stable blood sugar levels over time. Similarly, finger millet has demonstrated hypoglycemic, hypocholesterolemic, nephroprotective, and anti-cataractogenic effects, attributed to its phenolic compounds, which exhibit significant α-glucosidase and pancreatic amylase inhibition.[30] These compounds, including gallic, ferulic, and cinnamic acids, effectively prevent cataractogenesis and provide antioxidative benefits.[31] Finger millet also improves antioxidant status and promotes wound healing, as observed in diabetic rat models.[32]

Barnyard millet has shown particular promise for type 2 diabetics, with dehulled varieties exhibiting a GI as low as 41.7 after heat treatment.[33] Similarly, proso millet improves glycemic responses and insulin sensitivity in obese, high-fat-fed diabetic mice. Collectively, these studies highlight the potential of millet grains in preventing and managing diabetes, emphasizing their multifaceted roles in modulating glucose metabolism, improving lipid profiles, and offering antioxidative and anti-inflammatory benefits.

4.3.3 Antioxidant Effects

Millet grains, rich in phenolics and flavonoids, exhibit significant antioxidant properties by chelating metal ions, preventing radical formation, and protecting cells from free radical-induced damage, thereby enhancing the endogenous antioxidant system.[15] Various millet varieties, including kodo, finger, little, foxtail, barnyard, and great millets, as well as their white varieties, have demonstrated notable antioxidant activity through their DPPH reduction capacity and ferric-reducing potential.[34] These properties position millet grains as potential anti-aging agents and protectors against metabolic syndromes.[35] The phenolic content in millet seed coats, as studied in finger millet, underscores their free radical scavenging potential.[36] Regular consumption of whole grains, including millets, is associated with reduced risk of chronic disorders due to the presence of antioxidants like phenolics and flavonoids.[37] Research has further highlighted the antioxidative benefits of millets. For instance, phenolic extracts from millets exhibit anticancer activity by neutralizing peroxyl and hydroxyl radicals.[18] Processing techniques such as roasting enhance the nutraceutical properties of little millet by increasing its phenolic content and antioxidant activity.[38] Millet-enriched diets have shown promise in mitigating hypertension and myocardial damage by reducing oxidative stress. Specific compounds in millets, such as quercetin, curcumin, ellagic acid, and catechins, contribute to detoxifying the body by neutralizing free radicals and promoting the excretion of toxins, particularly in the kidneys and liver.[39] The antioxidant capacities of millet varieties, including kodo, finger, foxtail, proso, and pearl millets, are attributed to both soluble and insoluble phenolic extracts.[40] Studies have identified over 50 phenolic compounds in millets, including phenolic acids, flavan-3-ol monomers, and flavonoids, using advanced analytical techniques such as HPLC and HPLC-MS.[41] Finger millet has shown higher radical scavenging activity compared to wheat and rice, while defatted foxtail millet protein hydrolysates have also exhibited antioxidant potential.[42] These findings establish millets as valuable functional food ingredients and natural antioxidants, promoting health and reducing disease risk.

4.3.4 Anti-microbial Effects

Millet grains contain a variety of secondary metabolites, including bioactive compounds such as phenolics and flavonoids, which exhibit a wide range of biological properties, particularly antibacterial and antifungal activities.[43] Protein extracts from pearl millet demonstrated superior antifungal activity compared to other millet varieties, inhibiting pathogens such as *Rhizoctonia solani*, *Macrophomina phaseolina*, and *Fusarium oxysporum*.[44] Phenolic and flavonoid compounds in finger millet inhibit the proliferation of various bacterial pathogens, including *Escherichia coli*, *Bacillus cereus*, *Listeria monocytogenes*, *Staphylococcus aureus*, and others.[45] The antimicrobial potential of finger millet's phenolic compounds has been recognized as a promising alternative in the treatment of bacterial and fungal infections.[36] The mechanism behind the loss of fungal functionality when exposed to phenolic extracts from the seed coat of finger millet shows that these extracts act as free radicals, oxidizing microbial membranes and inactivating fungal proteins.[46] While research on the anti-inflammatory potential of millet grains is limited, finger millet has been studied for its bioactive compounds and their effects on the dermal wound healing process in diabetes-induced rats, particularly in modulating oxidative stress and inflammation.[47] Phenolic acids from finger millet fractions, particularly from the seed coat, exhibited higher antimicrobial activity than whole flour extracts, suggesting the potential of using finger millet seed coat as a natural antioxidant and food preservative.[36] Finger millet's phenolic content also positively affects its malt quality by reducing fungal load during germination, with high-phenol varieties outperforming low-phenol ones.[46] Moreover, Xu *et al.* (2011) isolated a potent antifungal peptide from foxtail millet seeds. These findings underscore the potential of millet-derived bioactive compounds as natural alternatives for food preservation and therapeutic use, though further studies are needed to fully validate their antimicrobial effects.[43]

4.3.5 Anti-cancer Effects

Millet grains, rich in phenolic acids, tannins, and phytate, are recognized for their health benefits, including cancer prevention. These nutrients, while often labeled as "antinutrients", have been shown to reduce the risk of colon and breast cancer in animal studies.[48] Additionally, research has highlighted the protective role of fiber in millet and sorghum against esophageal cancer, with lower incidences observed in populations consuming these grains compared to those consuming wheat or maize.[49] Recent studies suggest that increasing fiber intake, specifically over 30 grams per day, may reduce the risk of breast cancer in women by more than 50%. Millet phenolics, in particular, have demonstrated potential in preventing the initiation and progression of cancer *in vitro*.[50] Sorghum, known for its anti-carcinogenic properties, has been well-studied for its health benefits. Both *in vivo* and *in vitro* research show that sorghum consumption positively impacts cancer prevention. The polyphenols and tannins in sorghum exhibit anti-mutagenic and anti-carcinogenic effects, acting against human melanoma cells and promoting melanogenic activity.[51,52] Additionally, procyanidin extracts from sorghum may induce cytochrome P-450, a protein involved in converting certain promutagens into mutagenic derivatives.[53] Epidemiological studies, including one conducted in Sachxi Province, China, revealed lower incidences of esophageal cancer in populations that consumed sorghum. Over six years, researchers found that sorghum consumption was associated with a reduced mortality rate from esophageal cancer compared to wheat and maize.[49]

4.3.6 Gut Protective Effects

Regulating the digestive process plays a crucial role in improving nutrient retention and reducing the risk of serious gastrointestinal conditions such as gastric ulcers and colon cancer. The fiber content in millets aids in alleviating digestive disorders like constipation, excessive gas, bloating, and cramping. Additionally, millets can be beneficial for individuals with celiac disease, an immune-mediated enteropathy triggered by the ingestion of gluten in genetically susceptible individuals.[54] A gluten-free diet, which is essential for managing celiac disease, has a significant impact on food consumption, particularly in the grain food group. Replacing gluten-containing cereals like wheat, barley, and rye with gluten-free grains such as rice, corn, sorghum, millet, amaranth, buckwheat, quinoa, and wild rice can help individuals adhere to a gluten-free diet.[55] Given that millets are naturally gluten-free, they hold great potential in the production of gluten-free foods and beverages, catering to the growing demand from individuals with celiac disease.[56] As the demand for healthier, gluten-free options rises, millets, along with oats and sorghum, have carved a niche in the market for non-wheat-based cereal products.[57] Celiac disease, once considered a rare disorder mostly affecting European children, is now recognized as one of the most common lifelong disorders globally, affecting a wide range of populations.[54] In this context, millet grains and their derivatives not only show promise in cancer prevention but also have the potential to support the production of gluten-free products for those with celiac disease.

4.4 Millet Protein Digestibility

Millets are a rich source of plant-based proteins, making them a valuable dietary choice, particularly for vegans. However, their protein digestibility poses challenges. Studies by Mertz *et al.* (1984) show that millets have lower protein digestibility compared to other grains, which is concerning in regions where millets are a primary protein source.[58] This issue is worsened by certain processing methods, such as hydrothermal treatments, which further reduce protein digestibility in specific millet varieties.[58] Despite this, millets are considered an excellent source of plant protein, containing minimal saturated fats compared to animal proteins. The presence of antinutrients, which inhibit protein digestibility, is a key factor in these challenges.[59] Simple processing techniques like dehulling, milling, soaking, and heating have been found to reduce antinutrients and improve protein digestibility. Table 4.2 shows the effect of different processing methods on millet protein digestibility. For instance, alkaline cooking, fermentation, germination, and popping have all shown improvements in protein quality, particularly in foxtail millet. Germination of foxtail millet, for example, led to an increase in protein content, and similar improvements were observed in pearl millet cultivars such as Gadarif and Gazeera. The fermentation of millet flour has been shown to enhance protein digestibility, as it helps

Table 4.2 Effect of processing methods on millet protein digestibility.

Processing method	Millet type	Key findings	Ref.
Soaking	Foxtail millet	Increased IVPD value from 62.3% to 76%. Combined with dehulling and cooking, IVPD rose to 83%.	73
Germination	Barnyard millet	Tannin levels reduced significantly from 1.594 to 0.657 mg per 100 g.	74
	Finger millet	Phytic acid dropped by 51.11% after 48 hours of germination at 25 °C.	75
	Foxtail millet	After 4 days, tannin reduced from 0.33 to 0.02 $mg\,g^{-1}$; saponin dropped from 0.39 to 0.002 $mg\,g^{-1}$; phytic acid declined from 6.40 to 2.67 $mg\,g^{-1}$.	76
	Proso millet	After 4 days, tannin decreased from 0.51 to 0.05 $mg\,g^{-1}$; saponin dropped from 0.44 to 0.007 $mg\,g^{-1}$; phytic acid reduced from 8.09 to 4.86 $mg\,g^{-1}$.	
Cooking	Foxtail millet	IVPD values exceeded 80% after cooking.	73
	Proso millet	Pan-cooked, microwave-cooked, and uncooked proso millet had starch digestibility values of 57.18%, 42.20%, and 43.73%, respectively.	77
	Little millet	Raw, pan-cooked, and microwave-cooked small millet showed starch digestibility values of 41.96%, 29.17%, and 30.09%, respectively.	
Fermentation	Finger millet	Fermentation at 27 °C for 16 hours reduced phytic acid by 56.63%. Combining germination and fermentation resulted in a 59.57% phytic acid reduction.	75

degrade antinutritional factors like phytates and converts insoluble proteins to soluble forms. Soaking and malting pearl millet also led to increased protein content. These findings suggest that processing methods such as soaking, malting, germination, and fermentation can significantly enhance protein digestibility, making millets a better protein source, particularly in regions facing protein–energy malnutrition.[8]

Decortication, which removes a portion of the outer husk, bran, and germ, has also been shown to improve protein digestibility by reducing antinutrients. Dehulling of pearl millet up to 17.5% increased protein content, but further dehulling led to a decrease in protein. Similarly, milling pearl millet with bran-rich grains showed the highest protein digestibility. In addition to improving protein digestibility, these processes offer a sustainable approach to increasing millet's nutritional value, making them accessible for domestic use, particularly in underdeveloped regions. However, despite these improvements, plant storage proteins in millets still have lower digestibility than animal proteins, due to factors such as protease inhibitors, low protein solubility, and rigid cell walls. These characteristics hinder enzymatic digestion, particularly in millets like sorghum, which experience structural changes that further reduce digestibility.[60] While some studies have reported a slight increase in protein digestibility after cooking, others found no improvement or even a reduction after hydrothermal treatments.[61] In particular, the presence of polyphenolic compounds like tannins can bind to proteins and further lower digestibility, especially in varieties with higher tannin levels, such as kodo millet. However, millets with intermediate tannin levels still maintain a relatively high protein digestibility. Further research is needed to refine processing methods to enhance the digestibility of millet proteins, particularly for varieties like proso millet, which show significant decreases in digestibility after certain treatments. Overall, while millets are nutritionally valuable, their lower protein digestibility poses a challenge, especially in regions where they are a primary protein source. However, processing methods such as dehulling, germination, fermentation, and soaking offer practical solutions to improve their digestibility, making millets a more effective tool in addressing protein–energy malnutrition. These methods can be applied at the domestic level, providing an accessible and sustainable approach to enhancing the nutritional quality of millet-based products.[62]

4.5 Effect of Millets on Changes in Gut Microbiome

The intricate ecosystem of the human gut microbiome, comprising trillions of microorganisms, profoundly influences various aspects of health and disease.[63] Recent scientific endeavors have shed light on the pivotal role of diet in shaping the composition and function of the gut microbiota, offering promising avenues for preventive and therapeutic interventions. Among the diverse array of dietary options, millets have emerged as noteworthy candidates, celebrated for their nutritional richness and potential health benefits.

A list of studies on the effect of millet on gut microbiome changes is shown in Table 4.3. The gut microbiota primarily consists of *Firmicutes* and *Bacteroidetes*, with their ratio significantly influencing metabolic health. Fu *et al.* observed that the *Bacteroidetes* (B)/*Firmicutes* (F) ratio correlates positively with plasma glucose levels. A millet-enriched diet in diabetic models exhibited changes in this ratio, with the millet consumption group showing higher abundances of beneficial bacteria such as *Blautia*, *Akkermansia*, and *Odoribacter*, which are associated with anti-diabetic effects. Although the changes in the B/F ratio were not statistically significant, the enriched presence of specific genera underscores the potential modulatory effect of millets on gut health.[64] The effects of millet consumption on changes in gut microbiome and the implications on health are presented in Figure 4.1. The schematic illustrates the effects of millet consumption on gut microbiome composition and metabolic pathways. Millets contain bioactive components such as resistant starch, dietary fiber, and peptides that promote the growth of beneficial bacteria (*Lactobacillus, Bifidobacterium, Akkermansia muciniphila, Faecalibacterium, Roseburia, Prevotella, Enterococcus faecium*) while reducing pathogenic bacteria (*e.g.*, *Clostridium, Enterococcus, Streptococcus, Proteobacteria, Enterobacteriaceae*). These changes enhance intestinal integrity, motility, and immunity, primarily through the production of short-chain fatty acids (SCFAs) like acetate, propionate, and butyrate. The interaction of SCFAs with GPR41 and GPR43 plays a pivotal role in promoting metabolic and immune health. SCFAs like acetate, propionate, and butyrate activate these receptors to enhance energy homeostasis by regulating gut motility, fat storage, and glucose metabolism. GPR41 activation stimulates the release of peptide YY (PYY), a gut hormone that slows gastric emptying and reduces appetite, contributing to weight management and improved satiety. SCFAs also improve gut barrier integrity by strengthening tight junctions and reducing systemic inflammation caused by lipopolysaccharides (LPS). GPR43 activation modulates immune responses, promoting anti-inflammatory cytokines and preventing chronic inflammation. Additionally, SCFAs regulate insulin sensitivity, lipid metabolism, and histone deacetylase (HDAC) inhibition, which supports anti-carcinogenic effects and tumor suppression. These pathways collectively improve metabolic health, reduce obesity and diabetes risk, and support immune balance.[65] Dietary interventions, particularly those involving millets, have been linked to a reduction in inflammation. Molla reported higher abundances of *Bacteroidetes* in casein-based diets, associated with chronic liver diseases and fibrosis. Conversely, millet-based diets increased the abundance of *Lactobacillus* and *Allobaculum*, which contributes to higher butyric acid levels in the gut. Butyric acid plays a critical role in maintaining gut health, reducing inflammation, and supporting the integrity of the gut barrier.[66] The global burden of diabetes has spurred research into dietary strategies, with millets emerging as effective agents. Millet proteins and fibers modulate gut microbiota and metabolic regulation. Wang *et al.* demonstrated that heat-treated foxtail millet protein (HMP) improves

Table 4.3 Effect of millet consumption on change in gut microbiome.

Millet	Disease	Duration	Gut microbiome changes	Conclusion	Ref.
Quinic acid (QA) derived from millet	HFD-induced Alzheimer's disease in mice		Improved gut microbiota composition; increased tryptophan metabolites (IAA, KYNA), which correlated with reduced inflammation and Alzheimer's indicators.	QA counteracts HFD-induced brain oxidative stress and neuroinflammation by regulating gut microbial metabolites.	78
Foxtail millet (*Setaria italica* L.) bran	DSS-induced colitis mice		Restored gut microbiota composition; increased commensal bacteria (*Lachnospiraceae, Rikenellaceae*); elevated SCFA levels.	BPIS ameliorates DSS-induced colitis by preventing intestinal barrier damage and promoting gut microbiota health.	79
Heat-treated foxtail millet protein (HMP)	Prediabetic mice		Reduced *Dubosiella* and *Marvinbryantia*; increased *Lactobacillus*, *Bifidobacterium*, and *Erysipelotrichaceae*.	HMP improves gut microbiota composition and serum metabolic profiles, contributing to its hypoglycemic potential in prediabetes.	67
Millet and other African grain-based diets	Effects of sub-Saharan African diets on metabolic dysfunction, stunted growth, and microbiome-associated diseases in humans		Promoted gut lining integrity; modulated microbial abundance and inflammatory mediators; reduced *Enterobacteriaceae* taxa linked to inflammation.	Fiber-rich millet-based diets could act as biotherapeutics to downregulate inflammatory signals and improve immune responses.	80
Raw and cooked foxtail millet protein isolates	Streptozotocin (STZ)-induced diabetic mice	5 weeks	Both raw and cooked isolates mitigated diabetes-induced gut dysbiosis. Cooked isolate further enhanced GLP-1R/PI3K/AKT signaling and improved lipid profiles.	Both raw and cooked foxtail millet protein isolates exert hypoglycemic effects by modulating gut microbiota and glucose homeostasis pathways.	81

Oats, tartary buckwheat, foxtail millet	High-fat diet (HFD) fed rats	12 weeks	Increased *Lactobacillus* and *Romboutsia*; elevated butyrate level by 2.16-fold with oats supplementation.	Oats and tartary buckwheat modulate gut microbiota, improve lipid metabolism, and reduce oxidative stress and inflammation in HFD-fed rats.	82
Millet (MSB+), corn-soy-blend (CSB+)	Malnourished pigs (pure maize diet model)	3 weeks	MSB+ elevated *Bacteroidetes*, *Firmicutes*, and *Lachnospira* spp., reduced Proteobacteria; strong correlation between *Faecalibacterium* and *Lachnospira* spp. abundance and growth parameters.	Millet-based nutritional supplementation restored gut microbiome diversity and alleviated malnutrition symptoms in pigs.	83
Millet (Vitexin)	DSS-induced colitis mice		Vitexin altered gut microbiota by decreasing harmful bacteria (*Veillonella*, *Terrisporobacter*, *Klebsiella*, *Paeniclostridium*) and increasing beneficial bacteria (*Parabacteroides*, *Flavonifractor*, *Blautia*).	Vitexin alleviates colitis by regulating gut microbiota, improving intestinal barrier, and reducing inflammation.	84
Cooked foxtail millet (PCFM)	High-fat diet + STZ-induced diabetic mice	5 weeks	PCFM alleviated diabetes-related gut microbiota dysbiosis, improving the microbiome composition.	PCFM ameliorates glucose metabolism disorders and exhibits hypoglycemic effects by modulating gut microbiota and serum metabolomics.	64
Millet shell polyphenols (MSPs)	$ApoE^{-/-}$ mice fed with high-fat diet		MSPs enriched *Oscillospira* and *Ruminococcus*, and reduced *Allobaculum* in $ApoE^{-/-}$ mice fed a high-fat diet.	MSPs inhibit atherosclerotic plaque development, regulate gut barrier integrity, and modulate gut microbiota, showing cardiovascular protective effects.	85
Foxtail millet (bound polyphenols, BPIS)	High-fat diet (HFD)-induced non-alcoholic fatty liver disease (NAFLD) mice	12 weeks	BPIS remodeled gut microbiota structure, including changes at the phylum and genus levels, towards a healthier profile.	BPIS from foxtail millet protects against NAFLD by restoring gut microbiota and upregulating tight junction proteins.	86

Table 4.3 *(Continued)*

Millet	Disease	Duration	Gut microbiome changes	Conclusion	Ref.
Foxtail millet (Bowman–Birk type trypsin inhibitor, FMB-BBTI)	Apo $E^{-/-}$ atherosclerotic mice		Increased *Firmicutes* at the phylum level and significant enrichment of *Lactobacillus* at the genus level.	FMB-BBTI from foxtail millet bran restrains AS progression by remodeling gut microbiota, contributing to its anti-AS activity.	87
Mixture of millet, maize, oat, soybean, and purple potato	HFD-induced obese mice		Increased relative abundance of *Lactobacillus* and *Bifidobacterium*, enhanced diversity of gut microbiota, promoted SCFA release.	Mixture of coarse cereals could prevent HFD-induced obesity by modulating gut microbiota and enhancing SCFA production.	88
Pearl millet (*Pennisetum glaucum*)	Male Wistar rate		Increased *Bacteroidetes* and decreased *Firmicutes* in gut microbiota.	PMAX (arabinoxylan from pearl millet bran) modulates gut microbiota, showing potential as a nutraceutical.	89
Foxtail millet	Colorectal cancer mice		Increased abundance of *Bifidobacterium* and *Bacteroidales*_S24-7.	Foxtail millet reduced tumor burden, disease activity, and inflammation in CAC mice by modulating gut microbiota.	90
Foxtail millet (FM)	High-fat diet and streptozotocin-induced diabetic rats		Increased abundance of *Lactobacillus* and *Ruminococcus*_2, negatively correlated with fasting glucose (FG) and 2-h glucose.	FM supplementation improves hypoglycemia by modulating gut microbiota and activating PI3K/AKT signaling.	91
Kodo millet (whole grain and bran)	Diet-induced obese mice	16 weeks	Increased abundance of *Lactobacillus* sp., *Bifidobacteria*, *Akkermansia*, and *Roseburia* spp. Increased short-chain fatty acid and acetate levels.	Kodo millet supplementation improves glucose tolerance, lipid profile, and promotes beneficial gut bacteria.	92

Millet bran–insoluble dietary fiber (MBIDF)	DSS-induced colitis mice		Inhibited growth of *Muribaculaceae*_unclassified, *Bacteroides*, *Parasutterella*; promoted growth of *Candidatus Saccharimonas* and *Clostridia*_UCG-014. Increased *Lachnospiraceae*_NK4A136.	MBIDF alleviated DSS-induced chronic colitis by reducing inflammation and improving gut barrier function.	93
Extract of millet bran protein (EMBP)	HFD-induced metabolic syndrome in mice		Improved gut microbiome dysbiosis; upregulated *Bacteroides acidifaciens*; increased gamma-aminobutyric acid (GABA).	EMBP suppresses MetS by remodeling gut microbiota and enhancing NRF2-dependent gut barrier function, making it a candidate for food–medicine dual use.	94
Foxtail millet (*Setaria italica*) porridge	Diphenoxylate-induced constipation in mice	14 days	Increased *Bifidobacterium* and *Lactobacillus*; decreased *Escherichia coli*, *Enterococcus*, and *Bacteroides*.	Millet porridge accelerates intestinal motility, changes proportions of intestinal flora, and exerts a potent prebiotic effect.	95
Millet ethanol extract and whole grain millet	HFD-induced hyperlipidemia		Increased SCFAs (acetate, propionate, butyrate); decreased *Firmicutes*; increased *Actinobacteria*.	Millet exhibits hypolipidemic effects by regulating lipid metabolism-related gene expression (SREBP-1C, FAS, HMGCR) and gut microbiota composition.	96
Millet bran polysaccharide (MBP)	HFD + STZ induced diabetes in mice	4 weeks	Increased levels of beneficial bacteria; decreased harmful bacteria; improved intestinal microbial community.	MBP alleviates T2DM symptoms, reduces fasting blood glucose, improves glucose tolerance, exhibits anti-inflammatory effects, and modulates gut microbiota.	97

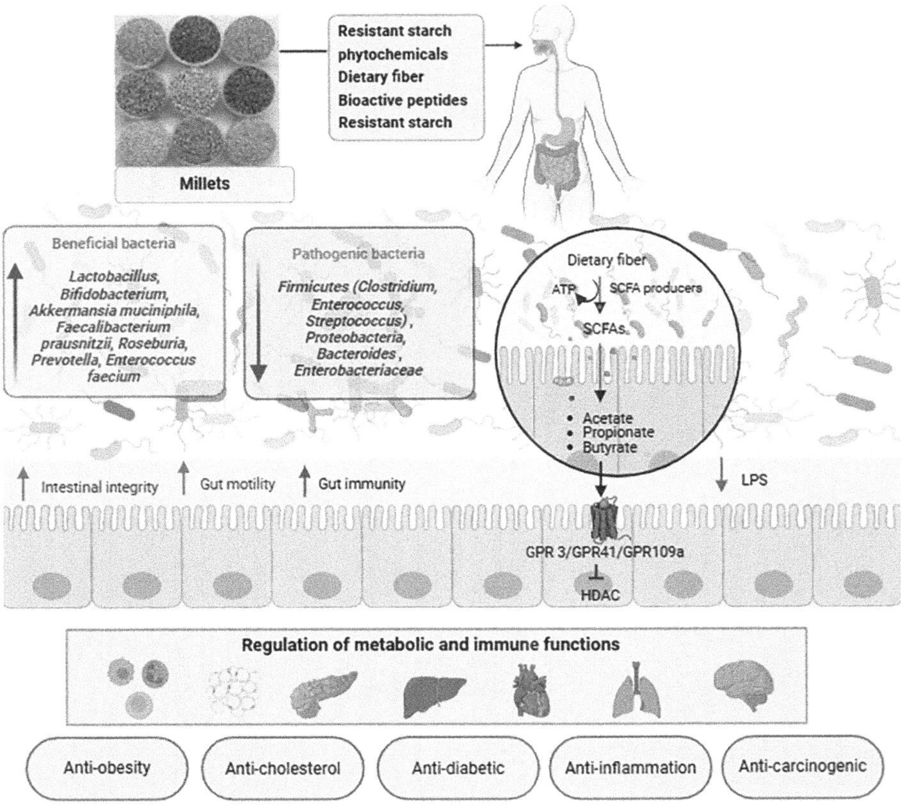

Figure 4.1 The effect of millet consumption on gut microbiome changes.

glucose metabolism, insulin sensitivity, and gut microbiota composition. Similarly, Fu *et al.* found that millet protein isolates can regulate intracellular signaling pathways associated with glucose homeostasis. These findings highlight the potential of millets in diabetes management through gut microbiota modulation.[67] Malnutrition exacerbates gut dysbiosis and immune dysfunction. Millet-based nutritional supplementation has proven effective in restoring microbial balance. Li *et al.* investigated millet-soy-blend (MSB+) supplementation in malnourished pigs, observing improved gut microbiota diversity, colonization by beneficial bacterial species, and better growth parameters. This research underscores the therapeutic potential of millets in addressing malnutrition-related gut dysbiosis.[68] Millets, enriched with bioactive compounds, contribute to cardiovascular protection. Li *et al.* demonstrated that millet shell polyphenols (MSPs) inhibit atherosclerotic plaque development by modulating inflammatory pathways and enriching beneficial bacterial taxa. These findings suggest the potential of millets as dietary interventions for cardiovascular health through gut microbiota modulation.[69] Colorectal cancer (CRC) remains a significant global health challenge. Emerging evidence suggests that millets may confer protective effects against CRC. Zhang *et al.* observed that millet consumption suppresses tumor burden and modulates inflammatory signaling pathways in colitis-associated CRC models. The observed changes in gut microbiota composition, such as increased *Bifidobacteria* and *Lactobacilli*, further highlight the anti-inflammatory and gut microbiota-modulatory properties of millets.[70] Millets contain polyphenols and non-digestible carbohydrates with prebiotic properties. These compounds support beneficial bacterial growth, inhibit pathogens, and produce short-chain fatty acids (SCFAs) like butyrate. SCFAs enhance gut barrier integrity, reduce inflammation, and regulate immune responses. Additionally, millets influence the secretion of gut hormones, such as glucagon-like peptide-1 (GLP-1) and peptide YY (PYY), which regulate appetite and glucose homeostasis. These combined effects contribute to metabolic health and overall well-being.[2] Innovative applications of millets include their integration into probiotic-enriched foods. Subbaiyan *et al.* incorporated *Lactobacillus acidophilus* into traditional millet-based laddus, demonstrating enhanced gut health benefits. Such approaches offer convenient and culturally relevant dietary options for promoting gut health.[71] The interaction between millets and the gut microbiome is multifaceted, involving fermentation, SCFA production, prebiotic effects, gut barrier modulation, and hormonal regulation. Millets hold immense potential as dietary interventions for managing metabolic disorders, enhancing immune function, and preventing chronic diseases. Future research should focus on clinical trials to validate these findings and elucidate the molecular mechanisms underlying millet's effects on gut microbiota and health.

4.6 Conclusion

In conclusion, millets represent a valuable nutritional resource, offering significant health benefits through their diverse composition and positive effects on the gut microbiome. These ancient grains are rich in proteins, dietary fiber, polyphenols, and various bioactive compounds, making them not only a source of essential nutrients but also a functional food with therapeutic potential. The nutritional value of millets extends beyond their macronutrient content, as their bioactive compounds, including phenolic acids and flavonoids, exhibit antioxidant, anti-inflammatory, and anti-diabetic properties. These benefits are particularly notable in the context of gut health and disease prevention. The digestibility of millet protein, while influenced by factors such as processing techniques, plays a key role in its impact on gut microbiota composition and overall health. Millets have been shown to contribute to the production of short-chain fatty acids (SCFAs), such as butyrate, which help maintain gut barrier integrity, reduce inflammation, and support the growth of beneficial gut bacteria like *Akkermansia* and *Blautia*. These bacteria are essential for maintaining a balanced gut microbiome, which, in turn, influences metabolic regulation, immune function, and protection against chronic diseases. Millet consumption has demonstrated positive outcomes in managing metabolic disorders, particularly diabetes, through its low glycemic index and ability to modulate gut microbiota. Millet proteins and fibers have been linked to improved glucose metabolism, insulin sensitivity, and reductions in fasting blood glucose levels. By supporting a healthy gut microbiome, millets help regulate blood sugar levels, potentially lowering the risk of type 2 diabetes. Moreover, their polyphenolic compounds have shown anti-inflammatory effects, making them useful in preventing conditions such as inflammatory bowel disease (IBD) and colorectal cancer. Furthermore, millet-based interventions have shown promise in addressing cardiovascular health and obesity. Studies have highlighted millet's ability to modulate gut microbiota in ways that reduce inflammation, support lipid metabolism, and prevent atherosclerosis. Millets also aid in weight management by promoting satiety, due to their high fiber content, which influences gut hormone secretion and slows digestion, leading to better appetite control. Although millets contain some antinutritional factors such as phytates and tannins, which can hinder nutrient absorption, proper processing techniques like fermentation, soaking, and malting can reduce these compounds and enhance nutrient bioavailability. These processing methods also improve the digestibility of millet proteins, making their amino acids more accessible for absorption. In summary, millets offer a holistic approach to health, bridging nutrition and disease prevention through their effects on the gut microbiome. Their potential to combat malnutrition, metabolic disorders, and gastrointestinal diseases makes them an essential component of a healthy diet. However, further research is necessary to fully understand the complex interactions between millet consumption, gut health, and disease outcomes. Continued exploration into millet-based dietary interventions could unlock new possibilities for promoting human health and well-being, especially in regions facing food insecurity and diet-related chronic diseases.

References

1. S. A. Kheya, *et al.*, Millets: The future crops for the tropics - Status, challenges and future prospects, *Heliyon*, 2023, **9**, e22123.
2. J. Jacob, *et al.*, The nutrition and therapeutic potential of millets: an updated narrative review, *Front. Nutr.*, 2024, **11**, 1346869.
3. S. B. Singh, *et al.*, From farm to gut: Unraveling the role of millets in promoting metabolic well-being *via* gut microbiota, *J. Drug Res. Ayurvedic Sci.*, 2023, **8**, S50–S54.
4. H. V. Sheethal, C. Baruah, K. Subhash, R. Ananthan and T. Longvah, Insights of Nutritional and Anti-nutritional Retention in Traditionally Processed Millets, *Front. Sustainable Food Syst.*, 2022, **5**, 735356.
5. G. Kaur, *et al.*, Exploring sustainable novel millet protein: A look at the future foods through innovative processing, *Future Foods*, 2024, **9**, 100367.
6. M. Conlon and A. Bird, The Impact of Diet and Lifestyle on Gut Microbiota and Human Health, *Nutrients*, 2014, **7**, 17–44.

7. J. Tan *et al.* The Role of Short-Chain Fatty Acids in Health and Disease, in *Advances in Immunology*, Elsevier, 2014, vol. 121, pp. 91–119.
8. N. A. N. Gowda, *et al.*, Modern Processing of Indian Millets: A Perspective on Changes in Nutritional Properties, *Foods*, 2022, **11**, 499.
9. Y. Xiong, P. Zhang, R. D. Warner and Z. Fang, Sorghum Grain: From Genotype, Nutrition, and Phenolic Profile to Its Health Benefits and Food Applications, *Compr. Rev. Food Sci. Food Saf.*, 2019, **18**, 2025–2046.
10. B. N. Vinutha, S. C. Meti & S. Rawat, Finger Millet: Biology, Functional Potential and Sustainable Utilization: Major Millets, in *Millets: The Multi-Cereal Paradigm for Food Sustainability*, ed. M. Thakur, Springer Nature Switzerland, Cham, 2024, pp. 75–88.
11. S. Rani, R. Singh, R. Sehrawat, B. P. Kaur and A. Upadhyay, Pearl millet processing: a review, *Nutr. Food Sci.*, 2018, **48**, 30–44.
12. N. Sharma and K. Niranjan, Foxtail millet: Properties, processing, health benefits, and uses, *Food Rev. Int.*, 2018, **34**, 329–363.
13. K. Srivastava and P. Vijayakumar, Comparative Nutritional Benefits of Kodo and Little Millets against other Cereals and their Role in Dairy Production, *J. Adv. Biol. Biotechnol.*, 2024, **27**, 1228–1244.
14. R. Krishnan, U. Dharmaraj, R. Sai Manohar and N. G. Malleshi, Quality characteristics of biscuits prepared from finger millet seed coat based composite flour, *Food Chem.*, 2011, **129**, 499–506.
15. S. Nithiyanantham, *et al.*, Nutritional and functional roles of millets—A review, *J. Food Biochem.*, 2019, **43**, e12859.
16. S. Saleem, *et al.*, Millets as smart future food with essential phytonutrients for promoting health, *J. Food Compos. Anal.*, 2023, **124**, 105669.
17. A. Scalbert, C. Manach, C. Morand, C. Rémésy and L. Jiménez, Dietary Polyphenols and the Prevention of Diseases, *Crit. Rev. Food Sci. Nutr.*, 2005, **45**, 287–306.
18. A. Chandrasekara and F. Shahidi, Bioaccessibility and antioxidant potential of millet grain phenolics as affected by simulated *in vitro* digestion and microbial fermentation, *J. Funct. Foods*, 2012, **4**, 226–237.
19. A. Fardet, E. Rock and C. Rémésy, Is the *in vitro* antioxidant potential of whole-grain cereals and cereal products well reflected *in vivo*?, *J. Cereal Sci.*, 2008, **48**, 258–276.
20. M. S. Edge, J. M. Jones and L. Marquart, A New Life for Whole Grains, *J. Am. Diet. Assoc.*, 2005, **105**, 1856–1860.
21. A. Truswell, Cereal grains and coronary heart disease, *Eur. J. Clin. Nutr.*, 2002, **56**, 1–14.
22. USDA, *Nutrition and Your Health: Dietary Guidelines for Americans*, 2005.
23. K.-O. Park, Y. Ito, T. Nagasawa, M.-R. Choi and N. Nishizawa, Effects of Dietary Korean Proso-Millet Protein on Plasma Adiponectin, HDL Cholesterol, Insulin Levels, and Gene Expression in Obese Type 2 Diabetic Mice, *Biosci., Biotechnol., Biochem.*, 2008, **72**, 2918–2925.
24. T. P. Carr, *et al.*, Grain Sorghum Lipid Extract Reduces Cholesterol Absorption and Plasma Non-HDL Cholesterol Concentration in Hamsters, *J. Nutr.*, 2005, **135**, 2236–2240.

25. J. W. Anderson and T. J. Hanna, Whole grains and protection against coronary heart disease: what are the active components and mechanisms? *Am. J. Clin. Nutr.*, 1999, **70**, 307–308.
26. P. Lakshmi Kumari and S. Sumathi, Effect of consumption of finger millet on hyperglycemia in non-insulin dependent diabetes mellitus (NIDDM) subjects, *Plant Foods Hum. Nutr.*, 2002, **57**, 205–213.
27. Y.-Y. Choi, *et al.*, Effects of Dietary Protein of Korean Foxtail Millet on Plasma Adiponectin, HDL-Cholesterol, and Insulin Levels in Genetically Type 2 Diabetic Mice, *Biosci., Biotechnol., Biochem.*, 2005, **69**, 31–37.
28. P. Würsch and F. X. Pi-Sunyer, The Role of Viscous Soluble Fiber in the Metabolic Control of Diabetes: A review with special emphasis on cereals rich in β-glucan, *Diabetes Care*, 1997, **20**, 1774–1780.
29. M. P. R. Prasad, B. D. Rao, K. Kalpana, M. V. Rao and J. V. Patil, Glycaemic index and glycaemic load of sorghum products, *J. Sci. Food Agric.*, 2015, **95**, 1626–1630.
30. J. Pei, *et al.*, A Review of the Potential Consequences of Pearl Millet (Pennisetum glaucum) for Diabetes Mellitus and Other Biomedical Applications, *Nutrients*, 2022, **14**, 2932.
31. S. Chethan, S. M. Dharmesh and N. G. Malleshi, Inhibition of aldose reductase from cataracted eye lenses by finger millet (Eleusine coracana) polyphenols, *Bioorg. Med. Chem.*, 2008, **16**, 10085–10090.
32. S. A. S. Jayawardana, J. K. R. R. Samarasekera, G. H. C. M. Hettiarachchi and M. J. Gooneratne, Antidiabetic properties of finger millet (Eleusine coracana (L.) Gaertn.) varieties cultivated in Sri Lanka, *J. Herb. Med.*, 2022, **32**, 100534.
33. R. Ugare, B. Chimmad, R. Naik, P. Bharati and S. Itagi, Glycemic index and significance of barnyard millet (Echinochloa frumentacae) in type II diabetics, *J. Food Sci. Technol.*, 2014, **51**, 392–395.
34. D. Kumari, T. Madhujith and A. Chandrasekara, Comparison of phenolic content and antioxidant activities of millet varieties grown in different locations in Sri Lanka, *Food Sci. Nutr.*, 2017, **5**, 474–485.
35. A. Kumar, *et al.*, Nutritional Significance and Antioxidant-Mediated Antiaging Effects of Finger Millet: Molecular Insights and Prospects, *Front. Sustainable Food Syst.*, 2021, **5**, 684318.
36. V. Viswanath, A. Urooj and N. G. Malleshi, Evaluation of antioxidant and antimicrobial properties of finger millet polyphenols (Eleusine coracana), *Food Chem.*, 2009, **114**, 340–346.
37. S. Bellato, R. Ciccoritti, V. Del Frate, D. Sgrulletta and K. Carbone, Influence of genotype and environment on the content of 5-n alkylresorcinols, total phenols and on the antiradical activity of whole durum wheat grains, *J. Cereal Sci.*, 2013, **57**, 162–169.
38. S. R. Pradeep and M. Guha, Effect of processing methods on the nutraceutical and antioxidant properties of little millet (Panicum sumatrense) extracts, *Food Chem.*, 2011, **126**, 1643–1647.
39. R. Tsao, Chemistry and Biochemistry of Dietary Polyphenols, *Nutrients*, 2010, **2**, 1231–1246.

40. A. Chandrasekara and F. Shahidi, Content of Insoluble Bound Phenolics in Millets and Their Contribution to Antioxidant Capacity, *J. Agric. Food Chem.*, 2010, **58**, 6706–6714.
41. A. Chandrasekara and F. Shahidi, Bioactivities and Antiradical Properties of Millet Grains and Hulls, *J. Agric. Food Chem.*, 2011, **59**, 9563–9571.
42. M. dJ. Abedin, A. T. M. Abdullah, M. A. Satter and T. Farzana, Physical, functional, nutritional and antioxidant properties of foxtail millet in Bangladesh, *Heliyon*, 2022, **8**, e11186.
43. W. Xu, *et al.*, A novel antifungal peptide from foxtail millet seeds: A novel antifungal peptide, *J. Sci. Food Agric.*, 2011, **91**, 1630–1637.
44. R. Radhajeyalakshmi, K. Yamunarani, K. Seetharaman and R. Velazhahan, Existence of Thaumatin-like Proteins (TLPs) in Seeds of Cereals, *Acta Phytopathol. Entomol. Hung.*, 2003, **38**, 251–257.
45. S. Banerjee, Finger millet (Eleusine coracana) polyphenols: Investigation of their antioxidant capacity and antimicrobial activity, *Afr. J. Food Sci.*, 2012, **6**(13), 362–374.
46. M. Siwela, J. R. N. Taylor, W. A. J. De Milliano and K. G. Duodu, Influence of phenolics in finger millet on grain and malt fungal load, and malt quality, *Food Chem.*, 2010, **121**, 443–449.
47. N. S. Rajasekaran, M. Nithya, C. Rose and T. S. Chandra, The effect of finger millet feeding on the early responses during the process of wound healing in diabetic rats, *Biochim. Biophys. Acta*, 2004, **1689**, 190–201.
48. E. Graf, Antioxidant functions of phytic acid, *Free Radic. Biol. Med.*, 1990, **8**, 61–69.
49. S. J. van Rensburg, Epidemiologic and dietary evidence for a specific nutritional predisposition to esophageal cancer, *J. Natl. Cancer Inst.*, 1981, **67**, 243–251.
50. A. Chandrasekara and F. Shahidi, Antiproliferative potential and DNA scission inhibitory activity of phenolics from whole millet grains, *J. Funct. Foods*, 2011, **3**, 159–170.
51. H. R. Grimmer, V. Parbhoo and R. M. McGrath, Antimutagenicity of polyphenol-rich fractions from sorghum bicolor grain, *J. Sci. Food Agric.*, 1992, **59**, 251–256.
52. C. Gómez-Cordovés, B. Bartolomé, W. Vieira and V. M. Virador, Effects of Wine Phenolics and Sorghum Tannins on Tyrosinase Activity and Growth of Melanoma Cells, *J. Agric. Food Chem.*, 2001, **49**, 1620–1624.
53. V. Parbhoo, H. R. Grimmer, A. Cameron-Clarke and R. M. McGrath, Induction of cytochrome P-450 in rat liver by a polyphenol-rich extract from a bird-resistant sorghum grain, *J. Sci. Food Agric.*, 1995, **69**, 247–252.
54. C. Catassi and A. Fasano, Is this really celiac disease? Pitfalls in diagnosis, *Curr. Gastroenterol. Rep.*, 2008, **10**, 466–472.
55. I. Comino, *et al.*, The Gluten-Free Diet: Testing Alternative Cereals Tolerated by Celiac Patients, *Nutrients*, 2013, **5**, 4250–4268.
56. S. Rai, A. Kaur and C. S. Chopra, Gluten-Free Products for Celiac Susceptible People, *Front. Nutr.*, 2018, **5**, 116.

57. A. Angioloni and C. Collar, Effects of pressure treatment of hydrated oat, finger millet and sorghum flours on the quality and nutritional properties of composite wheat breads, *J. Cereal Sci.*, 2012, **56**, 713–719.
58. E. T. Mertz, *et al.*, Pepsin digestibility of proteins in sorghum and other major cereals, *Proc. Natl. Acad. Sci. U. S. A.*, 1984, **81**, 1–2.
59. V. Shukla, S. Srivastava, S. Singh, M. Mursal and S. Hussain, Unveiling the intricacies of phytate antinutrients in millets and their therapeutic implications in breast cancer, *Intell. Pharm.*, 2024, **2**, 516–527.
60. K. G. Duodu and N. Nwabisa, 5 Processing of Millet Foods and its Impact on Nutraceutical and Health-Promoting Properties, in *Nutriomics of Millet Crops*, 2023, pp. 99–123.
61. P. Gulati, *et al.*, Heating Reduces Proso Millet Protein Digestibility *via* Formation of Hydrophobic Aggregates, *J. Agric. Food Chem.*, 2017, **65**, 1952–1959.
62. F. S. Pushparaj and A. Urooj, Influence of Processing on Dietary Fiber, Tannin and in Vitro Protein Digestibility of Pearl Millet, *Food Nutr. Sci.*, 2011, **2**, 895–900.
63. C. M. Guinane and P. D. Cotter, Role of the gut microbiota in health and chronic gastrointestinal disease: understanding a hidden metabolic organ, *Ther. Adv. Gastroenterol.*, 2013, **6**, 295–308.
64. Y. Fu, *et al.*, Hypoglycemic Effect of Prolamin from Cooked Foxtail Millet (Setaria italic) on Streptozotocin-Induced Diabetic Mice, *Nutrients*, 2020, **12**, 3452.
65. R.-G. Xiong, *et al.*, Health Benefits and Side Effects of Short-Chain Fatty Acids, *Foods*, 2022, **11**, 2863.
66. M. M. Molla, *Effect of foxtail millet diet on liver injury and blood lipid profile induced by D-galactosamine in mice*, China Agricultural University (CAU), 2016.
67. H. Wang, *et al.*, Heat-treated foxtail millet protein delayed the development of pre-diabetes to diabetes in mice by altering gut microbiota and metabolomic profiles, *Food Funct.*, 2023, **14**, 4866–4880.
68. X. Li, *et al.*, Millet-based supplement restored gut microbial diversity of acute malnourished pigs, *PLoS One*, 2021, **16**, e0250423.
69. F. Liu, *et al.*, Millet shell polyphenols prevent atherosclerosis by protecting the gut barrier and remodeling the gut microbiota in $ApoE^{-/-}$ mice, *Food Funct.*, 2021, **12**, 7298–7309.
70. B. Zhang, *et al.*, Dietary Supplementation of Foxtail Millet Ameliorates Colitis-Associated Colorectal Cancer in Mice *via* Activation of Gut Receptors and Suppression of the STAT3 Pathway, *Nutrients*, 2020, **12**, 2367.
71. R. Subbaiyan, A. Ganesan, V. Varadharajan, P. R. Jeyachandran and H. Thangavel, Formulation and validation of probioticated foxtail millet laddu as a source of antioxidant for biological system using response surface methodology, *Braz. J. Microbiol.*, 2024, **55**, 647–661.
72. United States Department of Agriculture/Human Nutrition Information Service (USDA/HNIS), *Composition of Foods: Cereal Grains and Pasta*, 1984, pp. 8–20.

73. V. D. Pawar and G. M. Machewad, Processing of foxtail millet for improved nutrient availability, *J. Food Process. Preserv.*, 2006, **30**, 269–279.
74. S. Sharma, D. C. Saxena and C. S. Riar, Analysing the effect of germination on phenolics, dietary fibres, minerals and γ-amino butyric acid contents of barnyard millet (Echinochloa frumentaceae), *Food Biosci.*, 2016, **13**, 60–68.
75. S. O. Azeez, *et al.*, Impact of germination alone or in combination with solid-state fermentation on the physicochemical, antioxidant, *in vitro* digestibility, functional and thermal properties of brown finger millet flours, *LWT*, 2022, **154**, 112734.
76. K. Kumari, P. Kashyap and P. Chakrabarti, Germination and probiotic fermentation: a way to enhance nutritional and biochemical properties of cereals and millets, *Food Sci. Biotechnol.*, 2024, **33**, 505–518.
77. S. R. Kumar, M. B. Sadiq and A. K. Anal, Comparative study of physicochemical and functional properties of pan and microwave cooked underutilized millets (proso and little), *LWT*, 2020, **128**, 109465.
78. S. Li, *et al.*, Quinic acid alleviates high-fat diet-induced neuroinflammation by inhibiting DR3/IKK/NF-κB signaling via gut microbial tryptophan metabolites, *Gut Microbes*, 2024, **16**, 2374608.
79. R. Yang, *et al.*, Polyphenols from foxtail millet bran ameliorate DSS-induced colitis by remodeling gut microbiome, *Front. Nutr.*, 2022, **9**, 1030744.
80. P. Isibor, *et al.*, Significance of African Diets in Biotherapeutic Modulation of the Gut Microbiome, *Bioinform. Biol. Insights*, 2021, **15**, 11779322211012697.
81. Y. Fu, *et al.*, Protein Isolates from Raw and Cooked Foxtail Millet Attenuate Development of Type 2 Diabetes in Streptozotocin-Induced Diabetic Mice, *Mol. Nutr. Food Res.*, 2021, **65**, 2000365.
82. Y. Wang, *et al.*, Effects of Oats, Tartary Buckwheat, and Foxtail Millet Supplementation on Lipid Metabolism, Oxido-Inflammatory Responses, Gut Microbiota, and Colonic SCFA Composition in High-Fat Diet Fed Rats, *Nutrients*, 2022, **14**, 2760.
83. X. Li, *et al.*, Millet-based supplement restored gut microbial diversity of acute malnourished pigs, *PLoS One*, 2021, **16**, e0250423.
84. S. Li, *et al.*, Regulation of gut microbiota and alleviation of DSS-induced colitis by vitexin, *Eur. J. Nutr.*, 2023, **62**, 3433–3445.
85. F. Liu, *et al.*, Millet shell polyphenols prevent atherosclerosis by protecting the gut barrier and remodeling the gut microbiota in $ApoE^{-/-}$ mice, *Food Funct.*, 2021, **12**, 7298–7309.
86. I. Ghani, Y. An, Q. Qiao, S. He and Z. Li, Polyphenols from Foxtail Millet Improve Non-Alcoholic Fatty Liver Disease by Regulating Intestinal Microbiome in Mice, *Foods*, 2024, **13**, 1683.
87. S. Shan, *et al.*, Bowman-Birk Major Type Trypsin Inhibitor Derived from Foxtail Millet Bran Attenuate Atherosclerosis *via* Remodeling Gut Microbiota in $ApoE^{-/-}$ Mice, *J. Agric. Food Chem.*, 2022, **70**, 507–519.
88. Y. Ji, *et al.*, Dietary intake of mixture coarse cereals prevents obesity by altering the gut microbiota in high-fat diet fed mice, *Food Chem. Toxicol.*, 2021, **147**, 111901.

89. A. Singh, *et al.*, Arabinoxylan from pearl millet bran: Optimized extraction, structural characterization, and its bioactivities, *Int. J. Biol. Macromol.*, 2024, **279**, 135247.
90. B. Zhang, *et al.*, Dietary Supplementation of Foxtail Millet Ameliorates Colitis-Associated Colorectal Cancer in Mice *via* Activation of Gut Receptors and Suppression of the STAT3 Pathway, *Nutrients*, 2020, **12**, 2367.
91. X. Ren, *et al.*, Foxtail Millet Improves Blood Glucose Metabolism in Diabetic Rats through PI3K/AKT and NF-κB Signaling Pathways Mediated by Gut Microbiota, *Nutrients*, 2021, **13**, 1837.
92. S. M. Sarma, *et al.*, Kodo millet whole grain and bran supplementation prevents high-fat diet induced derangements in a lipid profile, inflammatory status and gut bacteria in mice, *Food Funct.*, 2017, **8**, 1174–1183.
93. M. Li, Q. Wang, M. Niu, H. Yang and S. Zhao, Protective effects of insoluble dietary fiber from cereal bran against DSS-induced chronic colitis in mice: From inflammatory responses, oxidative stress, intestinal barrier, and gut microbiota, *Int. J. Biol. Macromol.*, 2024, **283**, 137846.
94. S. Shan, *et al.*, Gut microbiota remodeling drived by dietary millet protein prevents the metabolic syndrome, *Food Sci. Hum. Wellness*, 2024, **13**, 1987–2001.
95. Y. Chen, R. Zhang, J. Xu and Q. Ren, Alteration of intestinal microflora by the intake of millet porridge improves gastrointestinal motility, *Front. Nutr.*, 2022, **9**, 965687.
96. S. Li, *et al.*, Effects of millet whole grain supplementation on the lipid profile and gut bacteria in rats fed with high-fat diet, *J. Funct. Foods*, 2019, **59**, 49–59.
97. J. Zhang, *et al.*, Antidiabetic Effect of Millet Bran Polysaccharides Partially Mediated *via* Changes in Gut Microbiome, *Foods*, 2022, **11**, 3406.

CHAPTER 5

Millet Proteins Beyond Nutrition: Functional, Structural and Engineering Attributes

ZEYNEP SALIHA GÜNEŞ,[a,b] RUKIYE GÜNDOGAN,[a,c] AYSUN YÜCETEPE[a,d] AND TUBA ESATBEYOGLU*[a]

[a] Department of Molecular Food Chemistry and Food Development, Institute of Food and One Health, Gottfried Wilhelm Leibniz University Hannover, Hannover, Germany; [b] Department of Food Engineering, Istanbul Sabahattin Zaim University, 34303 Halkali, Istanbul, Türkiye; [c] Department of Food Engineering, Faculty of Chemical and Metallurgical Engineering, Istanbul Technical University, Türkiye; [d] Department of Food Engineering, Faculty of Engineering, Aksaray University, Aksaray TR-68100, Türkiye
*Email: esatbeyoglu@foh.uni-hannover.de

5.1 Introduction

Global warming and climate change threaten agriculture, undermining food security and increasing malnutrition, especially hidden hunger in developing countries. Millets offer a sustainable solution for food security due to their resilience to extreme climates, drought, and pests. These crops require minimal water, have a shorter growing season, and are less prone to diseases, making them an environmentally friendly alternative. These seeds are ecologically significant, adapt well to mixed and intercropping systems, and withstand adverse environmental factors, including salinity and drought.[1]

Food Chemistry, Function and Analysis No. 48
Millet Proteins: Nutritional, Structural and Functional Attributes
Edited by Shweta Suri, Anuradha Dutta and Tarun Belwal

Published by the Royal Society of Chemistry, www.rsc.org

Rich in calcium, fiber, polyphenols, and protein, millets are gaining attention as a nutritious staple to address food shortages and nutritional deficiencies.[2,3] All warm-season cereal grains, apart from wheat and rice, are categorized as coarse cereals, which is the general term for millets. These grains, which have historically been utilized as animal and bird feed, are becoming known as cutting-edge human medicinal foods.[1,4] Vital minerals, vitamins, and beneficial secondary metabolites are abundant in millets. They have a mineral profile that is superior to those of the main cereals, such as rice and wheat, and they are naturally high in fiber and gluten-free. Notably, finger millet supports bone and dental health since it contains ten times more calcium than other grains.[5,6]

Major and minor millets are two types of millets that are grown as cereal crops or grains all over the world. The four major millets are finger millet (*Eleusine coracana*), foxtail millet (*Setaria italic*), proso or white millet (*Panicum miliaceum*), and pearl millet (*Pennisetum glaucum*). Browntop millet (*Urochloa ramose*), kodo millet (*Paspalum scrobiculatum*), Guinea millet (*Brachiaria deflexa*), barnyard millet (*Echinochloa* spp.), little millet (*Panicum sumatrense*), and sorghum (*Sorghum* spp.) are the minor millets.[7]

Millets have higher protein, lipid and fiber content than rice and wheat seeds and they also contain polyphenolic compounds.[1,8] Protein makes up around 10% of millet's total weight, making it the second most prevalent nutrient after starch.[1] Typical millet seed proteins contain fractions of albumin, globulin, cross-linked prolamin, and glutelin. In millet proteins, a high concentration of essential amino acids containing sulfur, like cysteine and methionine, is present.[7] Particularly, finger millet is quite valuable since it has a high methionine content but a low lysine content.[8] Millet seeds are a viable alternative protein source because they may be processed into protein isolates and concentrates with an appropriate amino acid profile.[9] Millet protein isolates and hydrolysates offer significant potential for applications in the food, pharmaceutical, and nutraceutical industries. Millet protein fractions showed outstanding nutritional and sensory qualities, and fortifying diets with complementary proteins or necessary amino acids can improve nutritional quality.[10,11] Several variables, including cultivar, growing environment, and farming methods, affect the millet proteins' concentration, content, and quality. Protein quality and quantity are also influenced by growing temperature, chemical inputs, water availability, and soil quality.[12]

The creation of functional foods and nutraceuticals can be aided by researching millet proteins to find those associated with these health-promoting qualities.[13]

Foxtail millet can reduce the risk of chronic conditions by improving cholesterol metabolism and blood glucose control. Its antioxidants, vitamins, minerals, and fibers help lower blood glucose, improve insulin sensitivity, and support a healthier lipid profile. The millet's slow-digesting carbohydrates and fatty acids maintain a low glycemic index, aiding diabetes and cholesterol management.[14] In addition, millets are known to provide health advantages, including cancer prevention, reduced risk of heart disease, and improved gastrointestinal health.[15] Bioactive peptides, derived from proteins, play a key role in defining the structural and functional properties of millet proteins. These peptides offer various health effects, including high biological value, low toxicity, and easy metabolism. They remain active after being released from proteins by proteolytic enzymes. Studies have shown that bioactive peptides from pearl millet exhibit antioxidant and therapeutic properties, with potential benefits for managing hypertension and cholesterol metabolism.[16]

Proteins perform different roles in food matrices, referred to as functional properties, which are influenced by factors such as pH, drying, heating, ionic strength, storage conditions, the presence of reducing agents, and physical, chemical, or enzymatic modifications. Various methods can modify proteins, and treatments such as microwave, ultrasonication, high pressure processing, and milling/decortication significantly impact millet proteins, leading to changes in their structure, stability, functionality, digestibility, and foaming ability.[16,17] Knowing the protein profile of millets provides important information about their nutritional worth, including the quality of the protein and the mix of amino acids. In brief, research on millet proteins is essential for maximizing their nutritional, agricultural, and economic potential and for helping to address global concerns, including sustainable agriculture, public health, and food security. This chapter will examine the wider potential of millet proteins, focusing on their structural, functional, and engineering properties, and their prospective role in promoting sustainable food systems.

5.2 Structural Attributes of Millet Proteins

Millets are a class of small-seeded cereal crops that are prized for both their substantial nutritional profile and their ability to withstand adverse environmental conditions. The functional characteristics, digestibility, and general applicability of millet proteins in food systems are significantly influenced by their structural characteristics. The structural characteristics of millet proteins will be examined in this section, with particular attention paid to their conformation and structure.

5.2.1 Protein Structure and Conformation

The linear arrangement of amino acids connected by peptide bonds is known as a protein's fundamental structure. Essential amino acids are abundant in millet proteins, albeit the exact makeup varies depending on the millet species.[18] Prolamins and glutelins, storage proteins with different amino acid sequences, are the most abundant proteins in millet. High quantities of proline and glutamine are characteristics of millets' prolamins, often referred to as "pennisetins" for pearl millet and "setarins" for foxtail millet. These prolamins are typically insoluble in water but soluble in alcohol–water mixes.[19] Glutelins make up a sizable amount of the total protein composition in millets and are more soluble in alkaline solutions. A study found that millet proteins' prolamin fraction, pennisetin, could be isolated in trace amounts without the need of a reducing agent.[20] The amount of isolated pearl millet proteins increased as the concentration of sodium metabisulfite increased. Interestingly, the isolated pennisetin yield from pearl millet flour dropped as the quantities of dithiothreitol and β-mercaptoethanol increased. The glutelin fraction had a primary band at 15 kDa and multiple minor bands at 11, 22, 24, 78, and 209 kDa, while the prolamin fractions revealed three strong peptide bands at 11, 14, and 24 kDa.[21] The 20 kDa band was also linked to prolamin fractions, and the main polypeptides in pearl millet proteins were found to be within the molecular weight range of 12–43 kDa.[22] The distribution of proteins in millet is likewise comparable to that of maize, especially in the form of genuine prolamins that dissolve in alcohol.[23]

The Dawn cultivar's total protein content is composed of 9.5% albumin, 4.1% globulin, 47.2% prolamin, and 39.1% glutelin. By contrast, the Plateau cultivar's total protein composition is 34.5% glutelin, 50.8% prolamin, 6.1% globulin, and 8.6% albumin. While the prolamin fractions showed the least distinguishable peak, the millet protein fractions showed a single enthalpy peak. All fractions in various cultivars have an average denaturation temperature of around 82.1 ± 3.5 °C.[21]

The distinctive folding patterns of prolamins are influenced by their high proline concentration, which also affects their structural stiffness. Additionally, millet proteins' nutritional value is impacted by their comparatively low quantities of tryptophan and lysine, which are abundant in cereal proteins. However, compared to many other cereals, millets have higher concentrations of sulfur-containing amino acids (such cysteine and methionine), which are crucial for the formation of disulfide bonds and the three-dimensional structure of proteins. Changes in the proportions of β-sheets and α-helices were the primary cause of functional property differences in millet flours during bioprocessing. The variety of millet crop affects the amounts of α-helix, β-sheet, and β-coil proteins. The secondary structure of millet proteins has been analyzed using methods including Fourier-transform infrared spectroscopy (FTIR) and circular dichroism (CD) spectroscopy, which have shown that β-sheet topologies predominate.[10,24]

Although the protein's stiffness and stability are enhanced by these β-sheet structures, their solubility and digestion may be restricted. Local folding of the polypeptide chain into regular structures like α-helices and β-sheets, supported by hydrogen bonds, is referred to as the secondary structure of proteins (Smetana and Misra, 2017).[25] Although α-helices are also found, β-sheet topologies predominate in millet proteins, especially in their prolamin and glutelin fractions.[26] Prolamins typically form longer β-sheet structures with less α-helical content because of their high proline concentration. Cereal storage proteins typically have this conformation, which produces a compact, folded structure that is less soluble in water.[27,28] α-Helices and β-sheets may be distributed more evenly in glutelins, which would explain their solubility in alkaline environments as well as their usefulness in gel formation and emulsification.[29]

Sharma *et al.*[30] investigated the secondary structures of millet proteins. Using Gaussian deconvolution and the second derivative of the FTIR spectra in the amide I region (1700–1600 cm^{-1}), the spectral regions identified were β-turns (1700–1680 cm^{-1}), α-helices (1660–1650 cm^{-1}), random coils (1640–1650 cm^{-1}), and β-sheets (1680–1670 cm^{-1} and 1640–1600 cm^{-1}).

The total three-dimensional folding of the protein, influenced by disulfide bond formation, van der Waals forces, hydrophobic interactions, and hydrogen bonding, is referred to as the tertiary structure.[27] Disulfide linkages between cysteine residues sustain the complex tertiary structures that are often formed by millet proteins, particularly the glutelins. According to Bean *et al.*,[31] these disulfide bridges are essential for preserving the protein's stability and resistance to denaturation in the presence of heat and stress. Quaternary structures, in which several polypeptide chains (subunits) come together to form bigger protein complexes, are seen in the storage proteins of millet. In the millet seed, for example, glutelins and prolamins may exist as oligomers, generating bigger complexes that support these proteins' storage role.[16]

5.2.2 Protein–Amino Acid Profile

The amino acid content of millet proteins contributes significantly to their excellent nutritional value and variety of functional qualities. Around the world, research has been carried out to determine the total millet protein content of various types of millet, with protein content ranging from 2.6% to 20.8%.[32] Due to their high protein content, millets have become promising functional foods when coupled with lysine-rich foods from the Fabaceae family, such as beans, peas, and soybeans. For instance, finger millet has roughly 9.8% protein and foxtail and pearl millets have roughly 13.6% crude protein, which is much greater than that of other cereals like wheat.[33] The amino acid composition of different types of millet seeds and proteins from studies in the literature are tabulated in Table 5.1. Differences in millet varieties and dehulling process have no significant impact on the amino acid contents. On the other hand, in other studies, compared to dehulled millet samples, which ranged from 7.74 to 11.49 g per 100 g, the proteins of millet cultivars show a greater concentration range for the total amino acids (6.90–14.30 g per 100 g).[15,35,36] The results also show that certain cultivars have significantly higher quantities of leucine and alanine, two important amino acids, than other gluten-free grains. The three most prevalent amino acids in millet grains are leucine (1.4 ± 0.2 g per 100 g), alanine (1.1 ± 0.2 g per 100 g), and glutamic acid/glutamine (2.1 ± 0.3 g per 100 g). The amount of methionine differed greatly throughout cultivars. Essential amino acids like lysine, methionine, and cysteine, which are frequently lacking in other cereals like rice and wheat, are found

Table 5.1 Amino acid profiles of different millet varieties.[a]

Crop	His	Lys	Leu	Isoleu	Phe	Val	Thr	Trp	Met	Met + Cys	Ref.
Pearl millet variety 1[b]	2.89	4.02	9.99	4.59	5.30	6.20	3.97	0.66	1.04	1.51	10
Pearl millet variety 2[b]	2.15	2.81	7.56	3.42	4.14	4.53	2.87	1.00	0.83	1.66	10
Finger millet variety 1[b]	2.93	4.57	9.14	4.39	5.30	6.45	4.39	0.55	1.83	2.47	10
Finger millet variety 2[b]	2.85	4.00	9.77	4.62	5.31	6.69	4.15	0.70	1.77	2.23	10
Foxtail millet[c]	0.18	0.18	1.25	0.35	0.50	0.48	0.32	nd	0.23	0.04	39
Finger millet[b]	2.07	2.15	10.65	3.29	4.90	5.00	3.11	nd	1.82	3.81	40
The mean values of 165 different pearl millets[b]	0.42	0.22	2.89	1.22	1.11	1.67	1.02	1.58	0.21	0.39	41
Proso millet variety 1[b]	1.22	1.90	2.60	1.27	2.06	1.99	1.24	nd	0.65	nd	42
Proso millet variety 2[b]	1.39	2.13	2.84	1.30	2.11	2.13	1.38	0.01	0.73	nd	42
Foxtail millet protein concentrate[c]	2.28	2.49	6.88	4.65	5.96	5.06	3.57	2.05	4.36	6.38	11
Pearl millet[c]	0.03	0.01	0.41	0.24	0.13	nd	0.04	nd	0.02	nd	43
Proso millet variety 1[d]	2.73	1.69	17.19	4.28	nd	4.67	6.83	1.33	nd	2.93	44
Proso millet variety 2[d]	2.84	1.35	18.49	4.48	nd	4.81	7.29	1.37	nd	3.14	44

[a] Phe (phenylalanine), Val (valine), Thr (threonine), Trp (tryptophan), Met (methionine), Met + Cys (methionine + cysteine), His (histidine), Lys (lysine), Leu (leucine), and Isoleu (isoleucine).
[b] g per 100 g protein.
[c] g per 100 g sample.
[d] $mg\,g^{-1}$; nd, no data available.

in several millet types, including pearl millet and finger millet.[15,35,36] According to Anitha *et al.*,[10] these amino acids improve the health benefits of millet, which include better protein digestion and balanced nutrition. The functional importance of pearl millet in human and animal diets is enhanced by its higher amount of lysine, threonine, methionine, and cysteine when compared to other grains like sorghum and corn.[37] Additionally, finger millet has higher levels of lysine, threonine, and valine than other millet types, making it well-balanced in essential amino acids.[37] The availability of necessary amino acids determines the quality of protein. Millet's amino acid profile is like that of wheat, barley, and rice, but has higher levels of lysine, threonine, methionine, and cysteine than sorghum and corn.[38] Pearl millet is also known for having a great balance of vital amino acids, which makes it a valuable protein and energy source for people. Pearl millet is more digestible than corn for amino acids such as arginine, threonine, valine, isoleucine, and leucine. The largest amounts of glutamic acid (21.4 and 18.9 g per 100 g) were found in foxtail millet protein concentrate and glutelin, followed by arginine (6.4 and 5.9 g per 100 g), leucine (6.9 and 7.2 g per 100 g), and alanine (8.8 and 8.1 g per 100 g) in that order.[11]

5.2.3 Structural Modifications

Factors such as protein–starch interactions and processing techniques impact the structural characteristics of millet proteins, specifically their digestibility and solubility, which can impact their use in food items.[45] The type, species, cultivar, and variety of millet can all have a substantial impact on the protein matrices, kinds, structures, and functions of millet proteins.[11] Like other plant-based proteins, the amino acid content of millet proteins, their arrangement along the polypeptide chain, and the secondary, tertiary, and quaternary structures that result have a substantial impact on their structural properties. Understanding the structure and conformation of millet proteins enables one to uncover their functional characteristics, including their capacity for foaming, emulsification, gelation, and water solubility. When assessing whether millet proteins are suitable for use in food formulations, particularly when producing plant-based food items, these capabilities are crucial.[34]

Hydrophobic interactions between non-polar amino acid residues affect the compact folding of millet proteins, especially the prolamins, which further decreases the solubility of proteins. The ability of millet proteins to unfold and refold in response to environmental factors (such as pH, temperature, and ionic strength) dictates their emulsifying and gel-forming capabilities, making their tertiary structure crucial to their functionality in food systems.[46,47]

The behavior of millet proteins during food processing, specifically in heat-induced gelation and the creation of protein networks in baked foods and extruded items, is influenced by their quaternary structure. The ability of millet proteins to produce viscoelastic gels, which are crucial for textural qualities in food applications, can be improved by their aggregation into bigger complexes.[48]

5.3 Millet Proteins: Beyond Nutrition

5.3.1 Health Benefits Beyond Nutrition

Millet plants are regarded as a suitable source of dietary proteins since they are hypoallergenic and gluten-free. The protein in millet seeds has a wide range of advantages for therapeutic effects in chronic diseases. Proso millet, for instance, has anti-proliferative, anti-hypertensive, and antioxidative properties in addition to improving lipid profiles and lowering cholesterol.[49] Conversely, kodo millet has a therapeutic effect on celiac disease, beriberi, post-menopausal women with high blood pressure, high cholesterol, and heart problems.[49] Comparably, finger millet protects against cardiovascular, gastrointestinal, and age-related diseases, as well as type II diabetes.[50]

5.3.1.1 Antioxidant Activity

Lipid oxidation has long been recognized as the primary process of food deterioration that impacts nutritional value and sensory appeal of food, particularly lipid-based products. Conversely, oxidative stress *in vivo* oxidizes DNA, cellular proteins, enzymes, and membrane lipids, resulting in destructive and fatal effects. As a result, it is believed to have destructive cellular effects linked to the pathophysiology of several health problems, such as atherosclerosis, inflammation, *etc.*[51]

Antioxidant peptides from finger millet protein hydrolysates have been described.[8] According to molecular docking analysis, the peptides' antioxidant activity resulted from the interaction of free radicals with serine and threonine residues.[8] Additionally, a fraction from foxtail millet protein hydrolysates, purified by chromatography, exhibited a closer antioxidant activity to that of α-tocopherol. However, it was lower than that of the synthetic antioxidant butylated hydroxytoluene.[52] In a study by Zhang *et al.*,[53] millet gliadin was subjected to high hydrostatic pressure (300 MPa) and enzymatic hydrolysis. According to the results of the study, the hydrolysates included certain amino acids, such as glycine, leucine, phenylalanine, and proline with high antioxidant activity.[53] In a study by Mudgil *et al.*,[54] pearl millet proteins were hydrolyzed with alcalase, bromelain, and chymotrypsin to investigate the bioactive properties of the millet protein hydrolysates. The protein hydrolysates prepared with alcalase showed the highest ABTS radical scavenging activity while chymotrypsin-generated protein hydrolysates demonstrated an important scavenging of DPPH free radicals. Moreover, hydrolysates prepared by bromelain hydrolysis displayed the highest ferric reducing antioxidant power.[54] The tyrosine/leucine-rich peptide fraction obtained from foxtail millet (*Setaria italica*) meal by fermentation with *Lactobacillus paracasei* Fn032 exhibited significant antioxidant activity.[55]

According to previous studies in the literature, it can be said that natural peptides found in finger millet are thought to be a promising source for functional food ingredients because of their strong antioxidant activity. For instance, the isolated peptide SDRLLGPNNQYLPK from pearl millet demonstrated great antioxidant activity.[56]

5.3.1.2 *Antidiabetic Effect*

The enzymes that hydrolyze carbohydrates, α-amylase and α-glucosidase, are in charge of breaking down starches in food and converting oligosaccharides into glucose, which causes a rise in postprandial glucose levels. Thus, inhibition of α-amylase and α-glucosidase activities is thought to be one of the main preventive measures to control type 2 diabetes out of all the existing antidiabetic therapy techniques.[57] Pearl millet protein hydrolysates showed α-glucosidase and α-amylase in contrast to protein isolates.[54] Peptide fractions were also demonstrated to have the ability to block α-amylase and α-glucosidase in millet protein gastrointestinal hydrolysates.[58]

The effects of protein isolates from cooked and raw foxtail millets on diabetic mice were examined in a study by Fu *et al.*[4] The results showed that the fasting blood glucose level was significantly lowered by the foxtail millet protein isolates. Problems with lipid metabolism brought on by diabetes can potentially hasten the development of other diseases. A notably lower level of triglycerides in the serum and liver indicated that the protein isolates from cooked foxtail millet corrected lipid metabolism abnormalities in diabetic mice.[4] According to a study by Mudgil *et al.*,[54] protein hydrolysates obtained from pearl millet proteins by hydrolysis with alcalase had higher *in vitro* antidiabetic potential and more inhibition against α-amylase, while hydrolysates treated with bromelain had the highest α-glucosidase inhibitory activity.

5.3.1.3 *Antimicrobial Effect*

Numerous complex and structurally diverse compounds can be found in plants and other natural sources. To overcome microbial resistance, there has been an increase in interest in the development of novel antimicrobial agents from natural plant sources in recent years.[59] For instance, the research on the antibacterial action of *Hevea brasiliensis* latex showed that the isolated protein rich in cysteines exhibited potent antimicrobial activity against various species of *Candida* spp.[60] Concentration-dependent inhibition zones were created in a study by Sitohy *et al.*[61] The authors investigated the antimicrobial effects of broad bean, soybean, and chickpea proteins against two pathogenic Gram-positive bacteria (*Bacillus subtilis* and *Staphylococcus aureus*), and two Gram-negative bacteria (*Pseudomonas aeroginosa* and *Escherichia coli*). The antifungal activity of peptides obtained from foxtail millet was demonstrated against a variety of fungal species such as *Fusarium oxysporum*, *Trichoderma viride*, *Botrytis cinerea*, and *Alternaria alternate*.[62] Similarly, proteins and peptides from finger millet, barnyard millet, and proso millet demonstrated antibacterial action against *Salmonella entrica* and *Pseudomonas aeruginosa*.[63] In a study by Amadou *et al.*,[64] the peptides from fermented foxtail millet inhibited *E. coli* ATCC 8099.

5.3.1.4 Angiotensin Converting Enzyme Inhibitory Activity

The primary independent cause of cardiovascular diseases, including peripheral arterial disease and coronary heart disease, is high blood pressure.[65] Angiotensin converting enzyme (ACE) is the main component in the renin–angiotensin system, which controls fluid volume in the body to control blood pressure. It changes the hormone angiotensin I into angiotensin II, which has the ability to constrict blood vessels. Thus, the use of various inhibitors to target ACE may have pleiotropic health effects in the management of hypertension, and angiotensin I-converting enzyme inhibitors account for more than one-sixth of all antihypertensive drugs prescribed worldwide.[16,66] ACE-inhibitory peptides found in food proteins have some advantages over chemically produced ACE inhibitors, including cost-effectiveness, availability from a variety of sources, safety, and customer acceptance. For instance, five novel ACE inhibitory peptides were isolated (QDPLFPL, FPGVSPF, SPAQLLPF, LVPYRP, and WYWPQ) from foxtail millet protein hydrolysates and these peptides were non-toxic and had a strong ACE inhibitory activity *in vitro*.[67] Similarly, the effect of extrusion and the fermentation process on ACE inhibitory activity in spontaneously hypertensive rats was investigated before protein extraction from foxtail millet and protein hydrolysis with digestive protease.[65] The results showed that raw and extruded hydrolysates had more effective antihypertensive activity in rats than fermented samples. On the other hand, ACE inhibition peptides with zinc-chelating activity from millet bran albumin hydrolysates were obtained. It was reported that only one peptide (GCHHY) among five peptides (VGSDGL, TGHGVA, GCHHY, LVEAVPRG and VLAQEE) had both ACE inhibition and zinc-chelating capacity.[68]

5.3.1.5 Protein Digestibility and Allergenicity

Protein availability can be determined *via* digestibility. Proteins with high digestibility offer more amino acids for absorption during proteolysis, making them more nutritionally valuable than proteins with lower digestibility.[16] The *in vitro* protein digestibility (IVPD) of various minor millet crops ranged from 72.0% to 75.4%.[69] The IVDP for finger millet and kodo millet was determined to be approximately 84% and 70%, respectively.[41] The IVDP of raw foxtail millet flour was 95.5%, 26.8%, and 92.95% with pepsin, trypsin, and papain enzyme systems, respectively.[70] The highest IVDP of pearl millet was determined to be 49.3%.[71]

Plant storage proteins are frequently less digestible compared to most animal proteins. This may be caused by a number of factors, including suppression of the digestion enzymes, lower enzyme accessibility due to rigid cell walls, and low protein solubility.[16] Moreover, tannins can reduce the IVDP since polyphenols bind to proteins.[16] Two pearl millet cultivars were spontaneously fermented at 30 °C for 36 h, resulting in a decrease in polyphenol and phytic acid content, whereas an increase in the IVDP was reported from ~60% to ~86% after fermentation.[72] Moreover, the IVDP of finger, foxtail and proso millet increased from 72.3%, 77.1% and 71.3%, to 85.5%, 91.6%, and 88.6%, respectively, after cooking processes.[73]

Protein allergenicity becomes a delicate matter when the isolated protein needs to be consumed as a source of protein in the diet. Some millets have demonstrated that their proteins are gluten free. Finger and foxtail millets showed gluten contents of 2.10% and 3.40%, respectively, whereas refined wheat had a gluten concentration of 27.7% (on a wet basis). Therefore, consuming millet proteins is advantageous for those suffering from celiac disease and for those who are sensitive to common proteins in wheat and dairy.[1] According to a mouse model study carried out by Bergamo *et al.*,[74] it is safe to use millet proteins in a celiac disease diet because they did not show any immune cross-reaction with gliadin epitopes associated with activation of celiac disease, which is a gluten-sensitive enteropathy with an immune basis.

5.4 Techno-functional Attributes of Millet Proteins

5.4.1 Solubility

Protein solubility is a crucial characteristic that influences a variety of functions, including foaming, gelling, and emulsification. Thus, food's flavor and texture are influenced by the solubility of proteins. When nitrogen fertilizer was applied to various varieties of proso millet, the protein solubility ranged from 9.94% to 20.34%.[42] A study conducted by Jhan *et al.*[22] investigated the effect of nanoreduction on the millet protein functionality and structure. Nanoprotein millet samples obtained by ultrasonication demonstrated enhanced functional properties, such as solubility (at pH 4, pH 7, and pH 10), emulsification, foaming, and reduced hydrophobicity, along with decreased turbidity. With a solubility range of 5.7% to 100%, the protein fractions showed maximum solubility at pH 9 and minimal solubility at pH $\leq$7.[21] In another study by the same author,[75] for every proso millet protein fraction considered, it was found that the solubility of ultrasonicated proso millet protein was significantly enhanced when compared to the native. As the ultrasound power and treatment duration increased, this increment was seen to rise. For both the prolamin and glutelin protein types, the ultrasound treatment had a significant impact on the increase in solubility, but the cultivar type only significantly affected the prolamin protein and not the glutelin protein.

Following hydrolysis, protein solubility of millet proteins is noticeably high across the broad pH range of 2–10. It was discovered that the millet protein isolates' solubility varied with pH, with the lowest solubility occurring at about pH 4. Hydrolysates showed higher solubility than isolates, potentially because of decreased surface hydrophobicity and molecular weight.[76] A study by Mohamed *et al.*[9] found that the pH-dependent protein solubility profiles of two distinct millet protein concentrate varieties were very similar to those of soy protein concentrate, showing a U-shaped curve with the highest solubility values at alkaline pH. All proteins were soluble (above 40%) at pH 2.0 in an acidic environment. Millet proteins dissolved between 50% and 80% at pH 6.0 or higher. For the majority of seed proteins, the solubility curve is common.

The solubility of protein can be enhanced by sprouting. Increased proteolytic activity brought on by sprouting breaks down millet storage proteins into shorter peptides and free amino acids, increasing the proteins' solubility in sprouted flours. The solubility of raw millet protein was found to be 31.6% for finger, 33.9% for little, 36.4% for proso, 33.17% for barnyard, 32.8% for kodo, and 35.9% for foxtail. Protein solubility increased significantly with increasing sprouting time and soaking time. Following 48 hours of sprouting, the highest solubility record was found for foxtail millet (48.6%), followed by proso millet (47.6%).[77]

5.4.2 Water and Oil Holding Capacity

The ability of a protein to absorb and retain water or oil inside a protein matrix in the face of gravity is known as the "interaction of water or oil with protein." Because they have an impact on the flavor and texture of food, interactions between proteins and water and oil are crucial in food systems. Protein shape, surface hydrophobicity/polarity, and amino acid composition are intrinsic characteristics that influence the ability of dietary protein to bind water or oil. When it comes to culinary applications, the ability of proteins to hold water or oil is more significant for food systems. In a study conducted by Wang *et al.* (2021), following hydrolysis, the water holding capacity (WHC) of various millet protein isolates rose noticeably. The range of WHC for millet proteins was between 59.2 and 64.4 mL g^{-1}.[42] The fact that the hydrolysis process strengthened the polar groups may be directly related to this event. The WHC of proso millet protein varied between 381.4% and 466.2% for four distinct isolates cultivated with nitrogen fertilizer.[42] The oil holding capacity (OHC) of the millet protein isolates in the same study was much higher than that of the millet hydrolysates, which ranged from 2.8 to 3.18 mL g^{-1}.[76] The production of lower molecular weight peptides upon hydrolysis results in peptides that have a smaller interface surface area to connect with a fat molecule. Since non-covalent bonds are directly responsible for fat absorption, the decrease in hydrophobicity and the release of ionizable groups of millet protein hydrolysates also causes decreased fat absorption.[78] According to a study by Mohamed *et al.*,[9] the WHC of two varieties of foxtail millet protein concentrate were 7 and 5 g g^{-1}, significantly higher than that of the soy protein hydrolysate. The OHCs of different millet protein concentrates were determined to be 5 and 6 g g^{-1} in the same study.

Moreover, the OHC of proso millet values ranged from 218.6% to 268.42%, and the OHC of proso millet protein was considerably enhanced by nitrogen fertilizer treatment.[42] After heating, the WHC of prolamins isolated from various foxtail millet proteins dropped dramatically to 0.5–0.9 $g\,g^{-1}$ from 1.9–2.3 $g\,g^{-1}$. The WHC of prolamins varied significantly across the different varieties ($p < 0.05$). In addition to substantial protein–protein interactions, heat denaturation caused the polypeptide chains to unfold and to change to a more disordered state. The WHC of polar amino acid groups may decrease as a result of all these factors.[79] High WHC and OHC are crucial for creating food systems like breads, cakes, muffins, sausages, batters, mayonnaise, and salad dressings.

5.4.3 Emulsification Ability and Stability

Once a protein can diffuse at the oil-in-water interface, its emulsion stability index (ESI) indicates how effectively it supports the interface. According to Arkarume *et al.*,[21] the greatest ESI recorded was about 60 min, whereas the majority of fractions showed an emulsifying activity index (EAI) $< 0.25\ m^2\,g^{-1}$. The prolamin fractions and pearl millet flour's functional characteristics were investigated. Although pennisetin had a poorer emulsion stability, it had comparable oil/water binding capabilities to pearl millet flour, as well as a higher foaming capacity and emulsifying activity index. The functional characteristics of pennisetin were mostly similar to or lesser than those of widely used proteins such as peanut, rice, sorghum, or wheat.[20] The highest foaming and emulsification capacity was reported at pH 10 for all millet varieties due to high solubility at alkaline pH. Additionally, the nanoreduction treatment enhanced stability by increasing the negative ion charge for different millet varieties. Scanning electron microscopy (SEM) also showed disruption in the protein microstructure following treatment.[22]

The study found that the protein isolates and hydrolysates of pearl, finger, and green sorghum millet had the highest emulsion activity index at pH 10 and the lowest at pH 4. Following hydrolysis, the emulsion activity of protein isolates was found to be greater in green sorghum millet ($92.66\ m^2\,g^{-1}$), finger millet ($93.49\ m^2\,g^{-1}$), and pearl millet ($91.88\ m^2\,g^{-1}$). This could be because hydrolysates are surface-active substances that encourage oil-in-water emulsion because of their hydrophobic and hydrophilic groups. Protein can swiftly diffuse and adsorb at the interface due to its enhanced solubility upon hydrolysis, increasing the emulsion activity of hydrolysates.[76] In another study, millet proteins were shown to have emulsifying activity of 65% and 74%, which was much higher than that of soy protein concentrate (53%). The high hydrophobic residues and surface tension of millet protein may be the cause of the strong emulsifying activity of millet protein concentrates.[9] Additionally, the emulsion capacity values of proso millet varied greatly, ranging from 45.3% to 53.9%. Proso millet proteins have an emulsion stability of above 93.7%. Because of the high concentration of hydrophobic amino acids, proteins treated with nitrogen fertilizer had a greater emulsion capacity than the control group, which did not receive fertilizers. The interaction of several hydrophobic amino acids improved the emulsifying performance of protein by increasing the binding ability of the emulsifier and oil droplet.[42] In a study, when compared to native protein fractions, ultrasound treatment for prolamin protein marginally raised the EAI at 5 min, but as treatment time rose to 10 min, the EAI was seen to decrease. Less aggregation may have allowed the protein to diffuse into the oil–water interface, which may have contributed to the rise in the EAI at the 5-minute treatment. On the other hand, the glutelin protein's EAI was shown to be lower than that of its natural protein. Both prolamin and glutelin showed notable alterations in EAI. The ultrasound treatment seemed to have little influence on the ESI of the glutelin protein, but at longer treatment times, it raised the ESI of the prolamin relative to the native protein fractions.[75]

5.4.4 Foaming Capacity

Foaming capacity refers to a protein's ability to incorporate and retain air within a liquid or semi-solid matrix, creating a dense, viscoelastic layer surrounding gas bubbles at the air–water interface.[21] Research has explored the foaming properties of foxtail millet protein, highlighting its promise as a versatile component in functional food applications requiring foam formation and stability.

The foaming features of white and yellow foxtail millet protein concentrates (WMPC and YMPC) have been compared to soy protein concentrate (SPC). WMPC exhibited the highest foam capacity (137 $g\,mL^{-1}$) and foam stability (73 $g\,mL^{-1}$), outperforming both YMPC (124–46 $g\,mL^{-1}$) and SPC (124–61 $g\,mL^{-1}$).[9] These results underscore the suitability of foxtail millet protein concentrates in food applications demanding good foam formation and stability.

Ultrasound treatment improves the foaming properties of proteins from foxtail millet, sorghum, and pearl millet. Nanoreduction *via* ultrasonication enhanced the proteins' ability to trap air and stabilize foams by reducing particle size and molecular weight, as confirmed by SDS-PAGE.[22] The structural modifications increased hydrophobicity and solubility, boosting surface activity and foam stability, making these treated proteins ideal for aerated food products.

The effects of enzymatic hydrolysis on defatted foxtail millet proteins results in significant improvements in foaming properties. The foaming capacities and stabilities of defatted foxtail millet protein hydrolysates (DFMPH) ranged from 185.33 to 137.80 $g\,mL^{-1}$, while those of defatted foxtail millet protein isolate (DFMPI) varied between 185.33 and 111.1 $g\,mL^{-1}$. Enzymatic hydrolysis was found to enhance the proteins' ability to form and stabilize foams significantly, further highlighting DFMPH as a promising functional ingredient for food products requiring superior foaming functionality.[80]

5.4.5 Gelling and Film Formation

Gelation, a key functional property, arises from protein and starch in ingredients. Heat-induced protein gelation occurs due to molecular unfolding and aggregation, while starch gelation results from granule swelling and hydration upon heating. The least gelation concentration (LGC) measures gelation potential, with lower LGC indicating better gelation.[81] The gelation properties of protein hydrolysates derived from pearl millet, finger millet, and green sorghum millet were evaluated after enzymatic hydrolysis.[76] Among these, pearl millet protein hydrolysate demonstrated the highest gelation capacity, achieving 7%. This enhancement indicates that enzymatic hydrolysis improves the gel-forming potential of millet proteins, making them suitable for food applications requiring gel-like textures.

Biofilms derived from renewable plant proteins offer a sustainable solution to reduce plastic use within the food sector and biomedical applications, mitigating climate impact. Millet proteins can be fabricated into films using wet solvent casting or dry methods like compression, molding, and extrusion to produce thermoplastics. Structural and functional enhancements can be achieved through plasticizers, heat, acid/alkaline treatments, cross-linking agents, enzymes, or by developing composite films with other polymers. These biodegradable films are ideal for packaging due to their mechanical strength and effective aroma barrier properties.[82]

Pennisetin, a prolamin derived from pearl millet, is a renewable substitute for petroleum-based plastics. Films were developed using ethanol solutions and plasticized with glycerol, a glycerol–citric acid blend, and a glycerol–lactic acid–polyethylene glycol combination. Among these, the glycerol–citric acid mixture improved tensile strength and barrier properties, likely due to alterations in the film's microstructure.[83] The results demonstrated that pennisetin films possess mechanical and barrier characteristics similar to other protein-based materials derived from cereals, making them a viable option for eco-friendly packaging.

5.5 Engineering Attributes of Millet Proteins

Millet proteins are emerging as valuable ingredients for sustainable, nutritious, and innovative food products, addressing the growing need for plant-based, gluten-free, and allergen-free alternatives. Their distinct textural and rheological properties support diverse applications, such as in meat substitutes, dairy alternatives, and gluten-free items. This section explores their engineering properties, processing advantages, and recent advances in product innovation across the food industry.

5.5.1 Application in Food Products

Millet proteins offer diverse functionalities, enabling their use as nutrients in nutritional supplements, food fortification, and as techno-functional aids in food processing. Additionally, they serve as biopolymers in the production of food, pharmaceutical, and healthcare products. Despite their benefits, millet proteins present some nutritional limitations, such as low levels of tryptophan and lysine, high leucine and isoleucine content, and partial digestibility. These can be addressed by combining millet proteins with other protein sources in food formulations.[1]

Recent studies have demonstrated their potential as functional food ingredients, particularly peptides derived from foxtail millet, which exhibit antioxidant capabilities by reducing reactive oxygen species and raising glutathione levels. Moreover, foxtail millet prolamin has shown promise in improving glucose homeostasis, regulating gut flora, and modifying serum metabolism, making it a valuable candidate for food products aimed at diabetes prevention.[48,84] Its versatility allows it to be used in a variety of dishes, such as casseroles, stews, and soups, where it can replace meat or beans to enhance nutritional value.

5.5.1.1 Meat Analogs

Meat production faces criticism for environmental and health concerns. To address this, meat analogs have been developed as sustainable and healthier alternatives. Although they make up only 1–2% of the meat market, they appeal to vegetarians, vegans, and those reducing meat for health, ethical, or religious reasons. Common meat analogs include tofu, tempeh, and seitan, made from plant-derived materials like soy, wheat gluten, and legumes.[85]

The worldwide growth in gluten sensitivity has driven demand for gluten-free products. Millets present a prospect to create gluten-free meat analogs for this growing consumer group. However, there is limited literature on creating innovative meat substitutes featuring millets as a key ingredient. Millet flours (sorghum, pearl millet, and finger millet) have been used as secondary ingredients in soy protein isolate blends to produce fibrous high-moisture meat analogues *via* extrusion.[86] Millets were added at 10%, 20%, and 30% levels with moisture contents of 60%, 65%, and 70%. Pearl millet achieved good texture at 30%, while sorghum and finger millet performed well at 20%. SPI-millet blends had lower hardness, chewiness, and tensile strength than SPI-gluten blends, resembling chicken texture. Millets also created unique fiber patterns, colors, and gluten-free options, emphasizing their potential for sustainable and diverse plant-based meat products.

In another study, researchers intended to produce plant-based burger patties featuring pea protein isolates (PPIs) and barnyard millet flour (BMF) *via* high-moisture extrusion cooking. The optimal mixture of 90 : 10 pea protein isolate to barnyard millet flour, processed at a screw speed of 80 rpm, produced extrudates with favorable characteristics, including moderate water absorption, high hardness, and strong adhesion. The resulting plant-based burger patties maintained good texture and protein content. When stored for 12 days under refrigeration, patties in vacuum packaging had less cooking loss, moisture content, color changes, and microbial growth compared to those in aerobic packaging. This demonstrated the prospects offered by PPIs and BMF as key ingredients in plant-based meat alternatives and the benefits of vacuum packaging for preserving quality during storage.[87]

5.5.1.2 Dairy Alternatives

Millet milk presents significant opportunities in the Indian food industry, particularly as a dairy option for individuals with lactose intolerance or milk allergies. It can be utilized in various products, including beverages, functional foods, ready-to-cook formulations, and even ice creams. Millet milk not only serves as a lactose-free and vegan-friendly option but also enhances the nutritional value of food products by enhancing antioxidants and dietary fiber. Studies suggest that fermented millet milk can produce appealing and healthy beverages suitable for plant-based diets.[88]

Millet milk has been blended with coconut milk to develop a nutritious, dairy-free alternative. Among various blends tested for sensory and nutritional attributes, the 50% millet milk and 50% coconut milk combination emerged as the most nutritionally rich and acceptable sensory experience. This blend demonstrates great potential as a flavorful and nutrient-dense option for those following plant-based diets or avoiding dairy.[89]

Encapsulation of probiotics has been studied as a solution to maintain their viability in food products during storage and digestion. *Lactobacillus casei* was encapsulated using sodium alginate, inulin, and materials such as skim milk, whey protein, soy protein, and flaxseed oil. Freeze-dried microcapsules were tested for encapsulation efficiency, probiotic survival, oxidative stability, and sensory properties in a millet yogurt formulation.[90] Encapsulation rates exceeded 90%, and co-microencapsulation with flaxseed oil notably improved probiotic viability under gastrointestinal conditions over a 28-day period at 4 °C, and in yogurt. This approach demonstrates the promise of sodium alginate–inulin matrices to enhance probiotic stability and functionality in millet-based dairy alternatives.

Additionally, the fermentation of millet was investigated using Fiti sachets, which included *Lactobacillus rhamnosus* GR-1 and *Streptococcus thermophilus* C106, both formerly employed in Africa to inhibit malnutrition.[91] The bacteria thrived in an 8% millet–water mixture, achieving probiotic counts exceeding 10^6 CFU per mL. Fermentation reduced phytic acid, improving micronutrient bioavailability. Sensory analysis favored millet fermented in milk, highlighting its suitability as a cost-effective, nutritious alternative when milk is unavailable, particularly in resource-constrained settings.

5.5.1.3 Gluten-free Products

Gluten, a key wheat protein, provides elasticity and viscosity to dough, making it essential for baking. It is composed of gliadin and glutenin, which significantly influence dough properties. However, among people diagnosed with celiac disease, an immune-based response triggered by gluten, gluten consumption damages the small intestine, leading to nutrient malabsorption and complications like osteoporosis and infertility. A permanent gluten-free diet remains the only therapeutic option, though the limited availability and variety of gluten-free products affect patients' quality of life. Despite these challenges, the gluten-free food industry has grown substantially in recent times., reflecting rising demand for innovative and nutritious gluten-free options.[92]

Millets, including pearl, proso, niger, and foxtail, are naturally gluten free and increasingly favored for developing foods for individuals with celiac disease. Their excellent digestibility and lack of allergenic potential make millets a potentially effective replacement for wheat. While gluten-free flours such as teff, chestnut, and soy protein have been shown to boost both nutritional value and sensory attributes of gluten-free bread, millet proteins remain underexplored in gluten-free food formulations. Research has predominantly focused on millet flours rather than proteins. Studies incorporating alternative proteins like pea, lupine, and soy have reported mixed results, with pea protein showing the most promise in enhancing the performance attributes of gluten-free products.[93]

Gluten-free cookies and pasta were developed using finger millet, pearl millet, soybean, and groundnut. These ingredients, rich in fiber, protein, fat, and minerals, improved both the nutritional profile and sensory attributes of the products. The optimal cookie formulation consisted of a 25% mix of all four ingredients, while the most preferred pasta blend contained 30% pearl millet, 35% finger millet, and 35% soybean flour. Fried pasta was favored over steamed pasta in sensory evaluations, although formulations with higher pearl millet content received slightly lower sensory scores. The inclusion of these gluten-free ingredients significantly increased fiber, protein, fat, and mineral content in comparison with the control samples.[94]

Underutilized cereals like sorghum, finger millet, and pearl millet can potentially be used as wheat alternatives in the African baking industry. Sourdough bread produced using these cereals and their blends exhibited protein content ranging from 16.29% to 39.26%, fat content from 14.02% to 18.80%, and carbohydrate content from 42.16% to 65.61%. The energy content of these gluten-free breads (405.99–446.39 kcal per 100 g) was higher than that of wheat bread (396.43 kcal). However, as the nutritional profiles were significantly enhanced, the breads were only moderately accepted by consumers due to texture issues and the preference for wheat-based products. This highlights the need for future studies to enhance the sensory qualities of millet-based gluten-free products and promote their adoption as viable alternatives to wheat.[95]

5.5.2 Texture and Rheological Properties

Research demonstrates that incorporating foxtail millet flour influences the texture and rheological properties of noodles. It enhances water absorption and shortens cooking time while reducing cooking yield and the swelling index compared to noodles made solely from wheat flour. These findings highlight foxtail millet's potential to enhance the functional characteristics of noodle formulations by enhancing water retention, solubility, and overall nutritional value.[96] In a study by Agrahar-Murugkar *et al.*, a gluten-free, eggless cake was created using a composite flour blend of finger millet, sprouted soy, and amaranth, designed for individuals with celiac disease.[97] The eggless cake (T2) was evaluated against a control cake (C) made with refined wheat flour and eggs, as well as another eggless cake (T1) prepared with composite flour comprising whole wheat, malted finger millet, sprouted soy, and amaranth. The T2 batter demonstrated comparable textural properties, flow behavior, and consistency indices to the control (C). Additionally, T2 exhibited a higher volume (454.4 cm^3) than T1 (437.1 cm^3) and shared similar textural qualities, such as springiness, resilience, and cohesiveness.

In a separate study, the researchers investigated the effects of substituting wheat flour with 30–50% native finger millet flour (NFMF) or germinated finger millet flour (GFMF) in soft dough biscuits. The rheological and baking characteristics of the flour blends were evaluated. As the replacement level increased, water absorption, dough stability, and extensibility declined, while the mixing tolerance index and dough hardness rose. Biscuit dough made with NFMF demonstrated greater hardness (125 to 234 N) compared to GFMF (118 to 126 N).[98]

5.5.3 Development of Novel Food Products: The Use of Millet Proteins in Innovative Food Products and Processing Techniques

5.5.3.1 Encapsulation and Delivery Systems for Bioactive Compounds

Polyphenols exhibit antioxidant, anti-carcinogenic, and anti-proliferative effects but face limitations in disease treatment due to their low water solubility, sensitivity to light, and poor bioavailability. To address these challenges, biopolymer-based nanoparticles (NPs) have been developed, offering advantages such as renewability, biocompatibility, and biodegradability.[99]

Prolamin, a protein abundant in nonpolar amino acids like proline, is of paramount importance in interacting with phytochemicals. It encapsulates bioactive compounds within its hydrophobic core, providing protection and enabling controlled release. Due to its biocompatibility, biodegradability, and strong affinity for bioactives, prolamin is widely used as a delivery system.[48] Studies comparing nanoparticles made from zein and foxtail millet prolamin, stabilized with sodium caseinate, have demonstrated differences in encapsulation efficiency and stability across four bioactives with varying polarities. Foxtail millet prolamin-based nanoparticles showed superior resveratrol encapsulation, attributed to their lower hydrophobic amino acid content. Additionally, co-encapsulation shifted curcumin into hydrophobic regions, enhancing long-term retention. These results highlight the impact of prolamin composition and polarity on stabilizing bioactives and highlight its potential in drug delivery.[99]

Curcumin-loaded nanoparticles were developed using foxtail millet prolamin, sodium caseinate, and curdlan sulfate. The nanoparticles, ranging from 84.7 to 136.4 nm in size with zeta potentials of −41.5 to −45.7 mV, demonstrated enhanced encapsulation efficiency (improving from 78.3% to 85.9%) and reduced curcumin release at physiological pH. These nanoparticles also exhibited improved stability and anticancer effects on MCF-7 cells, showcasing their potential for delivering hydrophobic bioactives.[100]

Proso millet protein was used as a delivery system for curcumin encapsulation. Proteins obtained *via* wet milling or ethanol formed spherical nanoparticles (250–350 nm) with encapsulation efficiencies reaching 78.9%.[101] Ethanol-extracted proteins showed superior performance, enhancing curcumin's stability at elevated temperatures while preserving its antioxidant properties. This suggests proso millet protein is an effective encapsulant for lipophilic bioactives, with potential applications in functional foods and nutraceuticals.

Crocin, a hydrophilic carotenoid from saffron, provides antioxidant, anticancer, and therapeutic benefits, but suffers from low stability and bioavailability due to sensitivity to pH, oxygen, light, and temperature.[102] Crocin-loaded nanoparticles were developed using proteins from sorghum, foxtail millet, and pearl millet through ultrasonication. Foxtail millet protein nanoparticles achieved the highest encapsulation efficiency (83.78%) and displayed controlled release under intestinal conditions, preserving crocin's bioactivity.[103] These nanoparticles exhibited antimicrobial, anticancer, and anti-hypertensive properties, indicating the effectiveness of millet-derived protein nanoparticles as delivery systems for crocin. These results suggest co-encapsulation of multiple bioactive compounds in millet protein-based nanoparticles is a promising area for upcoming studies, particularly for improving storage stability and bioavailability.

5.5.3.2 Innovative Processing Methods for Millet Proteins

One strategy to increase the use of protein in food production is to modify it using nonthermal methods like pulsed electric fields (PEF), high-pressure homogenization, and ultrasound.[104] High hydrostatic pressure (HHP), also known as ultra-high-pressure processing (UHPP), is a non-thermal technique used to extend the shelf life of food while minimizing physical damage. It works by applying pressures ranging from 100 to 1000 MPa and temperatures between 20–90 °C, helping preserve the food's taste and nutritional value. HHP has been widely researched for its impact on protein structure, denaturation, and changes in physicochemical properties, showing considerable potential in enhancing food safety and quality.[105] The impact of HHP on the structure and physicochemical properties of millet gliadin has been investigated.[106] The results showed that HHP treatment modified the spatial conformation of millet gliadin, enhancing its hydrophilicity. It also affected the surface charge distribution and increased protein aggregation, leading to larger particle sizes. Moreover, HHP treatment decreased the α-helix and random coil content, promoted the formation of disulfide bonds, and raised the denaturation temperature, contributing to a more stable protein structure. These findings indicate that HHP treatment can enhance millet gliadin's suitability for food applications by addressing its strong hydrophobicity.

PEF is an innovative, eco-friendly, non-thermal treatment method known for its energy efficiency and rapid application (in microseconds or milliseconds) under gentle environmental conditions, making it widely utilized for protein modification. It enhances protein flexibility, increases surface charge, and exposes more amino acid residues, improving protein–polyphenol affinity.[107] The impact of moderate-intensity pulsed electric field (MIPEF) processing on the structure and characteristics of foxtail millet prolamin has been reported. MIPEF treatment altered the protein's conformation, increased hydrophobicity and solubility in alcohol, and promoted the formation of larger aggregates.[28] It also reduced α-helix content, enhanced β-sheet content, and increased disulfide bond formation, resulting in a more ordered structure. These findings indicate that MIPEF improves characteristics of millet prolamin, making it well-suited for food applications, medicine, and biodegradable materials.

Ultrasound technology, which uses mechanical waves above 16 kHz, is classified into low-intensity (16–100 kHz) and high-intensity (100 kHz–1 MHz) applications. High-intensity ultrasound modifies food properties through cavitation, creating severe temperature conditions (5000 K), pressures (1000 atm), and turbulence, while low-intensity ultrasound is primarily used to analyze food characteristics such as sugar content and acidity. Moreover, ultrasound can produce reactive free radicals, facilitating chemical reactions in treated systems.[108] Researchers investigated the effect of high-power ultrasound (20 kHz) applied at different intensities and durations on the functional characteristics of millet protein concentrate. The findings demonstrated that ultrasound treatment notably enhanced solubility, emulsifying activity index, and foaming properties, with higher intensities showing improved outcomes. Specifically, treatment at 73.95 W cm^{-2} for 12.5 min increased solubility, raised the negative surface charge by 36%, and decreased molecular weight.[17] Foxtail millet protein, rich in essential amino acids, primarily consists of prolamin (31.10%) and glutelin (34.69%). High-intensity ultrasound (HIU) was applied at amplitudes of 5–10% and durations of 5–20 min to enhance the functional and physico-chemical characteristics of its concentrate and fractions, significantly improving solubility, particle size, zeta potential, and surface hydrophobicity.[11]

Cold plasma (CP) treatment offers a versatile approach for food applications, including modifying macromolecules (starch, protein, cellulose), enhancing seed germination, decontaminating microbes, and extracting bioactive compounds. As a non-thermal technology, CP has a partially ionized gas composed of free radicals, photons, ozone, energetic ions, and free electrons in both ground and excited states.[109] The protein concentrate from sprouted foxtail millet was exposed to dielectric barrier discharge plasma at 1 kV and 2 kV for durations of 5, 15, and 25 min, leading to significant changes in physicochemical properties, including particle size, zeta potential, and surface hydrophobicity.[110] Functional properties like solubility, emulsifying capabilities, and foaming capacities were also enhanced ($p \leq 0.05$). Electrophoretic profiles indicated minimal alterations; however, more intense bands emerged under stronger plasma treatments. Altered proteins offer potential as fortifiers, emulsifiers, stabilizers, and texture modifiers in plant-derived meat alternatives. Even so, there is still limited research regarding the effects of cold plasma processing on millet protein. Further research is required to reach conclusive findings.

Microwave technology is an electro-thermal process that transforms electrical energy into heat, operating within a frequency spectrum of 300 MHz to 300 GHz, with food-related applications typically using frequencies between 915 MHz and 2.45 GHz. This method is more energy-efficient and faster than traditional techniques like open-pan boiling, which rely on indirect heat transfer.[111] The effects of microwave (MW) and hot air-assisted radiofrequency (HARF) treatments were investigated on foxtail millet (FM) flour and its protein concentrate (FMP). HARF treatment at 15% moisture resulted in the lowest lipase activity (10.86%), while MW processing at 786.80 W for 6 min reduced lipase activity to 0.27%.[112] Both methods lowered peak viscosities, enhanced emulsifying and absorption properties, but reduced foaming ability and protein purity. These changes were attributed to decreased crystallinity and structural alterations. Table 5.2 highlights recent studies exploring diverse modification approaches and their influence on millet protein characteristics.

5.6 Challenges and Future Perspectives

Millet protein holds great promise as a plant-based protein source because of its agronomic benefits; however, its utilization is limited by challenges such as low digestibility, insufficient essential amino acids, and anti-nutritional factors. Techniques like roasting, fermentation, and germination improve digestibility and nutrient availability by reducing tannins and phytic acid. Despite these advances, millet protein continues to face limitations, including low solubility, undesirable sensory attributes, and restricted functional properties, which impact its broader application in food products.[114]

A commercially scalable process is crucial for isolating complete proteins while retaining their nutritional and functional properties. This requires the use of food-grade additives or isolation techniques, elimination of anti-nutritional factors, and optimization of desirable properties, such as solubility, water/oil absorption capacity, emulsification ability, foaming,

Table 5.2 Recent research on the effects of modification techniques on millet protein properties.

Millet protein	Treatment	Key findings	Ref.
Millet gliadin	High hydrostatic pressure (HHP)	HHP treatment of millet gliadin improved its hydrophilicity, surface charge, aggregation, particle size, and food suitability by altering conformation, reducing α-helix content, forming disulfide bonds, and raising denaturation temperature.	106
Millet gliadin	High hydrostatic pressure	Millet gliadin treated with 300 MPa HHP yielded an antioxidant peptide with DPPH, hydroxyl, and superoxide anion scavenging activities of 53.96%, 31.30%, and 52.73%, respectively, highlighting its capability as a natural antioxidant.	53
Foxtail millet prolamin	Pulsed electric field	Moderate-intensity pulsed electric field (MIPEF) treatment enhanced foxtail millet prolamin's hydrophobicity, solubility, and stability, while altering its structure and increasing β-sheet content.	28
Foxtail millet protein and fractions (prolamin and glutelin)	High-intensity ultrasound (HIU)	HIU treatment improved the functional and physicochemical properties of foxtail millet protein, including solubility, particle size, zeta potential, and surface hydrophobicity, enhancing its potential applications in the food industry.	11
Proso millet (*Panicum miliaceum* L.) protein	High-power ultrasound	The application of high-power ultrasound (20 kHz) to millet protein concentrates improved solubility, emulsifying, and foaming features.	17
Proso millet prolamin and glutelin proteins	High-power ultrasound	High-power ultrasound (29.29, 38.08, and 52.72 W) for 5–10 min improved solubility (up to 235%) and digestibility (28%–58%). The treatment enhanced protein functionality, especially with longer treatment times.	75
Albumin, globulin, and glutelin from white finger millet	Ultrasound	Ultrasound treatment (20 kHz, 100–300 W, 12 min) improved the solubility, emulsifying ability, and foaming ability of albumin, globulin, and glutelin proteins. from white finger millet by reducing particle size and increasing surface hydrophobicity.	113

Table 5.2 (*Continued*)

Millet protein	Treatment	Key findings	Ref.
Sprouted foxtail millet protein	Cold plasma	Foxtail millet protein was treated with dielectric barrier discharge plasma, leading to notable alterations in physicochemical characteristics and improved functional attributes ($p \leq 0.05$). Electrophoretic profiles showed intense bands with stronger treatments.	110
Foxtail millet (FM) flour and its protein-rich concentrate (FMP)	Microwave (MW) and hot air-assisted radiofrequency (HARF)	MW and HARF treatments on foxtail millet flour improved emulsifying and absorption properties but reduced viscosity, foaming, and protein purity, with structural changes confirmed by FTIR and XRD.	112

and filming properties. For food and therapeutic applications, achieving maximum protein concentration is essential.[1]

Acidic conditions are less effective for extracting millet protein as they do not efficiently break down cell walls. In contrast, alkali-based extraction technology, which dissolves cell walls and disrupts disulfide crosslinking bonds within the protein matrix, has proven more effective in promoting millet protein solubilization.[115] These advancements are critical for addressing existing limitations and unlocking the full potential of millet proteins for diverse applications.

A few obstacles must be overcome before the food or pharmaceutical industries may use novel ACE-inhibitory peptides. These obstacles include those related to the peptides' bioavailability, *in vivo* safety, and stability under various processing conditions. For instance, the critical importance of the amino acid sequence, particularly C-terminal tripeptide, in the physiological actions of ACE-inhibitory peptides has been established. However, stomach and intestinal proteases pepsin, trypsin, and dipeptidase can break down the active sequence of ACE-inhibitory peptides, rendering the peptides ineffective *in vivo*.[66,116] Investigating ACE-inhibitory peptides' physicochemical characteristics, safety, and stability under various food processing settings, as well as during gastrointestinal digestion, is a crucial issue.

Millets, despite their health benefits, face declining demand due to urbanization, changing diets, and inadequate processing methods. Solutions lie in adopting novel technologies like fortification, encapsulation, 3D printing, nanotechnology, and cold plasma, alongside enhanced digitalization and research efforts. Modern techniques like germination, roasting, and extrusion enhance millet's properties and shelf life, while emerging technologies like ultrasound and microwave processing show potential despite challenges with standardization and skilled labor. Future priorities include training, interdisciplinary research, and reviving traditional recipes to boost millet's global reach and sustainability.[117] Improved processing technologies could enable millets to become a crucial component in value-added and health-oriented food products, leading to increased demand from urban populations and emerging markets.[15]

5.7 Conclusion

Developing millet crops is a highly effective strategy to ensure food security in developing countries with cereal-based diets, where protein–energy malnutrition is exacerbated by limited agricultural resources and insufficient food production. This chapter highlights the role of millet as a protein source, and the importance of functional properties and nutritional and health attributes of millet protein. Despite its imbalanced amino acid profile and moderately reduced *in vitro* protein digestibility, millet protein serves as a viable alternative to animal protein, offering health benefits against chronic diseases. Furthermore, protein concentrates or isolates with enhanced functionality and processed millet flours could play a significant role in functional food applications.

To meet the growing demand for protein-rich foods, a comprehensive evaluation and refinement of millet-derived protein content can benefit both the food industry and household consumption. For the successful commercialization of millet proteins or millet-based products, it is imperative to address existing research gaps. These include the development of sustainable and cost-effective production methods, detailed toxicological assessments, an exploration of the nutritional and health benefits of isolated millet proteins, their functional applications in innovative or improved food systems, and a deeper understanding of consumer awareness, preferences, and sensory perceptions related to millet protein consumption.

References

1. N. Sachdev, S. Goomer, L. R. Singh, V. M. Pathak, D. Aggarwal and R. K. Chowhan, Current status of millet seed proteins and its applications: a comprehensive review, *Appl. Food Res.*, 2023, **3**(1), 100288.
2. S. A. Kheya, S. K. Talukder, P. Datta, S. Yeasmin, Md. H. Rashid, A. K. Hasan, Md. P. Anwar, A. K. M. A. Islam and A. K. M. M. Islam, Millets: The future crops for the tropics - Status, challenges and future prospects, *Heliyon*, 2023, **9**(11), e22123.
3. R. Samtani, S. S. Mishra and S. B. Neogi, Millets: Small grains, big impact in climate action, *J. Clim. Change Health*, 2024, **20**, 100345.
4. Y. Fu, R. Yin, Z. Liu, Y. Niu, E. Guo and R. Cheng, *et al.*, Hypoglycemic effect of prolamin from cooked foxtail millet (*Setaria italic*) on streptozotocin-induced diabetic mice, *Nutrients*, 2020, **12**(11), 3452.

5. S. Nithiyanantham, P. Kalaiselvi, M. F. Mahomoodally, G. Zengin, A. Abirami and G. Srinivasan, Nutritional and functional roles of millets—A review, *J. Food Biochem.*, 2019, **43**(7), e12859.
6. T. Maharajan, S. Antony Ceasar, T. P. Ajeesh Krishna and S. Ignacimuthu, Finger millet [*Eleusine coracana* (L.) Gaertn]: an orphan crop with a potential to alleviate the calcium deficiency in the semi-arid tropics of Asia and Africa, *Front. Sustainable Food Syst.*, 2021, **5**, 684447.
7. A. Majid and P. Priyadarshini CG, Millet derived bioactive peptides: A review on their functional properties and health benefits, *Crit. Rev. Food Sci. Nutr.*, 2020, **60**(19), 3342–3351.
8. H. Agrawal, R. Joshi and M. Gupta, Purification, identification and characterization of two novel antioxidant peptides from finger millet (Eleusine coracana) protein hydrolysate, *Food Res. Int.*, 2019, **120**, 697–707.
9. T. K. Mohamed, K. Zhu, A. Issoufou, T. Fatmata and H. Zhou, Functionality, in vitro digestibility and physicochemical properties of two varieties of defatted foxtail millet protein concentrates, *Int. J. Mol. Sci.*, 2009, **10**(12), 5224–5238.
10. S. Anitha, M. Govindaraj and J. Kane-Potaka, Balanced amino acid and higher micronutrients in millets complements legumes for improved human dietary nutrition, *Cereal Chem.*, 2020, **97**(1), 74–84.
11. N. Sharma, J. K. Sahu, A. Choudhary, M. Meenu and V. Bansal, High intensity ultrasound (HIU)-induced functionalization of foxtail millet protein and its fractions, *Food Hydrocolloids*, 2023, **134**, 108083.
12. K. S. Poutanen, A. O. Kårlund, C. Gómez-Gallego, D. P. Johansson, N. M. Scheers and I. M. Marklinder, *et al.*, Grains–a major source of sustainable protein for health, *Nutr. Rev.*, 2022, **80**(6), 1648–1663.
13. S. A. Ceasar, S. Prabhu and H. T. Ebeed, Protein research in millets: current status and way forward, *Planta*, 2024, **260**(2), 43.
14. N. Sharma and K. Niranjan, Foxtail millet: Properties, processing, health benefits, and uses, *Food Rev. Int.*, 2018, **34**(4), 329–363.
15. A. S. M. Saleh, Q. Zhang, J. Chen and Q. Shen, Millet Grains: Nutritional Quality, Processing, and Potential Health Benefits, *Compr. Rev. Food Sci. Food Saf.*, 2013, **12**(3), 281–295.
16. S. P. Bangar, S. Suri, S. Malakar, N. Sharma and W. S. Whiteside, Influence of processing techniques on the protein quality of major and minor millet crops: A review, *J. Food Process. Preserv.*, 2022, **46**(12), e17042.
17. B. Nazari, M. A. Mohammadifar, S. Shojaee-Aliabadi, E. Feizollahi and L. Mirmoghtadaie, Effect of ultrasound treatments on functional properties and structure of millet protein concentrate, *Ultrason. Sonochem.*, 2018, **41**, 382–388.
18. S. Singh, D. Yadav, M. Beckmann, A. Naveen, P. I. Gangashetty, L. A. Mur and R. S. Yadav, Variation in protein and amino acids in global collection of pearl millet (*Pennisetum glaucum*) germplasm, *J. Food Compos. Anal.*, 2024, **134**, 106557.

19. A. Leite, G. C. Neto, A. L. Vettore, J. A. Yunes and P. Arruda, The prolamins of sorghum, Coix and millets, in *Seed proteins*, 1999, pp. 141–157.
20. R. Bibi, H. Mokrane, K. Khaladi and H. Amoura, Improvement of pearl millet (*Pennisetum glaucum* (L.) R. BR) prolamin extractability: Chromatographic separation, characterization and functional properties, *J. Microbiol. Biotechnol. Food Sci.*, 2021, **11**(2), e3674.
21. F. Akharume, D. Santra and A. Adedeji, Physicochemical and functional properties of proso millet storage protein fractions, *Food Hydrocolloids*, 2020, **108**, 105497.
22. F. Jhan, A. Gani, N. Noor and A. Shah, Nanoreduction of Millet Proteins: Effect on Structural and Functional Properties, *ACS Food Sci. Technol.*, 2021, **1**(8), 1418–1427.
23. G. Ejeta, M. M. Hassen and E. T. Mertz, In vitro digestibility and amino acid composition of pearl millet (*Pennisetum typhoides*) and other cereals, *Proc. Natl. Acad. Sci. U. S. A.*, 1987, **84**(17), 6016–6019.
24. M. K. Tripathi, D. Mohapatra, R. S. Jadam, S. Pandey, V. Singh, V. Kumar and A. Kumar, Nutritional composition of millets, in *Millets and millet technology*, 2021, pp. 101–119.
25. J. H. C. Smetana and G. Misra, Principles of protein structure and function, in *Introduction to Biomolecular Structure and Biophysics: Basics of Biophysics*, Springer Singapore, Singapore, 2017, pp. 1–32.
26. S. Sharma, R. Sharma and B. Singh, Influence of bioprocessing treatments on phytochemical and functional properties, *in vitro* digestibility, protein secondary structure and morphological characteristics of Indian barnyard millet flour, *Int. J. Food Sci. Technol.*, 2022, **57**(8), 4744–4753.
27. F. Yildiz, 3 Amino Acids, Oligopeptides, Polypeptides, and Proteins, in *Advances in food biochemistry*, 2010, p. 51.
28. Y. Zhang, X. Jing, Z. Chen and X. Wang, Effects of moderate-intensity pulsed electric field on the structure and physicochemical properties of foxtail millet (*Setaria italica*) prolamin, *Cereal Chem.*, 2023, **100**, 360–370.
29. N. A. Mir, M. Bharadwaj, B. Yousuf, K. Gul, C. S. Riar and S. Singh, Food Biopolymers: Structural, Functional, and Nutraceutical Properties: Food Proteins: An Overview, in *Food biopolymers: Structural, functional and nutraceutical properties*, 2021, pp. 211–229.
30. R. Sharma, S. Sharma and B. Singh, Modulation in the bio-functional & technological characteristics, *in vitro* digestibility, structural and molecular interactions during bioprocessing of proso millet (*Panicum miliaceum* L.), *J. Food Compos. Anal.*, 2022, **107**, 104372.
31. S. R. Bean, L. Zhu, B. M. Smith, J. D. Wilson, B. P. Ioerger and M. Tilley, Starch and protein chemistry and functional properties, *Sorghum and millets*, 2019, pp. 131–170.
32. S. Anitha, A. Rajendran, R. Botha, C. Baruah, P. Mer and J. Sebastian, *et al.*, Variation in the nutrient content of different genotypes and varieties of millets, studied globally: a systematic review, *Front. Sustainable Food Syst.*, 2024, **8**, 1324046.

33. P. Singh and R. S. Raghuvanshi, Finger millet for food and nutritional security, *Afr. J. Food Sci.*, 2012, **6**(4), 77–84.
34. V. Wiedemair, S. Scholl-Bürgi, D. Karall and C. W. Huck, Amino Acid Profiles and Compositions of Different Cultivars of Panicum miliaceum L, *Chromatographia*, 2020, **83**, 829–837.
35. E. S. Sarita and E. Singh, Potential of millets: nutrients composition and health benefits, *J. Sci. Innov. Res.*, 2016, **5**(2), 46–50.
36. J. Jacob, V. Krishnan, C. Antony, M. Bhavyasri, C. Aruna and K. Mishra, *et al.*, The nutrition and therapeutic potential of millets: an updated narrative review, *Front. Nutr.*, 2024, **11**, 1346869.
37. G. J. F. C. Ravindran, Studies on millets: Proximate composition, mineral composition, and phytate and oxalate contents, *Food Chem.*, 1991, **39**(1), 99–107.
38. O. Adeola and J. I. Orban, Chemical composition and nutrient digestibility of pearl millet (*Pennisetum glaucum*) fed to growing pigs, *J. Cereal Sci.*, 1995, **22**(2), 177–184.
39. T. Wu, H. Li, J. Li and J. Hao, Nutrient composition of germinated foxtail millet flour treated with mixed salt solution and slightly acidic electrolyzed water, *Foods*, 2022, **12**(1), 75.
40. S. S. Audu, M. O. Aremu, B. W. Tukura and A. I. Ambo, Chemical composition of finger millet (*Eleusine coracana*) flour, no. October, 2018.
41. P. Singh, R. Singh, V. Bhadauria and H. Singh, The functionality and extraction of protein from sorghum, finger millet, and Kodo millet: a review, *Int. J. Food Sci. Technol.*, 2024, **59**(1), 512–521.
42. H. Wang, D. Li, C. Wan, Y. Luo, Q. Yang, X. Gao and B. Feng, Improving the functionality of proso millet protein and its potential as a functional food ingredient by applying nitrogen fertiliser, *Foods*, 2021, **10**(6), 1332.
43. A. A. M. Nour, A. R. Mohamed, O. Q. Adiamo and E. E. Babiker, Changes in protein nutritional quality as affected by processing of millet supplemented with Moringa seed flour, *J. Saudi Soc. Agric. Sci.*, 2018, **17**(3), 275–281.
44. R. Shen, Y. Ma, L. Jiang, J. Dong, Y. Zhu and G. Ren, Chemical composition, antioxidant, and antiproliferative activities of nine Chinese proso millet varieties, *Food Agric. Immunol.*, 2018, **29**(1), 625–637.
45. V. Kumar, M. Yadav, S. K. Awala, J. S. Valombola, M. S. Saxena, F. Ahmad and S. C. Saxena, Millets: a nutritional powerhouse for ensuring food security, *Planta*, 2024, **260**(4), 101.
46. T. P. Shukla and W. Bushuk, Cereal proteins: Chemistry and food applications, *Crit. Rev. Food Sci. Nutr.*, 1975, **6**(1), 1–75.
47. *Pseudocereals and less common cereals: grain properties and utilization potential*, ed. P. S. Belton and J. R. Taylor, Springer Science & Business Media, 2002.
48. W. Zhang, G. Zhang, W. Liang, J. Tian, S. Sun and X. Zhang, *et al.*, Structure, Functional Properties, and Applications of Foxtail Millet Prolamin: A Review, *Biomolecules*, 2024, **14**(8), 913.

49. A. K. Pandey and S. Verma, Protein hydrolysates of millets: an effective therapeutic to wide range of diseases, *Nutr. Food Sci.*, 2024, **54**(6), 1095–1105.
50. P. Asrani, A. Ali and K. Tiwari, Millets as an alternative diet for gluten-sensitive individuals: A critical review on nutritional components, sensitivities and popularity of wheat and millets among consumers, *Food Rev. Int.*, 2023, **39**(6), 3370–3399.
51. F. Shahidi and Y. Zhong, Measurement of antioxidant activity, *J. Funct. Foods*, 2015, **18**, 757–781.
52. T. K. Mohamed, A. Issoufou and H. Zhou, Antioxidant activity of fractionated foxtail millet protein hydrolysate, *Int. Food Res. J.*, 2012, **19**(1), 207–213.
53. Y. Zhang, X. Jing, Z. Chen and X. Wang, Purification and identification of antioxidant peptides from millet gliadin treated with high hydrostatic pressure, *LWT*, 2022, **164**, 113654.
54. P. Mudgil, M. Alblooshi, B. P. Singh, A. R. Devarajan and S. Maqsood, Pearl millet protein hydrolysates exhibiting effective in-vitro antioxidant, antidiabetic and anti-lipidemic properties as potential functional food ingredient, *Int. J. Food Sci. Technol.*, 2023, **58**(6), 3264–3272.
55. I. Amadou, G. W. Le, T. Amza, J. Sun and Y. H. Shi, Purification and characterization of foxtail millet-derived peptides with antioxidant and antimicrobial activities, *Food Res. Int.*, 2013, **51**(1), 422–428.
56. H. Agrawal, R. Joshi and M. Gupta, Isolation, purification and characterization of antioxidative peptide of pearl millet (*Pennisetum glaucum*) protein hydrolysate, *Food Chem.*, 2016, **204**, 365–372.
57. F. K. Ofosu, F. Elahi, E. B. M. Daliri, R. Chelliah, H. J. Ham and J. H. Kim, *et al.*, Phenolic profile, antioxidant, and antidiabetic potential exerted by millet grain varieties, *Antioxidants*, 2020, **9**(3), 254.
58. M. Karaś, A. Jakubczyk, U. Szymanowska, K. Jęderka, S. Lewicki and U. Zøotek, Different temperature treatments of millet grains affect the biological activity of protein hydrolyzates and peptide fractions, *Nutrients*, 2019, **11**, 550.
59. M. Balouiri, M. Sadiki and S. K. Ibnsouda, Methods for *in vitro* evaluating antimicrobial activity: A review, *J. Pharm. Anal.*, 2016, **6**(2), 71–79.
60. K. Kanokwiroon, R. Teanpaisan, D. Wititsuwannakul, A. B. Hooper and R. Wititsuwannakul, Antimicrobial activity of a protein purified from the latex of Hevea brasiliensis on oral microorganisms, *Mycoses*, 2008, **51**(4), 301–307.
61. M. Sitohy and A. Osman, Antimicrobial activity of native and esterified legume proteins against Gram-negative and Gram-positive bacteria, *Food Chem.*, 2010, **120**(1), 66–73.
62. W. Xu, L. Wei, W. Qu, Z. Liang, J. Wang, X. Peng, Y. Zhang and K. Huang, A novel antifungal peptide from foxtail millet seeds, *J. Sci. Food Agric.*, 2011, **91**(9), 1630–1637.

63. A. Bisht, M. Thapliyal and A. Singh, Screening and isolation of antibacterial proteins/peptides from seeds of millets, *Int. J. Curr. Pharm. Res.*, 2016, **8**(3), 96–99.
64. I. Amadou, G. W. Le and Y. H. Shi, Evaluation of antimicrobial, antioxidant activities, and nutritional values of fermented foxtail millet extracts by Lactobacillus paracasei Fn032, *Int. J. Food Prop.*, 2013, **16**(6), 1179–1190.
65. J. Chen, W. Duan, X. Ren, C. Wang, Z. Pan, X. Diao and Q. Shen, Effect of foxtail millet protein hydrolysates on lowering blood pressure in spontaneously hypertensive rats, *Eur. J. Nutr.*, 2017, **56**, 2129–2138.
66. Y. Zheng, X. Wang, M. Guo, X. Yan, Y. Zhuang, Y. Sun and J. Li, Two novel antihypertensive peptides identified in millet bran glutelin-2 hydrolysates: Purification, *in silico* characterization, molecular docking with ace and stability in various food processing conditions, *Foods*, 2022, **11**(9), 1355.
67. Y. Zhu, C. Chen, Z. Dai, H. Wang, Y. Zhang and Q. Zhao, *et al.*, Identification, screening and molecular mechanisms of natural stable angiotensin-converting enzyme (ACE) inhibitory peptides from foxtail millet protein hydrolysates: a combined *in silico* and *in vitro* study, *Food Funct.*, 2024, **15**(15), 7782–7793.
68. Y. Zheng, J. Ma, Y. Guo, Y. Zhuang, Z. Yang and Z. Zhu, *et al.*, In silico screening and characterization, inhibition mechanism on ACE, and stability of antihypertensive peptides with Zn-chelating capacity identified from millet bran albumin hydrolysates, *Food Bioscience*, 2023, **56**, 103419.
69. B. Sharma and H. S. Gujral, Influence of nutritional and antinutritional components on dough rheology and *in vitro* protein & starch digestibility of minor millets, *Food Chem.*, 2019, **299**, 125115.
70. P. V. Monteiro, D. H. Gopal, T. K. Virupaksha and G. Ramachandra, Chemical composition and *in vitro* protein digestibility of Italian millet (*Setaria italica*), *Food Chem.*, 1988, **29**, 19–26.
71. F. S. Pushparaj and A. Urooj, Influence of processing on dietary fiber, tannin and *in vitro* protein digestibility of pearl millet, *Food Nutr. Sci.*, 2011, **2**(8), 895–904.
72. S. H. Elyas, A. H. El Tinay, N. E. Yousif and E. A. Elsheikh, Effect of natural fermentation on nutritive value and *in vitro* protein digestibility of pearl millet, *Food Chem.*, 2002, **78**(1), 75–79.
73. G. Ravindran, Seed protein of millets: amino acid composition, proteinase inhibitorsand *in-vitro* protein digestibility, *Food Chem.*, 1992, **44**(1), 13–17.
74. P. Bergamo, F. Maurano, G. Mazzarella, G. Iaquinto, I. Vocca and A. R. Rivelli, *et al.*, Immunological evaluation of the alcohol-soluble protein fraction from gluten-free grains in relation to celiac disease, *Mol. Nutr. Food Res.*, 2011, **55**(8), 1266–1270.
75. F. Akharume and A. Adedeji, Effects of High-power Ultrasound on the *in vitro* digestibility, Physicochemical and Functional Properties of

Proso Millet Prolamin and Glutelin Proteins, *J. Food Meas. Charact.*, 2023, **17**(1), 178–186.
76. H. Agrawal, R. Joshi and M. Gupta, Functional and nutritional characterization of *in vitro* enzymatic hydrolyzed millets proteins, *Cereal Chem.*, 2020, **97**(6), 1313–1323.
77. B. Sharma and H. S. Gujral, Modifying the dough mixing behavior, protein & starch digestibility and antinutritional profile of minor millets by sprouting, *Int. J. Biol. Macromol.*, 2020, **153**, 962–970.
78. A. G. Wouters, I. Rombouts, E. Fierens, K. Brijs and J. A. Delcour, Relevance of the functional properties of enzymatic plant protein hydrolysates in food systems, *Compr. Rev. Food Sci. Food Saf.*, 2016, **15**(4), 786–800.
79. F. Zhang, Y. Fu, Z. Liu and Q. Shen, Comparison of the characteristics of prolamins among foxtail millet varieties with different palatability: Structural, morphological, and physicochemical properties, *Int. J. Biol. Macromol.*, 2021, **186**, 194–205.
80. M. T. Kamara, I. Amadou, F. Tarawalie and Z. Huiming, Effect of enzymatic hydrolysis on the functional properties of foxtail millet (*Setaria italica* L.) proteins, *Int. J. Food Sci. Technol.*, 2010, **45**(6), 1175–1183.
81. D. Dey, G. Singh, P. Ishwarya and A. Mateen, Functionality and Extrusion Processing of Millets—A Review, *Food Rev. Int.*, 2024, **40**(10), 3605–3633.
82. S. J. Calva-Estrada, M. Jiménez-Fernández and E. Lugo-Cervantes, Protein-Based Films: Advances in the Development of Biomaterials Applicable to Food Packaging, *Food Rev. Int.*, 2019, **11**(2), 78–92.
83. T. Gillgren, M.-V. Faye and M. Stading, Mechanical and Barrier Properties of Films from Millet Protein Pennisetin, *Food Biophys.*, 2011, **6**(4), 474–480.
84. N. Sharma, J. K. Sahu, V. Bansal, O. J. Esua, S. Rana and A. Bhardwaj, *et al.*, Trends in millet and pseudomillet proteins-Characterization, processing and food applications, *Food Res. Int.*, 2023, **164**, 112310.
85. O. P. Malav, S. Talukder, P. Gokulakrishnan and S. Chand, Meat Analog: A Review, *Crit. Rev. Food Sci. Nutr.*, 2015, **55**(9), 1241–1245.
86. A. Mateen and G. Singh, Evaluating the potential of millets as blend components with soy protein isolate in a high moisture extrusion system for improved texture, structure, and colour properties of meat analogues, *Food Res. Int.*, 2023, **173**, 113395.
87. C. W. X. Neo, Y. H. How, J. Kong, R. A. Talib and L. P. Pui, Development of plant-based burger patties with pea protein isolate and barnyard millet flour and its storage stability in aerobic and vacuum packaging, *J. Food Saf.*, 2024, **44**, 13134.
88. S. Pandey and A. Singh, Millet Milk: A Potential Cattle Milk Alternative—Extraction and Value Addition, *J. Food Qual.*, 2024, **2024**(1), 3964937.
89. J. D. Sunny, P. Kundu and R. Mehla, Preparation and physicochemical evaluation of dairy free alternative based on coconut and millet milk, *Int. J. Pharm. Biol. Sci.*, 2019, **9**, 565–572.

90. K. J. E. Kouamé, A. F. M. Bora, X. Li, Y. Sun, S. Tian, M. Hussain, L. Liu and I. Coulibaly, Development and characterization of probiotic (co)encapsulates in biopolymeric matrices and evaluation of survival in a millet yogurt formulation, *J. Food Sci.*, 2023, **88**(1), 9–27.
91. E. Stefano, J. White, S. Seney, S. Hekmat, T. McDowell, M. Sumarah and G. Reid, A Novel Millet-Based Probiotic Fermented Food for the Developing World, *Nutrients*, 2017, **9**(5), 529.
92. S. Bashir, M. Yaseen, V. Sharma, S. R. Purohit, S. Barak and D. Deepak Mudgil, Rheological and Textural Properties of Gluten Free Cookies based on Pearl Millet and Flaxseed, *Biointerface Res. Appl. Chem.*, 2020, **10**(5), 6565–6576.
93. S. Rai, A. Kaur and C. S. Chopra, Gluten-Free Products for Celiac Susceptible People, *Front. Nutr.*, 2018, **5**, 116.
94. R. Radhika, A. Virk, M. Kaur, P. Thakur, D. Chauhan, Q. U. E. H. Rizvi, S. Jan and K. Kumar, Development and nutritional evaluation of multigrain gluten free cookies and pasta products, *Curr. Res. Nutr. Food Sci.*, 2019, **7**(3), 842–853.
95. J. O. Adepehin, V. N. Enujiugha, A. A. Badejo, G. M. Young and D. A. Odeny, Physicochemical and sensory attributes of gluten-free sourdough breads produced from underutilised African cereal flours and flour blends, *Int. J. Food Sci. Technol.*, 2023, **58**, 493–501.
96. M. st Meherunnahar, M. M. Hoque, M. A. Satter, T. Ahmed, R. S. Chowdhury and S. Aziz, Effect of cooking characteristics, amino acid consistency, and functional properties of composite noodles made from foxtail millet, *Measurement: Food*, 2024, **13**, 100138.
97. D. Agrahar-Murugkar, A. Zaidi and S. Dwivedi, Development of gluten free eggless cake using gluten free composite flours made from sprouted and malted ingredients and its physical, nutritional, textural, rheological and sensory properties evaluation, *J. Food Sci. Technol.*, 2018, **55**(7), 2621–2630.
98. C. A. Shimray, S. Gupta and G. Venkateswara Rao, Effect of native and germinated finger millet flour on rheological and sensory characteristics of biscuits, *Int. J. Food Sci. Technol.*, 2012, **47**(11), 2413–2420.
99. X. Chen, Y.-C. Wu, Y. Liu, L.-H. Qian, Y.-H. Zhang and H.-J. Li, Single/co-encapsulation capacity and physicochemical stability of zein and foxtail millet prolamin nanoparticles, *Colloids Surf., B*, 2022, **217**, 112685.
100. X. Chen, C. Yu, J.-H. Wang, Y.-C. Wu, Y. Ma and H.-J. Li, Fabrication of curcumin–loaded foxtail millet prolamin–based nanoparticle: Impact of curdlan sulfate on the particle properties, *Colloids Surf., A*, 2023, **674**, 131893.
101. L. Wang, P. Gulati, D. Santra, D. Rose and Y. Zhang, Nanoparticles prepared by proso millet protein as novel curcumin delivery system, *Food Chem.*, 2018, **240**, 1039–1046.
102. A. M. Sánchez, M. Carmona, A. Zalacain, J. M. Carot, J. M. Jabaloyes and G. L. Alonso, Rapid Determination of Crocetin Esters and Picrocrocin from

Saffron Spice (*Crocus sativus* L.) Using UV–Visible Spectrophotometry for Quality Control, *J. Agric. Food Chem.*, 2008, **56**(9), 3167–3175.

103. F. Jhan, A. Gani, N. Noor, B. Ahmad Malla and B. A. Ashwar, Nano reduction coupled with encapsulation as a novel technique for utilising millet proteins as future foods, *Ultrason. Sonochem.*, 2022, **86**, 106006.
104. S. K. Vanga, J. Wang, S. Jayaram and V. Raghavan, Effects of pulsed electric fields and ultrasound processing on proteins and enzymes: A review, *Processes*, 2021, **9**(4), 722.
105. S. Suri, K. S. Balasaheb, D. K. Yadav, S. Malakar, P. Choudhary, A. Mohapatra and P. Dhurve, Overview of Millet proteins: Quality characteristics, effect of thermal/ non-thermal processing and applications, *Food Biosci.*, 2024, **57**, 103434.
106. Y. Zhang, X. Zhang, Z. Zhang, Z. Chen, X. Jing and X. Wang, Effect of high hydrostatic pressure treatment on the structure and physicochemical properties of millet gliadin, *LWT*, 2022, **154**, 112755.
107. Z.-L. Chen, Y. Li, J.-H. Wang, R. Wang, Y.-X. Teng, J.-W. Lin, X.-A. Zeng, M.-W. Woo, L. Wang and Z. Han, Pulsed electric field improves the EGCG binding ability of pea protein isolate unraveled by multi-spectroscopy and computer simulation, *Int. J. Biol. Macromol.*, 2023, **244**, 125082.
108. H. Hu, E. C. Y. Li-Chan, L. Wan, M. Tian and S. Pan, The effect of high intensity ultrasonic pre-treatment on the properties of soybean protein isolate gel induced by calcium sulfate, *Food Hydrocolloids*, 2013, **32**(2), 303–311.
109. A. Sarkar, T. Niranjan, G. Patel, A. Kheto, B. K. Tiwari and M. Dwivedi, Impact of cold plasma treatment on nutritional, antinutritional, functional, thermal, rheological, and structural properties of pearl millet flour, *J. Food Process Eng.*, 2023, **46**(5), e14317.
110. V. Monica, R. Anbarasan and R. Mahendran, Cold Plasma-Induced Changes in the Structural and Techno-functional Properties of Sprouted Foxtail Millet Protein Concentrate, *Food Bioprocess Technol.*, 2025, **18**(1), 850–867.
111. S. R. Kumar, M. B. Sadiq and A. K. Anal, Comparative study of physicochemical and functional properties of pan and microwave cooked underutilized millets (proso and little), *LWT*, 2020, **128**, 109465.
112. M. V. Rao, C. K. Sunil and N. Venkatachalapathy, Effect of microwave and hot air radiofrequency treatments on physicochemical and functional properties of foxtail millet flour and its protein isolate, *J. Cereal Sci.*, 2023, **114**, 103774.
113. E. Shylla, C. K. Sunil, A. Rawson and N. Venkatachalapathy, High-intensity ultrasound modification of techno-functional and structural properties of white finger millet protein fractions, *J. Food Sci.*, 2024, **89**, 8999–9014.
114. G. Kaur, S. Ahmadzadeh-Hashemi, S. Amir, Z. S. Khan, Z. Gulsunoglu-Konuskan, A. Karimidastjerd, S. Fayaz, M. S. Bhat, S. Rustagi, A. E.-D. A. Bekhit and T. Aijaz, Exploring sustainable novel millet protein: A look

at the future foods through innovative processing, *Future Foods*, 2024, **9**, 100367.
115. M. Contreras, A. Lama-Muñoz, J. M. Gutiérrez-Pérez, F. Espínola, M. Moya and E. Castro, Protein extraction from agri-food residues for integration in biorefinery: Potential techniques and current status, *Bioresour. Technol.*, 2019, **280**, 459–477.
116. M. Mirzaei, S. Mirdamadi, M. Safavi and N. Soleymanzadeh, The stability of antioxidant and ACE-inhibitory peptides as influenced by peptide sequences, *LWT – Food Sci. Technol.*, 2020, **130**, 109710.
117. A. Mundassery, J. Ramaswamy, T. Natarajan, S. Haridas and P. Nedungadi, Modern and conventional processing technologies and their impact on the quality of different millets, *Food Sci. Biotechnol.*, 2024, **33**(11), 2441–2460.

CHAPTER 6

Medicinal Properties of Millet Proteins

PRAVEEN KUMAR DIKKALA,[a] NIKHIL SWARAJ ARIGE,[a] GAYATRI DANGETI,[a] VIHARIKA KONDRU,[a] KANDI SRIDHAR*[b] AND MINAXI SHARMA*[c]

[a] Department of Food Technology, Koneru Lakshmaiah Education Foundation Deemed to be University, Vaddeswaram 522502, India; [b] Department of Food Technology, Karpagam Academy of Higher Education (Deemed to be University), Coimbatore 641021, India; [c] Research Centre for Life Science and Healthcare, Nottingham Ningbo China Beacons of Excellence Research and Innovation Institute (CBI), University of Nottingham Ningbo China, Ningbo, China
*Emails: sridhar4647@gmail.com; Minaxi.Sharma@nottingham.edu.cn

6.1 Introduction

Millets are the one of the earliest foods humans ever consumed and could be a game changer because of their remarkable traits, which include climate resilience, low input requirements, and a rich nutritional profile. Millets are frequently disregarded in favor of major staple crops like wheat and rice due to reasons such as urbanization and industrialization. With the capacity to flourish in arid, hot, and humid climate zones, where wheat and rice production is extremely unlikely, millets are a set of small seeds derived from annual plants that are extensively grown in semi-arid and dry land parts of the world.[1,2] They are essential to the ecosystem, work well in mixed and intercropping systems, are resilient to extreme weather conditions like drought and salt, require low inputs and low production costs, and give

Food Chemistry, Function and Analysis No. 48
Millet Proteins: Nutritional, Structural and Functional Attributes
Edited by Shweta Suri, Anuradha Dutta and Tarun Belwal

Published by the Royal Society of Chemistry, www.rsc.org

farmers a means of subsistence and food security. As millets require less synthetic fertilizers, herbicides, and insecticides, they are beneficial to both humans and environment.[3–5] Millets are easy to cultivate, renewable, less expensive and a regular source of protein consumed around the world as compared to animal-based foods consumed for proteins.[2,6] Improving protein cultivation benefits millets by preserving structure, synthesizing enzymes, performing photosynthesis, improving transportation and immunity, and serving as a nutrient reserve for seed growth and maturity. The millet market has been increasing because of the health benefits, pharmaceutical benefits and food security associated with millet.[5,7] By the end of 2025, the estimated worth is $12 billion globally.[8] This chapter aims to explore the development of various methods for the extraction, concentration, and characterization of proteins from millets. It will also examine the different bioactive peptides derived from millet proteins, highlighting their potential therapeutic applications. Additionally, the chapter will focus on analyzing the role of millet peptides in functional and therapeutic foods, emphasizing their health benefits and practical uses. By covering protein extraction techniques, characterization methods, and applications in pharmaceuticals and diverse food products, this chapter will serve as a valuable resource for entrepreneurs, scientists, and dietitians, facilitating a deeper understanding of how millet protein components can be effectively utilized in food and medicinal product development.

6.2 Types and Nutritional Value

To maintain a healthy lifestyle, it is vital to select functional foods based on their biological and physiological features. Various scientific investigations on millets show promising results in their use for a variety of nutraceutical foods.[9] Around 38% of the world's population lives in drylands, where climate change and human activity impact agricultural productivity. Farmers who understand climatic conditions and soil profiles may be able to pick crops more effectively. As millets are gluten free, they have received interest from food processing firms for the production of recipes based on their flour.[9,10] Millets are traditionally used to feed animals and birds but are now becoming more important as innovative therapeutic foods for humans. They are sometimes grouped together under the term 'coarse cereals', which encompasses all heated, seasoned cereal seeds apart from wheat and rice.[11] After starch, protein is the second most prevalent nutrient, comprising around 10% of millet's total weight.[5,12] Similar to rice and wheat, the protein content of different millet seed species ranges from 7.52% to 12.1%.[5,13] The species, cultivar, anatomical distribution, growing circumstances, and agricultural practices—such as soil quality, water content, chemical inputs supplied to the crop, and growth temperatures—all affect the quality and quantity of the millet proteins.[5,6] Compared to wheat and rice seeds, millets have a lower net carbohydrate content but a greater content of protein, dietary fiber, essential fatty acids, vitamins (primarily vitamin B complex), calcium, iron, magnesium, and phosphorus, as well as other trace elements. The seeds are inexpensive when compared to popular staple cereals.[14] They also include a number of beneficial polyphenolic substances.[4] Due to their nutraceutical properties, millets are involved in combating many diseases like blood pressure, thyroid, diabetes, cardiovascular disease, and celiac disease by keeping blood pressure and blood sugar levels stable.[15]

Because of their different nutritional properties, the Food and Agriculture Organization of the United Nations (FAO-UN) designated 2023 as the International Year of Millets due to the nutraceutical advantages of these plants, which are referred to as "smart crops". They promote the nutrition and health of the expanding population and a sustainable agricultural system that supports environmentally sustainable growth.[16]

Amongst the millets, finger (*Eleusine coracana*), foxtail (*Setaria italica*), proso (*Panicum miliaceum*), little (*Panicum sumantrense*), barnyard (*Echinochloa utilis*), and kodo millet (*Paspalum scrobiculatum*) are classified as minor or small millets because of their size. The major millets are sorghum (*Sorghum vulgare*) and pearl millet (*Pennisetum glaucum*).[17]

Millets are considered as "nutri cereals", as they have significant quantities of proteins, including phytosterols, essential fatty acids, dietary fiber and amino acids. Different processed products, including bread, biscuits, puff, flakes, snack items, traditional foods (such as tuwo (a Nigerian delicacy), fulanis, hausas (millet ball-Africa), gruel-pap, ogi, and akamu), wine, beer, spirits, burukutu, palmvine, pito, and ogogoro (local alcoholic beverages), and kunu (non-alcoholic beverage), all contain millet flour.[18]

6.3 Importance of Millets in Diet and Health

Millets are nutritionally comparable to, or perhaps better than, many important cereal grains, due to their low glycemic index, high fiber content, high antioxidant content, gluten-free protein content and hypoallergic properties (Table 6.1). The carbohydrate content of millet grains is around 70%, which includes soluble carbohydrates and dietary fiber. The majority of millet polysaccharides are made up of amylopectin and amylose (70–80%). They are higher in polyphenols (0.2–0.5%), as well as dietary fiber. The antioxidant activity of millets is primarily provided by polyphenols, tannins, and phytates, which also play a role in controlling the ageing process.[19]

Even though millets have an appropriate nutritional profile, they lack some essential amino acids like lysine, threonine, and tryptophan. The protein profile of millets offers a wide variety of advantages for treating chronic illnesses in people.[12,20–23] They are found in the unique storage vacuoles known as protein bodies, which are crucial for food, medicine and nutrition. Millets have "staple status" in the human diet and their production ensures global food security.[24]

Table 6.1 Types of millet and their health benefits.

Millet type	Features	Ref.
Foxtail millet	Lowers the chance of colon cancer. Lowers cholesterol and has anti-diabetic properties. Reduces the harm that ethanol causes to the liver.	25–34
Finger millet	In diabetic rats, it promotes wound healing and guards against tissue damage. It contributes to cancer prevention.	
Pearl millet	Because it is gluten free, it helps prevent celiac disease in people. Stimulates the immune system to stop the pathogenicity caused by Shigella in the mouse model.	
Proso millet	Lacks gluten and can prevent celiac disease in people. Because of its low glycemic index, it helps lower the risk of type 2 diabetes in people.	
Kodo millet	It has antioxidant properties, lowers the glycemic index, and prevents diabetes in the female human model.	
Barnyard millet	Causes the HT-29 human colon cancer cell line to undergo apoptosis, which inhibits the growth of cancer. Its phenolic concentration prevents the glycation and glycoxidation of proteins, which is a major factor in the development of diabetes.	
Little millet	Protects against contemporary metabolic diseases because polyphenols are present.	

6.4 Methodologies for Protein Characterization in Millets

6.4.1 Techniques for Protein Fractionation

Protein fractionation is a critical technique used to separate and isolate proteins from complex mixtures based on their unique physical and chemical properties (Figure 6.1). This process is achieved through methods like electrophoresis, precipitation, and chromatography, each using different characteristics based on the size, charge, and protein solubility.[5] Green technologies like sonication and microwaves are used to isolate specific protein fractions from millet grains. Isolated fractions exhibit physicochemical properties that affect solubility, emulsification properties and nutritional composition. Protein fractions from millet grains are suitable for diverse food and non-food applications.[35]

6.4.1.1 Solvent Extraction for Millet Protein Fractionation

Solvent extraction for protein fractionation relies on the wide solubility characteristics of different millet protein fractions, such as albumins, globulins, prolamins, and glutelin. The principles of solvent extraction involve the use of sequential extraction to isolate the fractions based on their

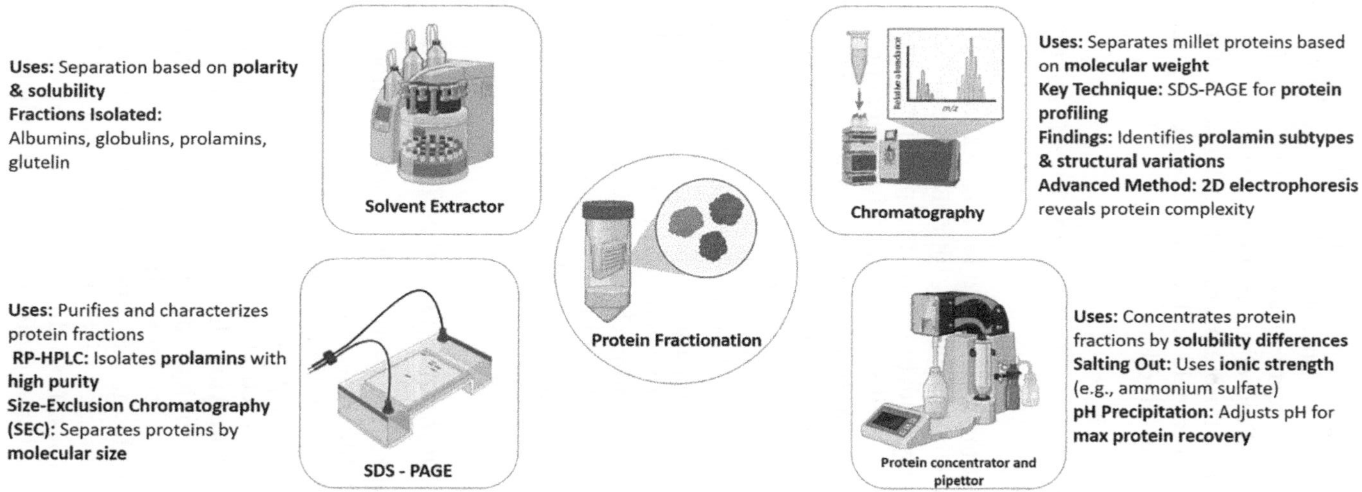

Figure 6.1 Techniques for protein fractionation.

polarity and ionic strength.[36,37] Ethanol-soluble proteins accounted for 40% of the total nitrogen, while SDS-soluble proteins comprised about 25%. Sequential extraction using 70% aqueous ethanol and sodium hydroxide can yield pennisetin with a protein content exceeding 90% from pearl millet flour, highlighting the effectiveness of solvent extraction in maximizing protein recovery. Marcellino *et al.* explored the optimization of extraction conditions for pearl millet proteins, reporting that a combination of solvents significantly improved the yield and purity of extracted proteins, particularly focusing on the prolamin fraction.[38]

6.4.1.2 Chromatographic Techniques for Millet Protein Fractionation

Chromatographic techniques are used for purification and detailed analysis of millet protein fractions following initial extraction. These techniques enhance the specificity and purity of the protein fractions, enabling researchers to investigate their structural and functional properties. Reversed-phase high-performance liquid chromatography (RP-HPLC) has been widely used to isolate and characterize prolamin fractions. For instance, Bibi *et al.* achieved high purity levels of prolamins from various pearl millet cultivars using RP-HPLC, uncovering valuable insights into their functional properties.[39] Similarly, Hong *et al.* demonstrated the utility of size-exclusion chromatography (SEC) in separating protein fractions from millet flour, providing a clear analysis of protein size distribution and associated functional attributes.[40] Furthermore, chromatographic methods have been used to reveal significant differences in functional properties among prolamin fractions, highlighting the potential of these techniques for characterizing millet proteins.[39]

6.4.1.3 Electrophoretic Analysis of Millet Proteins

Electrophoretic methods, particularly sodium dodecyl sulfate-polyacrylamide gel electrophoresis (SDS-PAGE), are crucial for analyzing molecular weight-based separation of millet proteins. SDS-PAGE has been instrumental in identifying protein profiles and structural characteristics of fractions. Zhang *et al.*[35] used SDS-PAGE to characterize prolamin fractions from foxtail millet, identifying distinct bands that correspond to various prolamin types. This technique provided essential insights into the structural attributes of these proteins. Tang *et al.*[41] conducted comparative studies of protein profiles across different pearl millet cultivars using SDS-PAGE, revealing significant variations that correlate with their nutritional and functional properties. Extending the scope of electrophoretic analysis, Manivannan, employed two-dimensional electrophoresis to explore the complexity of pearl millet protein extracts, uncovering their diversity and potential applications in food science and nutraceuticals.[42]

6.4.1.4 Precipitation Techniques for Millet Protein Enrichment

Precipitation methods offer an effective approach to concentrate specific protein fractions from millet based on solubility differences. These methods typically involve salt addition or pH adjustments to induce selective precipitation. The effectiveness of "salting out" techniques for isolating pearl millet proteins shows how controlled ionic strength adjustments can efficiently precipitate target proteins.[43] Similarly, Akharume *et al.* investigated pH-induced precipitation for prolamins from millet flour, identifying optimal pH conditions that maximize protein recovery.[44] Ammonium sulfate precipitation was utilized to isolate specific protein fractions from millet grains. Their work emphasized the method's potential to prepare concentrated protein extracts for subsequent functional and nutritional analysis.[45]

6.5 Analytical Methods for Protein Concentration and Composition

Spectrophotometric techniques, including UV-Vis and near-infrared spectroscopy (NIR), are vital for analyzing millet proteins for accurate quantification and characterization of their content, structure, and modifications. A study on *Panicum miliaceum* (proso millet) demonstrated the use of NIR spectroscopy for evaluating protein content, with handheld and benchtop spectrometers achieving high accuracy and minimal error for both intact and milled grains.[46] High-power ultrasound treatments on millet protein concentrate (MPC) enhanced functional properties such as solubility and emulsification, with spectrophotometric methods like FTIR and zeta potential used to identify structural modifications, including molecular weight reduction and increased negative surface charge.[47] Additionally, protein hydrolysates from pearl millet were analyzed using MALDI-TOF spectrometry, identifying antioxidant peptides and their activity against free radicals.[48] In finger millet, calcium-binding proteins crucial for seed development were characterized using nano-LC-MS, offering insights into their structure and functional roles.[49] Overall, these spectrophotometric methods are indispensable for advancing millet protein research, supporting their applications in nutrition and functional foods.

Sodium dodecyl sulfate-polyacrylamide gel electrophoresis (SDS-PAGE) is a powerful analytical technique for characterizing millet proteins, allowing for the determination of molecular weights, purity, and structural changes in protein fractions.

In a previously mentioned study, the effect of high-power ultrasound on millet protein concentrate (MPC) was investigated. SDS-PAGE was used to confirm structural changes, including molecular weight reduction and increased solubility, highlighting its value in characterization of protein fractions.[47] Similarly, SDS-PAGE has been used to analyze the impact of extrusion cooking on foxtail millet proteins. It demonstrated disulfide bond cross-linking among glutelin and setarin subunits, resulting in decreased protein solubility, providing critical information for understanding protein–protein interactions during processing.[50]

This technique also serves in proteomic studies. For example, it has been used in combination with mass spectrometry to profile protein distributions and identify proteoforms, providing in-depth insights into the molecular complexity of millet proteins.[51] Furthermore, SDS-PAGE aids in identifying calcium-binding proteins like calreticulin in finger millet, elucidating their role in calcium sequestration during seed development.[49]

6.6 Challenges and Innovations in Protein Analysis

Millet protein analysis faces two primary challenges: interference from complex matrix components and the lack of standardized methodologies. These barriers hinder consistent protein quantification and comparability across studies. However, innovations in extraction techniques and advanced analytical technologies have been developed to address these issues.

6.6.1 Complex Matrix Interference

Millet contains a variety of phytochemicals, including polyphenols, flavonoids, and tannins, which can interfere with protein quantification. These compounds may bind to proteins or mask them during analysis, leading to underestimation or inconsistent results. For instance, common protein assays like the Bradford or UV-based absorbance methods are sensitive to interference from these compounds, which compromises accuracy.[52] Advanced separation techniques, such as liquid chromatography (LC) and size exclusion chromatography (SEC), are increasingly used to mitigate these issues. Coupled with inductively coupled plasma mass spectrometry (ICP-MS), SEC enables precise quantification by targeting elements like sulfur, which are present in amino acids.[40]

The use of surfactant-assisted extraction has also been explored to improve protein recovery and reduce interference from phytochemicals. Surfactants enhance solubilization and separation of proteins, leading to more accurate quantification.[53] Despite these advancements, the choice of analytical technique remains critical, as no single method eliminates interference from the complex matrix.

6.6.2 Lack of Standardized Methods

Differences in sample preparation, enzymatic digestion, and analytical techniques across studies complicate comparisons and reduce reproducibility. This variability is particularly pronounced in proteomics, where mass spectrometry-based approaches dominate. For instance, isotope labelling and enzymatic digestion are widely used but lack universal guidelines for application and interpretation.[54]

Efforts to establish standardized workflows are ongoing. For example, the use of stable isotope-labeled (SIL) standards, such as AQUA peptides, provides consistent reference points for absolute quantification. These peptides mimic native protein digestion products and ensure accurate measurements across different conditions.[55] Additionally, computational normalization methods, such as MaxLFQ, have emerged as robust label-free approaches for quantification, offering high reproducibility and sensitivity.[56]

6.7 Innovations in Analytical Technologies

Mass spectrometry (MS) has revolutionized protein analysis with advancements like multiple-reaction monitoring (MRM) and high-resolution Orbitrap technology. These tools improve sensitivity and specificity in detecting low-abundance proteins, making them ideal for analyzing complex matrices like millet.[57] Label-free quantification methods further enhance proteomic analysis by eliminating the need for isotopic labels, simplifying workflows while maintaining high precision.[58]

Innovative extraction and detection methods, such as selective trapping techniques and post-column isotope dilution, have also been developed. These methods minimize matrix effects and enhance the accuracy of sulfur-based quantification for proteins.[59]

6.8 Bioactive Peptides Derived from Millet Proteins

6.8.1 Antibacterial Properties: Mechanisms and Examples

Bioactive peptides derived from millet proteins possess significant antibacterial properties, primarily through mechanisms such as bacterial cell wall disruption and inhibition of cell wall synthesis (Figure 6.2). Peptides, rich in cationic and amphiphilic properties, interact with negatively charged bacterial membranes, destabilizing their structural integrity and causing leakage of intracellular contents. Additionally, some peptides inhibit vital processes like cell wall synthesis, effectively hindering bacterial growth and survival. These mechanisms align with findings from studies on other food-derived bioactive peptides, underscoring their broad-spectrum potential against bacterial pathogens.[30]

Millet-derived peptides have demonstrated remarkable activity against common pathogens such as *Escherichia coli* and *Staphylococcus aureus*. These bacteria, often associated with severe infections, represent significant challenges in the context of rising antibiotic resistance. Millet peptides, by targeting essential bacterial structures and functions, provide a natural and effective solution to manage these infections.[30] Similar mechanisms have been observed in bioactive peptides from other sources, such as milk and fish proteins, which further validates the antimicrobial efficacy of millet peptides and supports their potential for broader applications.[42]

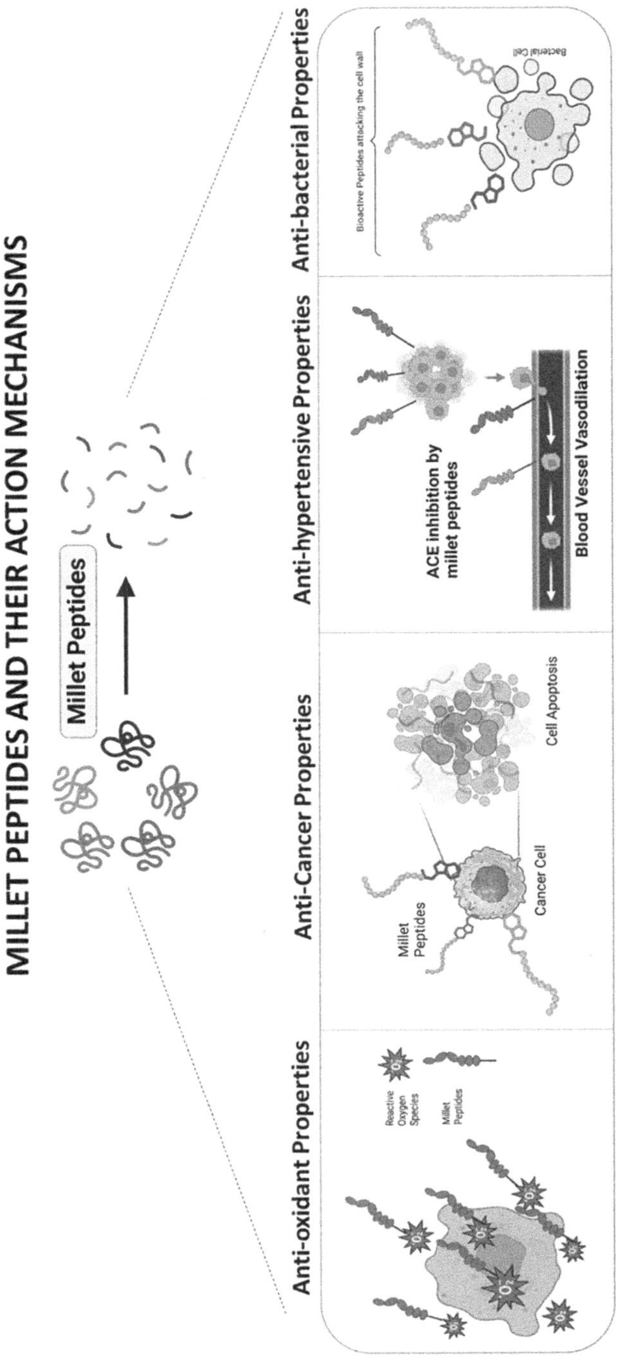

Figure 6.2 Millet peptides and their action mechanisms.

The application potential of millet-derived peptides is vast, ranging from food preservation, where they can prevent microbial contamination, to pharmaceuticals aimed at combating multidrug-resistant bacterial infections. Their natural origin, coupled with their multifunctional nature, makes them attractive alternatives to synthetic antibiotics, especially in addressing the global challenge of antibiotic resistance. However, further research into their bioavailability and *in vivo* efficacy is crucial to fully unlock their therapeutic and commercial potential.[30]

6.8.2 Antioxidant Activity: Evidence and Implications

Millet proteins are a promising source of bioactive peptides with potent antioxidant properties, capable of scavenging free radicals and mitigating oxidative stress (Figure 6.2). These peptides, often produced through enzymatic hydrolysis, donate hydrogen atoms or electrons to neutralize reactive oxygen species (ROS) and protect cellular structures from oxidative damage. For example, peptides derived from pearl millet were shown to neutralize DPPH and ABTS radicals effectively, inhibit lipid peroxidation, and chelate metal ions, showcasing their potential as natural antioxidants.[48]

Antioxidant activity in millet proteins is often enhanced by phenolic compounds, particularly in darker millet varieties, which are rich in flavonoids and phenolic acids. These compounds synergize with antioxidant peptides to enhance radical-scavenging activity and reduce lipid peroxidation. A review of cereal-derived peptides highlights the ability of such hydrolysates to donate protons or electrons, regulate oxidative pathways, and chelate metals, underscoring their broad applicability to reduce oxidative stress and its associated chronic diseases.[60]

Recent studies have identified specific antioxidant peptides from millet. Finger millet protein hydrolysates yielded peptides like TSSSLNMAVRGGLTR, which demonstrated strong radical-scavenging activity by interacting with hydroxyl and superoxide radicals. These peptides were shown to enhance endogenous antioxidant enzyme activity *in vitro*, such as superoxide dismutase (SOD) and glutathione peroxidase, offering significant protection against oxidative stress in human cells.[61]

Additional studies highlight the role of enzymatically hydrolyzed millet proteins in reducing oxidative stress markers, such as malondialdehyde, while promoting the activity of antioxidant enzymes. This supports their therapeutic potential in preventing oxidative stress-related diseases, including cardiovascular diseases, diabetes, and neurodegenerative disorders.[62]

Furthermore, millet-derived peptides have applications in functional food development. For example, peptides derived from other cereals and food proteins have demonstrated similar antioxidant capacities, such as those found in wheat and rice hydrolysates. These peptides not only scavenge radicals but also prevent the initiation of oxidative stress pathways, making them valuable for nutraceutical applications.[63]

In a broader context, millet-derived peptides contribute to reducing chronic inflammation caused by oxidative damage, offering protection against diseases like atherosclerosis, cancer, and aging-related disorders. These findings are supported by studies that link antioxidant activity to reduced inflammation and enhanced cellular resilience.[64]

6.8.3 Anti-hypertensive Effects: Peptide Functions and Clinical Relevance

Millet-derived peptides are increasingly recognized for their anti-hypertensive properties, primarily through the inhibition of angiotensin-converting enzyme (ACE), which plays a critical role in blood pressure regulation (Figure 6.2). By blocking the conversion of angiotensin I to the potent vasoconstrictor angiotensin II and preventing the degradation of bradykinin, a vasodilator, millet peptides effectively lower blood pressure. Additionally, these peptides influence other mechanisms, such as enhancing nitric oxide (NO) production, which promotes vascular relaxation.[65]

Studies have demonstrated the efficacy of millet-derived peptides in reducing hypertension. Enzymatically hydrolyzed millet proteins significantly reduced systolic blood pressure in hypertensive animal models without adverse effects on heart rate. These peptides, rich in amino acids like leucine and proline, have a high affinity for ACE, enhancing their inhibitory potential.[66] Similar ACE-inhibitory effects have been observed in peptides derived from other cereals. For instance, rice protein hydrolysates reduced systolic blood pressure in spontaneously hypertensive rats (SHRs), demonstrating a strong antihypertensive effect due to the inhibition of ACE.[67] Findings suggest that millet-derived peptides may offer comparable benefits, given millet's high protein content and distinct peptide profiles.

The clinical potential of antihypertensive peptides has been substantiated by long-term human studies. A randomized controlled trial using fermented milk enriched with ACE-inhibitory peptides demonstrated a significant reduction in systolic blood pressure by 6.7 mmHg and diastolic blood pressure by 3.6 mmHg after 21 weeks of daily supplementation.[68] Though such clinical studies on millet-derived peptides are limited, the evidence supports their potential as a dietary approach for managing hypertension. Moreover, millet-derived peptides may act beyond ACE inhibition. Emerging studies indicate their ability to modulate other pathways, including enhancing NO production, influencing endothelin signaling, and regulating oxidative stress, all contributing to improved vascular function.[69]

Research on other food-derived peptides also highlights potential applications. For instance, peptides from rice dregs demonstrated a reduction in systolic blood pressure by 26 mmHg in SHRs after long-term oral administration, showcasing strong antihypertensive effects without adverse side effects.[35] Additionally, peptides from rapeseed hydrolysates showed dual inhibition of ACE and renin, leading to significant blood pressure reductions in animal models, emphasizing the multifunctional nature of such peptides.[70]

6.8.4 Anticancer Potential: Mechanisms and Research Findings

Millet proteins and peptides have demonstrated diverse anticancer mechanisms in various studies. These mechanisms include apoptosis induction, inhibition of tumor growth factors, and modulation of cancer cell metabolism (Figure 6.2). For example, foxtail millet-derived peptides, such as the 35 kDa FMBP protein, have been shown to induce mitochondrial dysfunction and caspase-dependent apoptosis in colon cancer cells while sparing normal cells.[71] Additionally, millet bran polyphenols enhance chemotherapy efficacy in colorectal cancer by inhibiting multidrug resistance proteins and altering ganglioside metabolism [72]

Encapsulated polyphenol-rich extracts from pearl millet have shown anticancer activity in breast cancer models by inducing apoptotic and autophagic pathways, particularly in drug-resistant cancer cells. Peptides from finger millet also exhibit antioxidant and anticancer effects. Specific bioactive sequences identified include TSSSLNMAVRGGLTR and STTVGLGISMRSASVR, which demonstrate potent free radical scavenging and protective effects against oxidative stress.[61]

Moreover, millet protein hydrolysates generated under gastrointestinal conditions have been reported to yield peptides with strong anti-inflammatory and anticancer effects, including the inhibition of inflammatory markers such as TNF-α and interleukin-6 in cell models.[73] In addition, extracts from millet have shown cytotoxicity against liver cancer (HepG2) cells and antimicrobial properties, attributed to their rich polyphenolic content [74]

Finally, protein hydrolysates from various millet species have been highlighted for their nutraceutical potential, showcasing a range of anticancer, antioxidant, and anti-inflammatory activities, which underline the therapeutic promise of these bioactive compounds.[75]

6.8.5 Antidiabetic Properties: Impact on Glycemic Control and Mechanisms

Millet proteins have shown significant potential in improving glycemic control through mechanisms such as enhancing insulin sensitivity, reducing blood glucose levels, and modulating carbohydrate digestion (Figure 6.2). For instance, heat-treated foxtail millet protein (HMP) was found to lower fasting blood glucose, improve glucose tolerance, and alleviate insulin resistance in prediabetic mice, partly by altering gut microbiota composition.[76] Similarly, prolamin from foxtail millet demonstrated hypoglycemic effects by improving glucose metabolism and restoring gut microbiota balance in diabetic mice.[77]

In vitro studies further highlight the antidiabetic potential of millet grains. Finger millet showed strong inhibition of α-glucosidase and α-amylase enzymes, surpassing the effectiveness of standard drugs like acarbose in preventing glucose spikes.[78] Dipeptidyl peptidase-IV (DPP-IV) inhibitory peptides derived from millet protein also demonstrate promise as bioactive agents for diabetes management.[79]

Additionally, millet bran polysaccharides have been shown to reduce fasting blood glucose and improve glucose tolerance in diabetic models by positively influencing gut microbiota and reducing inflammation.[80] Dietary millet proteins like those from Japanese millet have also increased plasma adiponectin levels, improved lipid profiles, and reduced glucose levels in diabetic mice, further emphasizing their multifaceted role in managing diabetes.[81]

Moreover, finger millet-enriched probiotic fermented milk was found to reduce blood glucose levels and cholesterol significantly while improving antioxidant status and mitigating diabetic complications in rats.[82]

6.9 Therapeutic Potential and Applications

6.9.1 Use of Millet-derived Bioactive Peptides in Health Supplements

Millet-derived bioactive peptides are increasingly recognized for their potential health benefits and applications in health supplements. These peptides, released through enzymatic hydrolysis, fermentation, or gastrointestinal digestion, exhibit multifaceted health-promoting properties, including antioxidant, anti-inflammatory, antidiabetic, and antihypertensive effects. Their bioactivities make them valuable for managing chronic diseases, improving gut health, and enhancing immune responses.

Antioxidant peptides derived from pearl millet have shown the ability to scavenge free radicals and chelate metal ions, protecting cells from oxidative damage. A specific peptide sequence, SDRDLLGPNNQYLPK, isolated from pearl millet, demonstrated strong antioxidative activity, making it a promising ingredient for anti-aging and cellular protection supplements.[48] Additionally, millet bran peptides have been found to reduce inflammatory markers such as TNF-α and IL-1β in both *in vitro* and animal models. Their anti-inflammatory effects are mediated through pathways such as NF-κB and MAPK, alongside their ability to positively influence gut microbiota, further highlighting their potential in gut health-focused supplements.[20]

Millet-based peptides have also demonstrated significant antidiabetic effects. For example, dipeptidyl peptidase-IV (DPP-IV) inhibitory peptides derived from millet proteins improve glycemic control, which is a critical factor in diabetes management. These peptides work by modulating carbohydrate metabolism and enhancing insulin sensitivity, making them suitable candidates for inclusion in antidiabetic supplements.[79] Similarly, feruloylated xylo-oligosaccharides (FXOS) from millet exhibit antiglycation properties, protecting proteins from glycation-induced damage, which is vital in preventing complications related to diabetes and aging.[83]

Peptides can lower blood pressure and inhibit tumor progression, emphasizing their value in functional foods targeting cardiovascular health and cancer prevention.[30] Furthermore, millet peptides have been shown to enhance immune modulation, offering additional benefits for maintaining overall health.[82]

6.9.2 Potential for Developing Functional Foods

Millet proteins are increasingly recognized for their potential in functional food development, thanks to their rich nutritional composition, bioactive properties, and ability to address modern health challenges. These grains are an excellent source of proteins, essential amino acids, minerals, and bioactive compounds, including polyphenols and flavonoids, which exhibit antioxidant, antihypertensive, antidiabetic, and anticancer properties.

Processing methods such as sprouting, fermentation, and popping enhance the functional properties of millet grains, making them suitable for diverse food applications. Sprouting, for example, increases nutrient bioavailability, improves water absorption capacity, and enhances the digestibility of proteins and carbohydrates, thus boosting their functional properties.[84]

Foxtail millet protein isolates exhibit excellent emulsifying and foaming properties, surpassing those of soy protein in certain aspects. These characteristics make millet proteins suitable for use in emulsified products, bakery goods, and beverages.[85]

Additionally, low glycemic index and the ability to modulate glucose metabolism make them suitable for managing diabetes. Millet consumption has been linked to reduced risks of cardiovascular diseases and obesity, attributed to their fiber content, antioxidant activity, and hypolipidemic effects. Furthermore, millet-derived bioactive peptides exhibit antihypertensive, anti-inflammatory, and anticancer effects, supporting their use in functional foods aimed at disease prevention.[30]

The sustainability of millet cultivation adds another layer of value, as millets thrive in arid and semi-arid regions, requiring fewer resources compared to major cereals. This resilience positions them as a sustainable choice for addressing food security while offering health-promoting benefits.[86]

6.9.3 Future Directions in Research and Application

The therapeutic potential and applications of millet-derived bioactive peptides and proteins have gained significant attention; however, several avenues for future research and development remain unexplored.

6.9.3.1 Extraction and Characterization Techniques

Future research should focus on refining extraction, purification, and characterization methods for millet-derived bioactive peptides. Techniques such as high-pressure processing, ultrasound-assisted extraction, and enzymatic hydrolysis optimization can improve yield, functionality, and scalability.[86] Furthermore, advanced analytical tools like mass spectrometry and nuclear magnetic resonance (NMR) spectroscopy can help identify novel peptide sequences and unravel their structure–activity relationships, aiding in targeted applications.[48]

6.9.3.2 *Mechanisms of Action*

While existing studies highlight the bioactivities of millet peptides, further investigations are needed to elucidate their molecular mechanisms. For instance, understanding how millet peptides modulate signaling pathways like NF-κB, MAPK, and PI3K/Akt can provide deeper insights into their anti-inflammatory, antioxidant, and anticancer effects. Additionally, studies on gut microbiota modulation by millet peptides could clarify their role in gut–brain axis interactions and systemic health benefits.[82]

6.9.3.3 *Development of Targeted Functional Foods*

The application of millet peptides in creating condition-specific functional foods is a promising area. For example, formulating foods enriched with DPP-IV inhibitory peptides can cater to diabetic populations, while antioxidant peptides can be integrated into products targeting aging and oxidative stress-related diseases.[79] Combining millet peptides with other bioactive compounds such as probiotics, prebiotics, or plant-based polyphenols could enhance the synergistic health benefits of functional foods.

6.9.3.4 *Clinical Trials and Safety Evaluations*

Although *in vitro* and animal studies have demonstrated the bioactivities of millet peptides, human clinical trials are critical to confirm their efficacy and safety. Large-scale, randomized controlled trials should investigate their long-term health benefits, optimal dosages, and potential adverse effects. Furthermore, assessing bioavailability, stability, and metabolic pathways in humans is necessary to translate laboratory findings into practical applications.[30]

6.9.3.5 *Food Processing and Product Development*

Innovative processing techniques, such as encapsulation, nanotechnology, and extrusion, can enhance the stability and bioavailability of millet peptides in functional food formulations. Developing gluten-free, protein-rich, and low-glycemic index millet-based products for niche markets, such as athletes, aging populations, and individuals with dietary restrictions, presents a substantial market opportunity.[48] Additionally, leveraging millet's natural emulsifying and foaming properties can lead to innovative formulations in beverages and bakery products.[85]

6.9.3.6 Sustainability and Agronomic Research

Given millet's resilience to climate stress and low resource requirements, future research should explore strategies to promote its cultivation and integration into sustainable agriculture. Enhancing millet crop yields through genetic engineering, breeding programs, and soil management practices can ensure a consistent supply for both nutritional and functional food applications.[86] Exploring millet's role in carbon sequestration and water conservation can further support its position in sustainable food systems.

6.10 Conclusion

Millet, a coarse cereal seed and secondary commodity that farmers cultivate, remains underutilised. Despite their small size, millets are used worldwide for food and biodegradable products because of their low production costs and simplicity of growing, which can lead to a high protein output, particularly in warm and dry climates. They are rich in nutrients and contain valuable functional components, as previously mentioned in this chapter. Many dietary fiber elements of millets are non-starchy polysaccharides, which provide various health advantages, such as delayed nutrient absorption and blood lipid reduction. By controlling glucose homeostasis and preventing dyslipidemia, regular finger millet consumption as a meal or even as a snack aids in the management of diabetes and its consequences. However, with a good amount of protein, they are a wonderful substitute for animal-based proteins. They may be produced in huge quantities at a lesser cost, meeting the need for food, feed, and other purposes. Although a lot of research has been done on the proteins found in various millets, there aren't as many studies on these grains as there are on cereals like wheat, rice, and maize when it comes to cultivars, locations, agronomy, growing circumstances, and proteomics. For purification and fractionation of proteins and peptides, techniques such as electrophoresis, spectroscopic techniques, solvent extraction, NIR spectroscopy, MALDI-TOF spectrometry, and FTIR are used. Even with several approaches, millet protein analysis encounters two major challenges: interference from complex matrix components and a lack of standardized methodology. To overcome these challenges, novel extraction procedures and improved analytical technologies have been created. Future studies should look at how diverse physical, chemical, biological, and hurdle approaches affect the functional and structural qualities of millet proteins for a variety of applications and extraction procedures for different varieties of millets.

Declaration of Competing Interest

The authors declare that they have no known competing financial interests or personal relationships that could have appeared to influence the work reported in this chapter.

Data Availability

No data was used for the research described in this chapter.

Funding

The authors received no external/internal financial support for the research, authorship, and/or publication of this article.

References

1. P. Shah, A. Dhir, R. Joshi and N. Tripathy, *J. Bus. Res.*, 2023, **155**, 113372.
2. S. A. Kheya, S. K. Talukder, A. K. Hasan, P. Datta, S. Yeasmin, Md. H. Rashid, A. K. Hasan, Md. P. Anwar, A. K. M. A. Islam and A. K. M. M. Islam, *Heliyon*, 2023, **9**, e22123.
3. S. Maitra, *Crop Res.*, 2020, **55**, 57–63.
4. M. Tomar, R. Bhardwaj, M. Kumar, S. P. Singh, V. Krishnan, R. Kansal, R. Verma, V. K. Yadav, A. Dahuja, S. P. Ahlawat, J. Chand Rana, C. T. Satyavathi, S. Praveen and A. Sachdev, *LWT*, 2021, **149**, 111813.
5. N. Sachdev, S. Goomer, L. R. Singh, V. M. Pathak, D. Aggarwal and R. K. Chowhan, *Appl. Food Res.*, 2023, **3**, 100288.
6. K. S. Poutanen, A. O. Kårlund, C. Gómez-Gallego, D. P. Johansson, N. M. Scheers, I. M. Marklinder, A. K. Eriksen, P. C. Silventoinen, E. Nordlund, N. Sozer, K. J. Hanhineva, M. Kolehmainen and R. Landberg, *Nutr. Rev.*, 2022, **80**, 1648.
7. J. L. Ravi and S. S. Rana, *ACS Omega*, 2024, **9**, 38327.
8. S. Kumari, B. Singh and A. Kaur, *Food Chem. Adv.*, 2023, **3**, 100557.
9. A. Kaur, K. Kumar and H. S. Dhaliwal, *J. Food Process. Preserv.*, 2020, e14672.
10. K. S. Samuel and N. Peerkhan, *J. Food Sci. Technol.*, 2020, **57**, 3467–3473.
11. J. Fu, Y. Zhang, Y. Hu, Y. Zhao, Y. Tang and L. Zou, *Food Chem.*, 2020, **325**, 126761.
12. N. Sachdev, S. Goomer and L. R. Singh, *J. Sci. Food Agric.*, 2020, **101**, 831.
13. B. Sharma and H. S. Gujral, *Food Chem.*, 2019, **299**, 125115.
14. V. G. Renganathan, C. Vanniarajan, A. Karthikeyan and J. Ramalingam, *Front. Genet.*, 2020, **11**, 500.
15. K. Ambati and K. V. Sucharitha, *Int. J. Recent Sci. Res.*, 2019, **10**(7), 33943–33948.
16. J. Kane-Potaka, S. Anitha, T. W. Tsusaka, R. Botha, M. Budumuru, S. Upadhyay, P. Kumar, K. Mallesh, R. Hunasgi, A. K. Jalagam and S. Nedumaran, *Food Syst.*, 2021, **5**, 680777.

17. T. Maharajan, T. P. A. Krishna, N. M. Krishnakumar, M. Vetriventhan, H. Kudapa and S. A. Ceasar, *Agriculture*, 2024, **14**, 670.
18. C. R. Abah, C. N. Ishiwu, J. E. Obiegbuna and A. A. Oladejo, *Asian Food Sci. J.*, 2020, **14**, 9–19.
19. T. Tripathi and S. Vyas, *Int. J. Home Sci.*, 2023, **9**, 72–78.
20. R. He, M. Liu, Z. Zou, M. Wang, Z. Wang, X. Ju and G. Hao, *Food Funct.*, 2022, **13**, 1881–1889.
21. B. Zhang, Y. Xu, C. Zhao, Y. Zhang, H. Lv, X. Ji, J. Wang, W. Pang, X. Wang and S. Wang, *Food Funct.*, 2022, **13**, 2594–2605.
22. Y. Zheng, X. Wang, M. Guo, X. Yan, Y. Zhuang, Y. Sun and J. Li, *Foods*, 2022, **11**, 1355.
23. N. A. N. Gowda, K. Siliveru, P. V. V. Prasad, Y. Bhatt, B. P. Netravati and C. Gurikar, *Food*, 2022, **11**, 499.
24. W. Laskowski, H. Górska-Warsewicz, K. Rejman, M. Czeczotko and J. Zwolińska, *Nutrient*, 2019, **11**, 679.
25. X. Yang, S. Zhang, C. He, P. Xue, L. Zhang, Z. He, L. Zang, B. Feng, J. Sun and M. Zheng, *Mol. Cancer*, 2020, **19**, 46.
26. S. Vani, S. Shipra, S. Singh, M. d Mursal and S. Hussain, *Intell. Pharm.*, 2024, **2**, 516–527.
27. E. S. Sarita and E. Singh, *J. Sci. Innov. Res.*, 2016, **5**, 46–50.
28. O. M. Bello, A. B. Ogbesejana, A. Balkisu, M. Osibemhe, B. Musa and S. O. Oguntoye, *Clin. Compl. Med. Pharmacol.*, 2022, **2**, 100020.
29. L. Yousaf, D. Hou, H. Liaqat and Q. Shen, *Food Res. Int.*, 2021, **142**, 110197.
30. A. Majid and C. G. P. Priyadarshini, *Crit. Rev. Food Sci. Nutr.*, 2020, **60**, 3342–3351.
31. S. Das, R. Khound, M. Santra and D. K. Santra, *Agriculture*, 2019, **9**, 64.
32. V. Shukla, S. Srivastava, S. Singh, M. Mursal and S. Hussain, *Intell. Pharm.*, 2024, **2**, 516–527.
33. D. P. Ramadoss and N. Sivalingam, *Nutr. Cancer*, 2019, **72**, 1422–1437.
34. A. M. Anis and Y. N. Sreerama, *Food Chem.*, 2020, **315**, 126265.
35. W. Zhang, G. Zhang, W. Liang, J. Tian, S. Sun, X. Zhang, X. Lv, P. Guo, A. Qu and Z. Wu, *Biomolecules*, 2024, **14**, 913.
36. R. Shukla and M. Cheryan, *Ind. Crops Prod.*, 2001, **13**, 171–192.
37. A. V. Bailey and G. Sumrell, *US Pat.*, 4254022, 1980.
38. L. Marcellino, C. Junior and E. Gander, *Protein Pept. Lett.*, 2002, **9**, 237–244.
39. R. Bibi, H. Mokrane, K. Khaladi and H. Amoura, *J. Microbiol. Biotechnol. Food Sci.*, 2021, **11**, 1–7.
40. P. Hong, S. Koza and E. S. P. Bouvier, *J. Liq. Chromatogr. Relat. Technol.*, 2012, **35**, 2923–2950.
41. J. Tang, D. Yao, S. Xia, L. Cheong and M. Tu, *Food Chem. X*, 2024, **23**, 101540.
42. A. Manivannan, *Electron. J. Plant Breed.*, 2017, **8**, 371–378.
43. K. C. Duong-Ly and S. B. Gabelli, *Methods Enzymol.*, 2014, 85–94.
44. F. Akharume, D. Santra and A. Adedeji, *Physicochemical and Functional Properties of Proso Millet Storage Protein Fractions*, 2019.

45. P. Wingfield, *Curr. Protoc. Protein Sci.*, 2001, A.3F.1–A.3F.8.
46. V. Wiedemair, D. Mair, C. Held and C. W. Huck, *Talanta*, 2019, **205**, 120115.
47. B. Nazari, M. A. Mohammadifar, S. Shojaee-Aliabadi, E. Feizollahi and L. Mirmoghtadaie, *Ultrason. Sonochem.*, 2018, **41**, 382–388.
48. H. Agrawal, R. Joshi and M. Gupta, *Food Chem.*, 2016, **204**, 365–372.
49. M. Singh, M. Metwal, V. A. Kumar and A. Kumar, *J. Sci. Food Agric.*, 2016, **96**, 672–679.
50. X. Zhao, G. An, Z. Wang, Y. Wei and B. Zhang, *Adv. J. Food Sci. Technol.*, 2015, **7**, 522–529.
51. C. Chen, M. Wen and Y. Jin, *J. Proteome Res.*, 2022, **21**, 2311–2330.
52. S. Contreras-Martos, H. H. Nguyen, P. N. Nguyen, N. Hristozova, M. Macossay-Castillo, D. Kovacs, A. Bekesi, J. S. Oemig, D. Maes, K. Pauwels, P. Tompa and P. Lebrun, *Front. Mol. Biosci.*, 2018, **5**, 1–11.
53. X. Shen, S. Shen, J. Li, Q. Hu, L. Nie, C. Tu, X. Wang, B. Orsburn, J. Wang and J. Qu, *J. Proteome Res.*, 2017, **16**, 2445–2456.
54. F. Calderón-Celis, J. R. Encinar and A. Sanz-Medel, *Mass Spectrom. Rev.*, 2018, **37**, 715–737.
55. S. A. Gerber, J. Rush, O. Stemman, M. W. Kirschner and S. P. Gygi, *Proc. Natl. Acad. Sci. U. S. A.*, 2003, **100**, 6940–6945.
56. J. Cox, M. Y. Hein, C. A. Luber, I. Paron, N. Nagaraj and M. Mann, *Mol. Cell. Proteomics*, 2014, **13**, 2513–2526.
57. D. C. Liebler and L. J. Zimmerman, *Biochemistry*, 2013, **52**, 3797–3806.
58. N. M. Griffin, J. Yu, F. Long, P. Oh, S. Shore, Y. Li, J. A. Koziol and J. E. Schnitzer, *Nat. Biotechnol.*, 2010, **28**, 83–89.
59. M. Wang, W. Feng, W. Lu, B. Li, B. Wang, M. Zhu, Y. Wang, H. Yuan, Y. Zhao and Z. Chai, *Anal. Chem.*, 2007, **79**, 9128–9134.
60. R. Esfandi, M. E. Walters and A. Tsopmo, *Heliyon*, 2019, **5**, e01538.
61. H. Agrawal, R. Joshi and M. Gupta, *Food Res. Int.*, 2019, **120**, 697–707.
62. J. K. Willcox, S. L. Ash and G. L. Catignani, *Crit. Rev. Food Sci. Nutr.*, 2004, **44**, 275–295.
63. D. Agyei, M. K. Danquah, I. P. Sarethy and S. Pan, *in Free Radicals in Human Health and Disease*, Springer India, New Delhi, 2015, 417–430.
64. C. Torres-Fuentes, M. del M. Contreras, I. Recio, M. Alaiz and J. Vioque, *Food Chem.*, 2015, **180**, 194–202.
65. G. Herregods, J. Van Camp, N. Morel, B. Ghesquière, K. Gevaert, L. Vercruysse, S. Dierckx, E. Quanten and G. Smagghe, *J. Agric. Food Chem.*, 2011, **59**, 552–558.
66. F. Hong, L. Ming, S. Yi, L. Zhanxia, W. Yongquan and L. Chi, *Peptides*, 2008, **29**, 1062–1071.
67. Q. Chen, X. Phd, M. Fu, G. He, W. Wang, H. Zhang and H. Ruan, *Asia Pac. J. Clin. Nutr.*, 2007, **16**, 281–285.
68. L. Seppo, T. Jauhiainen, T. Poussa and R. Korpela, *Am. J. Clin. Nutr.*, 2003, **77**, 326–330.
69. P. Manzanares, J. B. Salom, A. García-Tejedor, R. Fernández-Musoles, P. Ruiz-Giménez and J. V. Gimeno-Alcañíz, *Food Funct.*, 2015, **6**, 2440–2452.

70. R. He, S. A. Malomo, A. T. Girgih, X. Ju and R. E. Aluko, *J. Agric. Food Chem.*, 2013, **61**, 8396–8402.
71. S. Shan, Z. Li, I. P. Newton, C. Zhao, Z. Li and M. Guo, *Toxicol. Lett.*, 2014, **227**, 129–138.
72. X. Zhang, S. Shan, J. Shi, H. Li and Z. Li, *Food Funct.*, 2021, **12**, 291–301.
73. S. Hu, J. Yuan, J. Gao, Y. Wu, X. Meng, P. Tong and H. Chen, *J. Agric. Food Chem.*, 2020, **68**, 9415–9426.
74. N. Singh, A. Sekhar and J. Abraham, *Pharma Innov. J.*, 2015, **3**, 82–86.
75. A. K. Pandey and S. Verma, *Nutr. Food Sci.*, 2024, **54**, 1095–1105.
76. H. Wang, Q. Shen, F. Zhang, Y. Fu, Y. Zhu, L. Zhao, C. Wang and Q. Zhao, *Food Funct.*, 2023, **14**, 4866–4880.
77. Y. Fu, R. Yin, Z. Liu, Y. Niu, E. Guo, R. Cheng, X. Diao, Y. Xue and Q. Shen, *Nutrients*, 2020, **12**, 3452.
78. F. K. Ofosu, F. Elahi, E. B.-M. Daliri, R. Chelliah, H. J. Ham, J.-H. Kim, S.-I. Han, J. H. Hur and D.-H. Oh, *Antioxidants*, 2020, **9**, 254.
79. H. Gu, J. Gao, Q. Shen, D. Gao, Q. Wang, M. Tangyu and X. Mao, *LWT*, 2021, **138**, 110587.
80. J. Zhang, W. Wang, D. Guo, B. Bai, T. Bo and S. Fan, *Foods*, 2022, **11**, 3406.
81. N. Nishizawa, T. Togawa, K.-O. Park, D. Sato, Y. Miyakoshi, K. Inagaki, N. Ohmori, Y. Ito and T. Nagasawa, *Biosci., Biotechnol., Biochem.*, 2009, **73**, 351–360.
82. S. Mudgal and J. K. Chaudhary, *Food Technol. Biotechnol.*, 2020, **58**, 192–202.
83. A. Singh and S. M. Eligar, *J. Food Meas. Charact.*, 2021, **15**, 5695–5706.
84. D. Panda, N. H. Sailaja, B. Padhan and K. Lenka, *Proc. Natl. Acad. Sci., India, Sect. B*, 2020, **90**, 79–86.
85. T. K. Mohamed, K. Zhu, A. Issoufou, T. Fatmata and H. Zhou, *Int. J. Mol. Sci.*, 2009, **10**, 5224–5238.
86. P. S. Belton and J. R. N. Taylor, *Trends Food Sci. Technol.*, 2004, **15**, 94–98.

CHAPTER 7

Millet Proteins: Extraction, Isolation and Purification

MOUMITA KARMAKAR,[a] ANKAN KHETO,[a] RACHNA SEHRAWAT,*[a] KHALID GUL[a] AND VEDA KRISHNAN[b]

[a] National Institute of Technology Rourkela, India,
Emails: moumita.krm19@gmail.com; ankan.kheto.711@gmail.com; fud.biopolymer@gmail.com; [b] ICAR-Indian Agricultural Research Institute, India, Email: vedabiochem@gmail.com
*Email: Sehrawatrachna2017@gmail.com

7.1 Introduction

The increasing global population can contribute to hunger, affecting more than 820 million people in the world. Food insecurity and malnutrition could be due to the unavailability of safe, nutritious food, along with inadequate quantity and quality of food, and the absence of protein in the daily diet. There is an increase in the demand for eco-friendly and sustainable sources of protein for the human diet. This demand can be met by substituting animal protein sources, like eggs, milk, meat and fish, for alternative plant-based protein sources.[1] The limited supply of animal protein, in addition to its cost, freshwater depletion, climate change and biodiversity loss, are drawbacks to the use of animal protein. Cereal crops may offer a solution to these problems as they are cost effective and renewable plant protein sources compared to the animal protein sources.[1] Cereal crops are staple foods, serve as a major component in the human diet and provide a key source of macro-nutrients, including proteins, thereby contributing to global food security. Millets are gaining popularity due to their nutritional, ecologic and economic benefits.[2,3] Millets are gluten free,

Food Chemistry, Function and Analysis No. 48
Millet Proteins: Nutritional, Structural and Functional Attributes
Edited by Shweta Suri, Anuradha Dutta and Tarun Belwal

Published by the Royal Society of Chemistry, www.rsc.org

drought resistant and require less water for cultivation compared to other cereals.[3,4] Millets belong to the *Panicoideae* grass subfamily and are commonly grown in arid and semi-arid regions.[5] Millet is the 6th highest yielding crop in the world and India is the largest producer of millet, followed by Niger and Sudan.[6] Based on the size of grains, millets are classified as major millets (sorghum, pearl and finger millet) and minor millets (foxtail, proso, kodo, barnyard and little millet).[7] Millets are rich in nutrients like carbohydrates (60–70%), protein (6–19%), dietary fiber (12–20%) and minerals (2–4%) and low in fat content (1.5–5%).[6] Moreover, millets are a suitable source of dietary protein due to their gluten-free, hypoallergenic properties, substantial quantities of amino acids (they are rich in cysteine and methionine) and can contribute to the reduction of chronic diseases.[8,9] An important consideration when using plant protein ingredients in food applications is the use of a suitable extraction technique. The choice of protein extraction method depends on the purpose of the final product, the accessible resources, and the technical ability of personnel.[10] Protein concentrates from an underutilized or new source, such as millets, can be used to enrich the food formulations for the development of healthy food.[11] According to Pojić *et al.*,[10] extraction techniques can be classified as wet or dry. In this chapter, various conventional techniques and non-conventional or thermal techniques are discussed. Isolated millet proteins with good nutritional properties can be used in the food and pharmaceutical industries.[12] Conventional protein extraction techniques sometimes result in a reduction of protein yield due to the denaturation of protein molecules during extraction. Various extraction parameters, such as extraction time, pH, type of solvent, and temperature, are responsible for this reduction in protein yield. However, non-conventional techniques can be adopted to reduce protein denaturation, resulting in higher protein extraction yield and better nutritional and functional properties. Here, different conventional (fractional, aqua, organic solvent, alkali-based) and non-conventional (ultrasound, high hydrostatic pressure, microwave-assisted, pulsed electric field) extraction techniques are discussed. After protein extraction, purification processes are required to obtain protein with high purity, enriched concentration, improved stability and reduced denaturation, and to remove toxins and reduce catalyst poisoning.[13] Different conventional purification techniques, like membrane filtration and ultrafiltration, along with several new technologies, like tangential flow filtration and three phase partitioning, can be used for the purification of protein. Moreover, for the separation of proteins based on their molecular weight, electrophoresis is used as a separation method. Different chromatographic separation methods are also used.

Commercial applications of millet protein require the isolation and concentration of protein. Scaling up the food and pharmaceutical applications of millet protein concentrates is influenced by various extraction techniques. Moreover, knowledge of millet proteins and process parameters are required for the selection of proper extraction and purification techniques.

7.2 Extraction Techniques for Millet Proteins

Demand for millet protein is increasing. However, the commercialization of millet proteins with favorable properties poses some difficulties. It is necessary to recognize the nature and functional properties of such proteins by extracting and purifying with several methods before their application in final products. Generally, several "wet isolation" extraction methods are proposed for the extraction of millet proteins.

7.2.1 Analytical Sequential Protein Fractionation Technique

The protein fractionation technique provides a comprehensive understanding of solubility properties, source variability, and protein–protein interactions during extraction, contributing to enhancement of processing strategies and commercial applicability.[14,15] The aim of performing fractionation of protein-rich millet flour is to isolate, identify and characterize different compositions of millet protein based on their cultivar, species, and growth conditions.[11] Generally, two methods are adopted for the fractionation of millet protein. One is the classical Osborne method, and the other is Landry and Moureaux's method.

7.2.1.1 Classical Osborne Method

The Osborne method is recognized as the most widely accepted fractionation method for the recovery of protein fractions like albumin, globulin, prolamin and glutenin, which are soluble in water, salt solutions, alcohol, acid and alkali solutions. Prolamin is highly hydrophobic in nature and contains a high proline and glutamine amino acid content. The albumin fraction is water soluble and easily extracted by distilled water followed by centrifugation to remove the insoluble fraction. After the extraction of the albumin fraction, the remaining solid is treated with 1 M NaCl solution for the dissolution of the globulin fraction. After removal of the globulin fraction, the prolamin fraction is extracted using 70% ethanol. After that, the remaining material is extracted with the help of 0.05 M NaOH solution for the dissolution of the glutenin fraction. After each extraction process, slurry (1 : 10 solid to liquid ratio) is centrifuged for the collection of protein-rich supernatant and the precipitate is resuspended in the subsequent solvent. This process is carried out 3 times for the better extraction of protein fractions.[16] As described by Akharume *et al.*,[17] albumin, globulin, prolamin and glutenin fractions were extracted from defatted proso millet flour using different solvent extraction methods. Figure 7.1 represents the extraction method of faba bean protein using the Osborne protein fractionation process.

7.2.1.2 Landry and Moureaux's Method

Landry and Moureaux's method was introduced by Landry and Moureaux in 1970 for the extraction of protein fractions from maize protein. This method

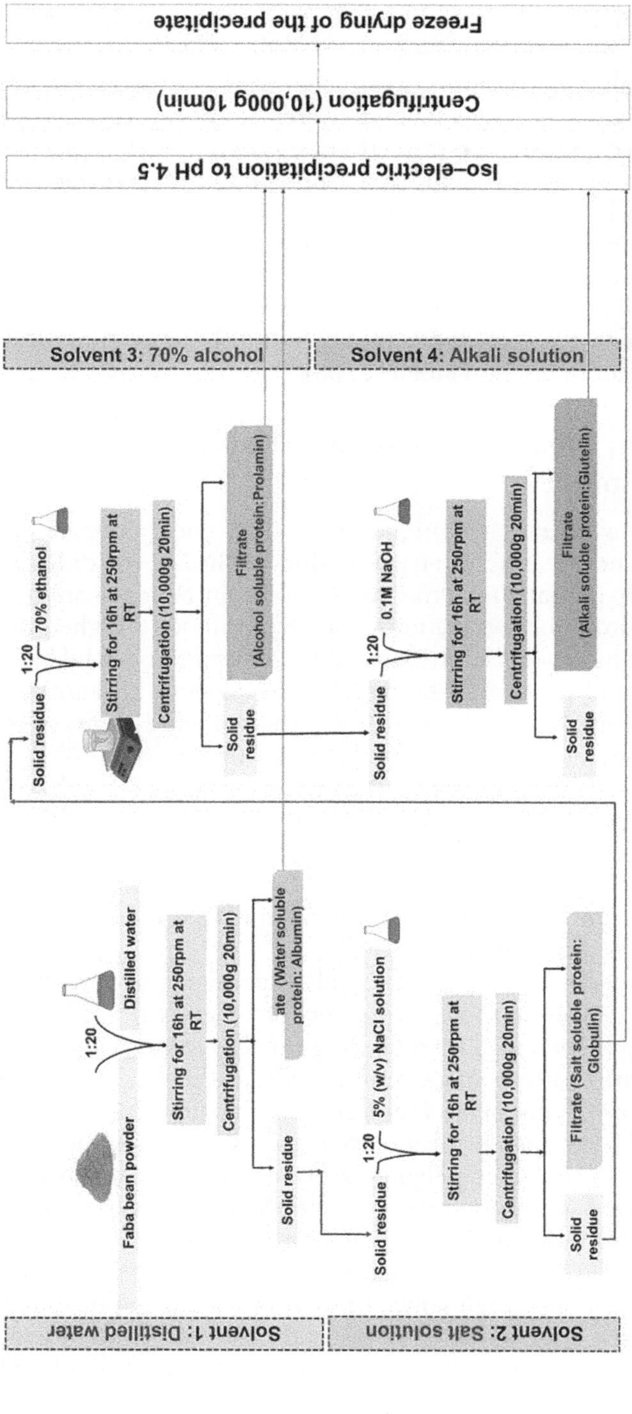

Figure 7.1 Example of Osborne extraction methods for faba bean protein. Reproduced from ref. 18, https://doi.org/10.1016/j.ifset.2023.103487, under the terms of the CC BY 4.0 license, https://creativecommons.org/licenses/by/4.0/.

is used to separate albumin, globulin, prolamin, prolamin-like, glutenin, and glutenin-like fractions. The true glutenin fraction couldn't be extracted by previous methods.[19] In this method, albumin, globulin and non-protein fractions are extracted using six solvents, followed by centrifugation at 4 °C. Initially, the albumin fraction is extracted with the use of deionized water and subsequently the pellet obtained after centrifugation is dissolved in 0.5 M NaCl solution for the extraction of globulins. 0.1 M borate buffer at pH 10.8 is used for the extraction of the glutenin-like fraction. 60% 2-propanol along with 1% dithiothreitol is required. Finally, for the removal of the cross-linked glutenin fraction, 0.1 M borate buffer along with 1% dithiothreitol and 1% sodium dodecyl sulfate is used at pH 10.8 at a temperature of 4 °C.[20] Generally, millet proteins are easily extracted by this method. Sorghum protein fractions are extracted using 0.5 M NaCl, 2-propanol, dithiothreitol (DTT), and sodium dodecyl sulfate (SDS).[20]

7.2.2 Aqua, Organic Solvent and Alkali Extraction Techniques

Conventional extraction techniques have high energy demands, are time consuming, and are less eco-friendly due to the use of different organic solvents, acids and alkalis. Proteins dissolved in aqueous solution display enhanced hydrophilic interactions, thereby promoting higher solubility in water. Thus, this process contributes to protein solubility while enabling the release of bound proteins. Bean *et al.* used aqueous ethanol as a protein extraction solvent, with 70% ethanol (v/v) and reducing agents (metabisulfite and glutathione), for the initial extraction of protein at 50 °C for 1 h. This was followed by centrifugation (4000 rpm for 10 min) of the solution to collect the supernatant, then dilution of the supernatant with 60% ethanol (v/v) for the precipitation of lipids. The sample was centrifuged (4000 rpm for 30 min at 0 °C) and the supernatant with 30% ethanol was chilled at 4 °C for 1 h and then centrifuged. Finally, the precipitate was collected and lyophilized to obtain the purified protein.[21]

Organic–aqueous extraction methods frequently use organic solvents and an aqueous solution for the recovery of proteins with higher solubility and stability. Proteins with lipid binding capacity, containing non-polar and polar side chains with aromatic amino acids, are easily dissolved in organic solvents (ethanol, butanol, acetone). However, an aqueous two-phase system is more suitable for the purification of proteins, cells, genetic materials and other organelles. Polymer–polymer and polymer–salt systems mostly use aqueous systems where the mixture of two polymers (*e.g.* polyethylene glycerol and dextran) results in the formation of aggregates and tends to separate into two different phases due to steric exclusion. Moreover, polymers with the addition of a high concentration of salts (*e.g.* phosphate, sulfate, citrate) captured large amounts of water for the extraction of protein. Additionally, partitioning between a two-phase system, with highly soluble and particulate matter on the top and hydrophobic, less polar matter on the bottom, was observed. This two-phase partitioning system depends on different parameters, including the hydrophobic properties of the phase system, electrical potential between phases, molecular size and bio affinity of proteins.[22] Sodium metabisulfite and NaOH were used in an aqueous ethanol solution by da Silva *et al.* for the extraction of protein. 70% ethanol (v/v) along with 0.35% NaOH (w/w) and 0.5% sodium metabisulfite (w/w) were mixed at 70 °C for 1 h. The solution was then centrifuged, and the supernatant removed and ethanol evaporated. Cold water (10 °C) was mixed with the solution in a 1 : 1 ratio and adjusted to pH 5 using 1 M HCl for the precipitation of protein. The protein was then lyophilized and defatted using hexane.[23]

Sometimes an acidic solution is used in the aqueous solution for the extraction of protein. Taylor *et al.* soaked a flour sample for 16 h with the addition of 0.5% aqueous metabisulfite and then the solution was centrifuged to collect the precipitate. 0.5% acetic acid was added to the pre-soaked flour and the solution was centrifuged. The supernatant was collected and mixed with chilled (10 °C) distilled water in a 1 : 1 ratio, followed by adjusting the pH to 5 for the precipitation of protein. Finally, the precipitated protein was recovered using centrifugation and the protein was lyophilized.[24]

Protein can also be extracted by subcritical water processing, which is an environmentally friendly, fast, inexpensive and efficient technique. Using this method, a sample was subjected to high temperature from 100 to 374 °C under pressure of 22 MPa. Under these conditions, a reduction in the viscosity and density of water and destruction of hydrogen bonding facilitated the solubilization of protein content.[25]

The alkali extraction method is the most commonly used conventional method of protein extraction and can result in a higher protein yield as compared to other organic extraction methods. The most commonly used alkalis are NaOH and KOH, which are used for the maintenance of basic pH, resulting in the breakdown of disulfide bonds present in the protein, enhancing protein solubility and yield. Any non-protein or insoluble components are removed by further centrifugation or filtration of the protein. The protein content that is solubilized in alkaline solution is then precipitated by changing the solution pH to the isoelectric point of the protein, reducing the net charge and electrostatic repulsion between the protein molecules. pH plays an important role in the extraction of protein due to the ionization of acidic and neutral amino acids at higher pH, resulting in higher solubility. Temperature is an important factor for stabilizing the structure of proteins, which is maintained by the covalent bonds between protein molecules. Other parameters, like sample to solvent ratio, alkali concentration, and extraction time, can be optimized for higher extraction yields.[26] However, higher alkalinity results in the aggregation, denaturation, cross-linking and hydrolysis of protein, which might be due to the exposure of non-polar and free-sulfhydryl groups on the surface of the protein causing an increase in hydrophobic attractions and disulfide bond formation.[27] The limitations of alkaline extraction have led to the use of non-thermal technologies like ultrasound, high-pressure processing, and microwave techniques, along with alkaline extraction, resulting in better protein extraction yields, reduced energy consumption and reduced processing time and cost. Aqua, organic solvent and alkali extraction techniques use salts, acids and alkali to extraction proteins. Some cereals are rich in the prolamin fraction and insoluble in aqueous solution but soluble in ethanol. Ethanol plays a key role in facilitating starch–protein separation by altering solvent characteristics, and 60 to 70% aqueous ethanol solutions are widely employed for protein extraction from sorghum, corn, and other cereals. Alcohol solutions combined with alkaline solutions are used in the enhancement of protein extraction from sorghum flour and corn.[28,29] Strong alkaline solutions result in the destruction of protein. Therefore, weak alkaline solution is preferred for the extraction of proteins, and various parameters like pH, temperature, and other factors enhance the extraction efficiency. According to a study by Alu'datt *et al.*,[30] protein recovery from barley flour was 60% with the use of NaOH solution. Moreover, extraction of the protein was higher in proso millet and pearl millet with the use of NaOH solution.[8,31] Additionally, organic solvents like acetone and hexane are used for the extraction of proteins, but excessive use of these solvents results in toxicity and degradation of proteins.

7.2.3 Enzymatic Extraction Technique

Enzymatic extraction is a green extraction technique for the economic recovery of highly nutritious protein. Enzymes like protease, cellulase, pectinase and others are used to extract proteins from plant residues. This technique can increase the amount of protein extracted, preserve nutritional value, and utilize waste. Rigid cell walls act as a barrier for the extraction of cellular components. Enzymatic extraction plays an important role in the enzymatic degradation of the components of cell walls for better extraction of protein. Generally, protease enhances the protein yield by unbinding the polysaccharide components from protein. Degradation of the cell wall promotes protein release, and subsequently protease enzymes fractionate high molecular weight proteins into smaller fragments, thereby resulting in more soluble proteins. To improve the catalytic efficiency of proteins, enzymes require an acidic or alkaline condition to facilitate interaction with amino acid side chains. Different parameters like the type of enzyme, concentration of enzyme, pH of solution, extraction temperature and particle size of extracted material are responsible for the effectiveness of protein extraction.[32] In this method, extraction parameters like pH and temperature vary based on the enzyme used. In a study by Sari *et al.*, enzyme-assisted protein extraction was performed by first preparing a solution of biomass in deionized water, then adjusting the pH and temperature. After 30 min, 1% enzyme was added and the reaction time was set to zero. After 2 h, 4% enzyme was added and samples were taken at 0, 2 and 3 h time intervals. Finally, the enzymatic reaction was stopped by heating the mixture at 90 °C for 10 min followed by centrifugation (5000 rpm, 10 min) to collect the supernatant. To check the feasibility of protein extraction at lower concentrations, 1% enzyme was added and samples were taken at different time intervals (0, 2, 3, 6 and 24 h).[33] According to a study by Wang *et al.*,[34] rice bran protein yield was increased from 34% to 74.6% with the use of phytase and xylanase enzymes at pH 5, 55 °C temperature and an incubation time of 2 h. Additionally, bi-enzymatic treatment resulted in 49% protein concentrate, whereas use of tri-enzymatic treatment had better protein recovery (78.3%).[35] However, the presence of enzymes may cause the enzymatic degradation of components of the cell wall, resulting in a barrier to extraction of proteins. Moreover, high enzyme cost, longer processing time, operational cost, pH and temperature requirements for optimal enzyme activation and prevention of protein denaturation are limitations of this technique.

7.2.4 Novel Extraction Techniques

To protect the environment, meet the demands of consumers, and enhance competition between industries to become more ecologic, innovative and economic, green extraction of new products is an innovative concept to meet these challenges.

7.2.4.1 Ultrasound-assisted Extraction

Ultrasound-assisted extraction is a sustainable technology that enhances the extraction efficiency, minimizes solvent requirements, and yields high purity products with lower processing time, energy consumption, and overall cost.[36] Ultrasound extraction uses the acoustic cavitation phenomenon, where bubble formation, growth and implosion results in the release of protein content to the extractive medium. Ultrasound waves disrupt the cell wall, minimize particle size to enhance the release of cellular components, and can enhance the efficiency of protein extraction after cell lysis.[37] During the ultrasound propagation through an alkaline solution, the liquid medium undergoes alternating compression and expansion, resulting in the localized negative pressure. The pressure generated results in the formation of cavitation bubbles. Subsequently, increasing bubble size followed by collapse releases high energy, resulting in the disruption of covalent bonds in the protein. Several mechanisms have been investigated for the extraction of plant protein using ultrasound. The impact of fragmentation due to the collision of inter-particles resulted in the collapse of bubbles created by cavitation. Due to the fragmentation mechanism, decreased particle size resulted in an increase of surface area of the particle followed by increased mass transfer and higher extraction yield.[38] In addition, the implosion of bubbles after cavitation improved the erosion on the surface of the plant material, resulting in the release of extraction medium. The sono-capillary effect mechanism refers to the intensity of liquid penetration into pores during sonication. Sonoporation refers to the generation of cell membrane pores for the release of cellular components under the application of higher and lower ultrasound frequency. During the implosion of bubbles created after cavitation, shear forces are generated in the solid materials. During ultrasound extraction, detexturation and destruction of plant materials is sometimes observed after the transfer of cell components to the solvent.[36] Various physical parameters like power, intensity, frequency of ultrasound, solvent, temperature, and dissolved gasses affect the extraction of protein.[36] For the extraction of isoflavone glucosides, fast and reliable ultrasound-assisted extraction gave higher quantitative recoveries at 60 °C after 20 min using 50% ethanol.[39] Ultrasound treatments with different times and intensities were used for the preparation of proso millet protein concentrate.[8] Furthermore, the efficiency of plant protein extraction was increased with the use of ultrasound combined with other conventional techniques. Ultrasound-assisted extraction treatment is a rapid, cost effective and eco-friendly technique that achieves high protein yield. Moreover, it enhances the functional properties and limits the degradation and destruction of extracted protein molecules. However, higher and prolonged sonication power might minimize the extracted protein yield. In addition, formation of disulfide bonds between cysteine residues results in the creation of protein aggregates during the generation of hydroxyl free radicals after ultrasound-assisted extraction. Similarly, acoustic agitation for a longer period of time after the cavitation of ultrasound-assisted treatment results in the denaturation and destruction of protein followed by reduction of protein yield.[40]

7.2.4.2 High Hydrostatic Pressure Extraction

High hydrostatic pressure as an extraction technique was first introduced in 2004 and was found to be a useful, economic technique with higher efficiency and lower energy requirements.[41] The high hydrostatic pressure extraction technique uses high pressure from 100 to 1000 MPa for a few seconds or minutes, which causes a large pressure difference between the interior and exterior of the plant cell wall, leading to increased permeability and increased dissolution of components from plant cells to the extraction solvent. Moreover, based on mass theory, the rate of mass transfer is equal to the pressure of mass transfer, and in high-pressure extraction treatments the pressure is increased from atmosphere to the preset pressure within a short period of time. Additionally, high pressure disrupts the cell wall, membrane and plant tissue, which enhances the transfer of cellular material to the solvents. At higher pressure, dissolution is based more on the phase behavior theory. High hydrostatic extraction is a promising extraction technique that can improve the extraction yield compared to other traditional techniques.[42,58] The high hydrostatic pressure technique follows Le Chatelier's principle, which states that changes in volume, temperature and concentration in a system result in an equilibrium shift.[58] Moreover, high pressure promotes phase transitions and molecular structural modifications, resulting in volume reduction and improved extraction efficiency. Additionally, high pressure breaks ionic bonds in molecules, which leads to a decrease in volume due to the action of electrostriction forces on water.

Subsequently, the disruption of molecular structures, cell wall, and cell membrane results in the release of intracellular constituents, thereby improving protein extraction efficiency.[43] Variable parameters, like raw materials, choice of solvents, process temperature, extraction pressure, use of additives, and solid : liquid ratio during the extraction process, affect the rate of extraction.[44] This extraction process has been used for the extraction of protein from oat bran[45] and rice bran,[46] and can be used in conjunction with other methods for better extraction.

7.2.4.3 Microwave-assisted Extraction

Microwave-assisted extraction was first introduced in 1986 by Ganzler and co-workers, where they observed that microwave-assisted extraction is more efficient, less time consuming and requires less energy compared to other conventional extraction techniques.[47] Microwaves are non-ionizing radiation that induce migration of ions and rotation of dipoles but do not change the molecular composition. The frequency range of the electromagnetic radiation lies between 300 MHz to 300 GHz, with a wavelength range of 1 mm to 1 m.[48] Rotation of dipoles and migration of ions induced by microwaves disrupts the cell wall by increasing the porosity of samples and simultaneously reducing the moisture content by increasing the pressure on the cell wall. Thus, it improves the infiltration of solvent into the plant cell and efficiently releases the protein content. Moreover, conversion of microwave energy to heat energy results in the evaporation of moisture, which generates high pressure on the cell wall.[49] 83% of protein was recovered from defatted rice bran meal using microwave-assisted extraction.[50] Moreover, a 1.54-fold increase in protein was extracted using microwave-assisted extraction as compared to conventional alkaline extraction.[51] This extraction technique is suitable for the extraction of protein from rice bran, wheat bran and other cereals, but extraction of millet protein using this technique has not been explored.

7.2.4.4 Extraction Using Pulsed Electric Field

The pulsed electric field technique uses short duration pulses (ms or μs) in the high electric field (10–80 $kV\,cm^{-1}$) generated between two electrodes, forming pores in the cell membrane through electroporation. The exposure of biological cell membranes to the higher electric field results in a rapid increase in electrical conductivity and permeability. This phenomenon is known as electroporation and was first introduced for non-excitable cells in 1967.[52,57] The transmembrane voltage caused by the strong electric field generates pores in the membrane. This electroporation can be reversible or irreversible. In reversible electroporation, pores reseal after the application of electric field, whereas in irreversible electrophoresis, pores do not reseal, or gradually reseal, resulting in the release of internal molecules.[53,57] The exposure of a cell to a strong electric field results in the formation of hydrophobic pores, promoting the exchange of molecular DNA and ions and subsequently increasing the extraction of protein.[54] The change from hydrophobic pores to hydrophilic pores can allow the movement of biomolecules throughout the cell.[54] It has been reported that protein can be extracted from olive kernels,[55] algae[56] and rice bran using this method. The pulsed electric field technique is a promising technique for the extraction of protein in its pure form, but the extraction yield is lower compared to other non-conventional methods. Moreover, there is a need for optimization of input factors during the extraction of protein.[40]

7.3 Extraction Techniques Used for Different Millet Proteins

7.3.1 Sorghum

The major protein fraction in sorghum millet is kafirin. For the extraction of kafirin, five extraction methods have been used: aqueous ethanol with NaOH and sodium metabisulfite, glacial acetic acid, aqueous ethanol with sodium metabisulfite, aqueous ethanol at acidic pH, and alkaline pH.[29] These different extraction methods are suitable for the preparation of different food products.[29] Kafirin is resistant to digestion and difficult to extract from sorghum millet.[59] To resolve this problem, different extraction methods like wet milling,[60] alkaline extraction,[61] sonication,[21] and glacial acetic acid extraction[24] were performed. Figure 7.2 represents commonly used wet milling processes for the isolation of important substances from sorghum.

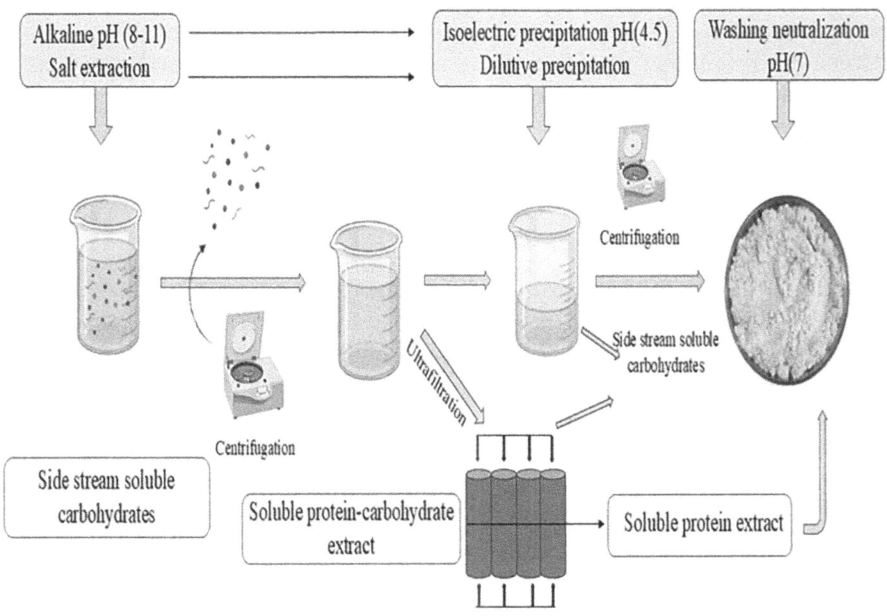

Figure 7.2 Isolation of valuable compounds from Sorghum. Reproduced from ref. 76 with permission from Elsevier, Copyright 2025.

7.3.2 Foxtail Millet

Foxtail millet protein was extracted using 1 M NaOH alkaline solution at pH 9.5 for 3 h at 30 °C.[62] As reported by Mohamed *et al.*,[62] 75% to 80% protein concentrate was extracted from foxtail millet with the potential of better functional properties for use as a functional food ingredient. Kasaoka *et al.*[63] reported that, foxtail millet protein was extracted using the heat stable α-amylase in an aqueous solution followed by heating in a water bath at 97 °C for 30 min, vacuum filtration, precipitation of solution using ethanol and finally drying of the precipitate. Finally, highly pure foxtail millet with 68.7% extracted protein concentrate was obtained with better functional and nutritional properties.[63]

7.3.3 Pearl Millet

Pearl millet is a staple cereal food grain with 20% protein concentrate but with limited lysine content. 70% aqueous isopropyl alcohol, sodium hydroxide and hydrochloric alcohol were used at different pH values for the extraction of protein concentrate from pearl millet.[64] Bailey and Sumrell[64] reported that 80% of protein was extracted from pearl millet for use as a protein ingredient in food. Protein from pearl millet was also extracted using alkaline solution (0.5 M NaOH) at pH 8.5 for 5 h for the preparation of hydrolysates and bioactive peptides for use as functional food ingredients.[31] Additionally, pearl millet protein was extracted in an aqueous solution using 0.3 M Tris HCl buffer, 10 mM 2-mercaptoethanol, and 2 M HCl at 20 °C with different pH conditions.[65] Pearl millet was solubilized by chymotrypsin or polymerized by transglutaminase, which improved the emulsifying capacity of the conjugate and enhanced the polymerized protein.[65]

7.3.4 Proso Millet

Proso millet contains a higher protein content (13.4%) compared to other cereals like wheat (10.5%) and rice (6.8% to 7.4%).[66] Protein from proso millet was extracted using 1 M NaOH solution at pH 9.5 with a sample to solution ratio of 1 : 4 for 1 h. The solution pH was then reduced to 4 for the precipitation of the solution, followed by the recovery of proso millet protein concentrate. Moreover, functional properties were improved after the application of ultrasound treatment to the proso millet protein concentrate.[8]

7.4 Purification Techniques for Millet Proteins

7.4.1 Ammonium Sulfate

Separation of different fractions of protein using ammonium sulfate salt was first introduced in 1988, where protein was precipitated after the addition of aluminium sulfate salt at a saturation limit.[67] This purification method is a batch purification method where quantities of sample were dissolved in a one-tenth saturated solution of ammonium sulfate. The solution was then filtered, followed by addition of more saturated ammonium sulfate solution (one-tenth) to make a final volume. The point at which the solution became turbid and the protein precipitated was observed.[67] Moreover, the extraction of protein and protein fractions using salt solution is described by Pool and Shooter (1955), where barley protein fractions were extracted using aluminium salt solution.

7.4.2 Tangential Flow Ultrafiltration

Tangential flow ultrafiltration (TFF) is a separation technique that removes low molecular weight particles for the purification of protein. TFF can separate samples in one of two ways: pore flow membrane and solution-diffusion membrane. Separation in the pore flow membrane occurs due to the flow driven by the pressure through the pores. Particles smaller than the pores can permeate easily, whereas large particles are filtered out. Separation through the solution-diffusion mechanism is based on the differences in the affinity and solubility of solutes.[68] Traditional TFF has been limited to the separation of solutes differing by about 10-fold in size. TFF is limited by the wide distribution of pore size, membrane fouling and concentration polarization. To overcome these limitations, high performance TFF has been used, where high resolution and high filtrate flux membranes resulted in maximum yield and purification of protein.[69]

7.4.3 Three-phase Partitioning

Three-phase partitioning (TPP) is a low-tech analytical technique for the precipitation of proteins and enzymes from aqueous solution using aqueous salt (ammonium sulfate) and aliphatic alcohol (methanol, ethanol, 1-propanol, 2-propanol, *t*-butanol). Aliphatic alcohol is completely soluble in water, but with the addition of ammonium salt, the solution was separated into a lower aqueous phase and a floating alcohol phase. Moreover, the presence of protein in the aqueous phase resulted in the formation of a third or intermediate phase between these two phases depending on the concentration of salt added.[70] This bi-separation technique is useful for the production of upstream and downstream proteins such as enzymes. Various parameters like concentration of salt, extract to alcohol ratio, pH, and temperature of the system are responsible for the efficiency of separation.[71] There are different types of TPP, including ultrasound-assisted, microwave-assisted and ionic liquid TPP, which have been modified to optimize extraction yield and purity of protein.

7.4.4 Other Methods

7.4.4.1 Dialysis, Ultrafiltration, and Adsorption

Dialysis is a purification method based on the concentration of protein molecules, which uses a semi-permeable membrane that protein molecules cannot pass through. This purification method can be combined with other methods, like centrifugal separation, salting out and other chromatographic techniques, for better purification of protein.[72] Ultrafiltration employs a membrane-based purification process in which proteins are concentrated and allows water and low-molecular weight solutes under pressure, thereby enabling protein concentration and purification.[73] This technique is mainly used for the desalination of protein, but it can separate protein based on molecular weight. The phenomenon of this technique involves concentration gradients across the membrane and adsorption on the membrane surface that reduces permeation flux, thereby affecting protein separation efficiency.[72]

7.5 Separation Techniques for Millet Proteins

7.5.1 Column Chromatography

Column chromatography is an absorption method where different loading conditions, such as pH, temperature, ionic strength, and buffer composition, are responsible for concentrating proteins. Hydrophobic interaction chromatography comprises a non-polar stationary phase and a polar mobile phase, where hydrophobic molecules bind to the stationary phase due to the van der Waals interactions during loading conditions. Hydrophobic interaction chromatography gives better chromatographic resolution if a high aluminium sulfate concentration is maintained in the process. In the first step, proteins undergo batch precipitation with 1.6 M aluminium sulfate and starting samples are stored in aluminium sulfate. During separation, samples are thawed and the aluminium sulfate concentration is adjusted before they are loaded in the hydrophobic interaction chromatographic column. Ion exchange (cation and anion) is the most common chromatographic technique and has been used for many years for the separation of protein. Mostly protein samples are loaded in the ion exchange column in low concentration salt buffer and at a pH that favors the protein. The goal of this chromatography is to bind the protein tightly at the head of the column. The buffer concentration should be low (1–10 mM) and the pH should be 1 unit below for that required for anion exchange and 1 unit above that required for cation exchange.[74] Gel filtration chromatography separates protein based on the size and shape of the protein. Compared to large proteins, small and compact proteins access more pores and elute slowly. Gel filtration chromatography is unable to concentrate protein, so protein samples should be concentrated first.[74] Affinity chromatography is a powerful chromatographic technique that exploits the interaction between protein and an immobilized ligand. This chromatographic technique is commonly used to elute reversibly bound substances like antigens and antibodies by lowering the pH to 3. High-performance liquid chromatography is used as a low-pressure liquid chromatography with excellent separation over a short time duration. This technique provides higher resolution than other low-pressure methods due to the greater recovery of enzymatic activity. Bibi *et al.*[75] reported that pearl millet protein was characterized by high performance liquid chromatography with a protein concentrate of 90.1% with high protein purity.

7.5.2 Electrophoresis

Electrophoresis is a separation technique where large voltage gradients are used to separate proteins based on their size, shape, charge, and charge/mass ratio. It is an analytical tool used to purify protein or estimate protein molecular weight. There are different types of analytical electrophoretic methods that are used for the separation of protein. The most common analytical electrophoretic method is sodium dodecyl sulfate-polyacrylamide gel electrophoresis (SDS-PAGE), where SDS is an anionic detergent that tightly binds proteins and denatures them at elevated temperature with the presence of β-mercaptoethanol. During electrophoresis, microorganisms and protein molecules are separated based on their molecular weight and purity. According to Akharume *et al.*,[17] SDS-PAGE analysis separated different fractions of proso millet protein with different molecular weights.

7.6 Conclusion

This chapter discussed different extraction, isolation, and purification techniques for millet protein. Advanced techniques are useful for the improvement of functional, physical, structural, and digestible properties of protein. Modified protein can be used as a functional food, including in confectionary food items and bakery products. However, there are limited studies on millet protein extraction, purification, and isolation. Further investigation is needed to elucidate its functional, physical, structural and digestible properties in applied food systems.

References

1. A. Kheto, R. Sehrawat, K. Gul and L. Kumar, *Food Chem.*, 2024, **444**, 138628.
2. A. Sarkar, T. Niranjan, G. Patel, A. Kheto, B. K. Tiwari and M. Dwivedi, *J. Food Process Eng.*, 2023, **46**(5), e14317.
3. W. Laskowski, H. Górska-Warsewicz, K. Rejman, M. Czeczotko and J. Zwolińska, *Nutrients*, 2019, **11**, 679.
4. P. Shah, A. Dhir, R. Joshi and N. Tripathy, *J. Bus. Res.*, 2023, **155**, 113372.
5. M. Chera, *Food Cult. Soc.*, 2017, **20**, 303–324.
6. H. M. Wani, P. Sharma, K. Gul, J. P. Bishnoi, I. A. Wani, S. Kothari and A. A. Wani, *Appl. Food Res.*, 2022, **2**, 100181.
7. P. Bora, S. Ragaee and M. Marcone, *Int. J. Food Sci. Nutr.*, 2019, **70**, 714–724.
8. B. Nazari, M. A. Mohammadifar, S. Shojaee-Aliabadi, E. Feizollahi and L. Mirmoghtadaie, *Ultrason. Sonochem.*, 2018, **41**, 382–388.

9. N. Sachdev, S. Goomer and L. R. Singh, *J. Sci. Food Agric.*, 2021, **101**, 831–842.
10. M. Pojić, A. Mišan and B. Tiwari, *Trends Food Sci. Technol.*, 2018, **75**, 93–104.
11. N. Sachdev, S. Goomer, L. R. Singh, V. M. Pathak, D. Aggarwal and R. K. Chowhan, *Appl. Food Res.*, 2023, **3**, 100288.
12. A. N. Sanjay, G. Basarkar and V. Buchake, *J. Pharmacogn. Phytochem.*, 2022, **11**, 177–185.
13. A. Saxena, B. P. Tripathi, M. Kumar and V. K. Shahi, *Adv. Colloid Interface Sci.*, 2009, **145**, 1–22.
14. N. Ghanghas, M. T. Mukilan, S. Sharma and P. K. Prabhakar, *Food Rev. Int.*, 2022, **38**, 354–383.
15. G. Jayaprakash, A. Bains, P. Chawla, M. Fogarasi and S. Fogarasi, *Polymers*, 2022, **14**, 3003.
16. J. S. Yang, F. F. Dias, T. T. K. Pham, D. Barile and J. M. de Moura Bell, *Food Hydrocolloids*, 2024, **146**, 109250.
17. F. Akharume, D. Santra and A. Adedeji, *Food Hydrocolloids*, 2020, **108**, 105497.
18. R. S. Das, X. Zhu, S. Hannon, E. Mullins, S. Alves, M. Garcia-Vaquero and B. K. Tiwari, *Innovative Food Sci. Emerging Technol.*, 2023, **89**, 103487.
19. R. Katoch and R. Katoch, *Anal. Tech. Biochem. Mol. Biol.*, 2011, 251–322.
20. A. E.-M. M. Afify, H. S. El-Beltagi, S. M. Abd El-Salam and A. A. Omran, *PLoS One*, 2012, 7, e31154.
21. S. Bean, B. Ioerger, S. Park and H. Singh, *Cereal Chem.*, 2006, **83**, 99–107.
22. J. A. Asenjo and B. A. Andrews, *J. Chromatogr. A*, 2011, **1218**, 8826–8835.
23. L. S. da Silva and J. R. Taylor, *Cereal Chem.*, 2004, **81**, 322–327.
24. J. Taylor, J. R. Taylor, M. F. Dutton and S. de Kock, *Cereal Chem.*, 2005, **82**, 485.
25. M. Álvarez-Viñas, P. Rodríguez-Seoane, N. Flórez-Fernández, M. D. Torres, B. Díaz-Reinoso, A. Moure and H. Domínguez, *Food Bioprocess Technol.*, 2021, **14**, 373–387.
26. C. Zhang, J. P. Sanders and M. E. Bruins, *Biomass Bioenergy*, 2014, **67**, 466–472.
27. M. Hadidi, F. Aghababaei and D. J. McClements, *Food Hydrocolloids*, 2023, 109132.
28. D. J. Myers, M. P. Hojilla-Evangelista and L. A. Johnson, *J. Am. Oil Chem. Soc.*, 1994, **71**, 1201–1204.
29. P. Pontieri, J. Troisi, S. R. Bean, M. Tilley, M. D. i Salvo, A. Boffa, D. Pignone, F. del Giudice, M. Aletta and P. Alifano, *Aust. J. Crop Sci.*, 2019, **13**, 1297–1304.
30. M. H. Alu'datt, T. Rababah, K. Ereifej, I. Alli, M. A. Alrababah, A. Almajwal, N. Masadeh and M. N. Alhamad, *Food Hydrocolloids*, 2012, **26**, 135–143.
31. H. Agrawal, R. Joshi and M. Gupta, *Food Chem.*, 2016, **204**, 365–372.
32. M. M. Poojary, V. Orlien, P. Passamonti and K. Olsen, *Food Chem.*, 2017, **234**, 236–244.

33. Y. W. Sari, M. E. Bruins and J. P. Sanders, *Ind. Crops Prod.*, 2013, **43**, 78–83.
34. M. Wang, N. Hettiarachchy, M. Qi, W. Burks and T. Siebenmorgen, *J. Agric. Food Chem.*, 1999, **47**, 411–416.
35. M. Houde, N. Khodaei, N. Benkerroum and S. Karboune, *Food Chem.*, 2018, **254**, 367–376.
36. V. Chawala, A. Kheto, L. Sharma and R. Sehrawat, in *Green Chemistry in Food Analysis*, Elsevier, 2024, DOI: 10.1016/B978-0-443-18957-9.00004-3.
37. A. L. Lupatini, L. de Oliveira Bispo, L. M. Colla, J. A. V. Costa, C. Canan and E. Colla, *Food Res. Int.*, 2017, **99**, 1028–1035.
38. A. Kheto, A. Chaudhari, S. Manikpuri, R. Sehrawat, K. Gul, L. Kumar and K. A. Khan, *LWT*, 2024, **213**, 117046.
39. M. A. Rostagno, M. Palma and C. G. Barroso, *J. Chromatogr. A*, 2003, **1012**, 119–128.
40. M. Kumar, M. Tomar, J. Potkule, R. Verma, S. Punia, A. Mahapatra, T. Belwal, A. Dahuja, S. Joshi and M. K. Berwal, *Food Hydrocolloids*, 2021, **115**, 106595.
41. Z. Shouqin, Z. Junjie and W. Changzhen, *Int. J. Pharm.*, 2004, **278**, 471–474.
42. H.-W. Huang, C.-P. Hsu, B. B. Yang and C.-Y. Wang, *Trends Food Sci. Technol.*, 2013, **33**, 54–62.
43. S. A. Khan, R. Aslam and H. A. Makroo, *J. Food Process Eng.*, 2019, **42**, e12896.
44. A. Mustafa and C. Turner, *Anal. Chim. Acta*, 2011, **703**, 8–18.
45. M. Cheng, X. Luo, Z. Chen, R. Wang, T. Wang, W. Feng, H. Zhang, J. He and C. Li, *Innovative Food Sci. Emerging Technol.*, 2022, **82**, 103204.
46. S. Tang, N. S. Hettiarachchy and T. H. Shellhammer, *J. Agric. Food Chem.*, 2002, **50**, 7444–7448.
47. K. Ganzler, A. Salgó and K. Valkó, *J. Chromatogr. A*, 1986, **371**, 299–306.
48. S. Manikpuri, A. Kheto, R. Sehrawat, K. Gul, W. Routray and L. Kumar, *J. Food Sci.*, 2024, **89**(4), 2188–2201.
49. A. Kheto, D. Joseph, M. Islam, S. Dhua, R. Das, Y. Kumar, R. Vashishth, V. S. Sharanagat, K. Kumar and P. K. Nema, *J. Food Process. Preserv.*, 2022, **46**(5), e16595.
50. M. Bhattacharya and M. Hanna, *J. Food Sci.*, 1988, **53**, 1230–1231.
51. S. Phongthai, S.-T. Lim and S. Rawdkuen, *J. Cereal Sci.*, 2016, **70**, 146–154.
52. A. Sale and W. Hamilton, *Biochim. Biophys. Acta, Gen. Subj.*, 1967, **148**, 781–788.
53. T. Kotnik, W. Frey, M. Sack, S. H. Meglič, M. Peterka and D. Miklavčič, *Trends Biotechnol.*, 2015, **33**, 480–488.
54. A. Golberg, M. Sack, J. Teissie, G. Pataro, U. Pliquett, G. Saulis, T. Stefan, D. Miklavcic, E. Vorobiev and W. Frey, *Biotechnol. Biofuels*, 2016, **9**, 1–22.
55. E. Roselló-Soto, F. J. Barba, O. Parniakov, C. M. Galanakis, N. Lebovka, N. Grimi and E. Vorobiev, *Food Bioprocess Technol.*, 2015, **8**, 885–894.

56. M. Coustets and J. Teissié, The Use of Pulsed Electric Fields for Protein Extraction from Nanochloropsis and Chlorella, in *1st World Congress on Electroporation and Pulsed Electric Fields in Biology, Medicine and Food & Environmental Technologies*, ed. T. Jarm and P. Kramar, IFMBE Proceedings, 2016, vol 53. Springer, Singapore, DOI: 10.1007/978-981-287-817-5_88.
57. A. Kheto, S. Manikpuri, A. Sarkar, R. Das, R. P. Bebartta, Y. Kumar, Y. Bist, R. Vashishth and R. Sehrawat, *Food Biosci.*, 2024, 104849.
58. H. Rostamabadi, A. C. Karaca, M. Nowacka, M. Z. Mulla, H. Al-Attar, K. Rathnakumar, B. G. Subasi, R. Sehrawat, A. Kheto and S. R. Falsafi, *Food Hydrocolloids*, 2023, 108375.
59. N. J. De Mesa-Stonestreet, S. Alavi and S. R. Bean, *J. Food Sci.*, 2010, **75**, R90–R104.
60. P. Yang and P. A. Seib, *Cereal Chem.*, 1996, **73**, 751–755.
61. Y. V. Wu, *J. Agric. Food Chem.*, 1978, **26**, 305–309.
62. T. K. Mohamed, K. Zhu, A. Issoufou, T. Fatmata and H. Zhou, *Int. J. Mol. Sci.*, 2009, **10**, 5224–5238.
63. S. Kasaoka, A. Oh-hashi, T. Morita and S. Kiriyama, *Nutr. Res.*, 1999, **19**, 899–910.
64. A. V. Bailey and G. Sumrell, *J. Agric. Food Chem.*, 1980, **28**, 475–477.
65. A. B. Hassan, G. A. Osman and E. E. Babiker, *Food Chem.*, 2007, **102**, 257–262.
66. M. Karmakar, A. Kheto, R. Sehrawat, Y. Kumar, K. Gul, W. Routray and L. Kumar, *Food Chem.*, 2024, 142383.
67. T. B. Osborne and I. F. Harris, *J. Am. Chem. Soc.*, 1903, **25**, 837–842.
68. P. Agrawal, K. Wilkstein, E. Guinn, M. Mason, C. I. Serrano Martinez and J. Saylae, *Org. Process Res. Dev.*, 2023, **27**, 571–591.
69. R. Van Reis, S. Gadam, L. N. Frautschy, S. Orlando, E. M. Goodrich, S. Saksena, R. Kuriyel, C. M. Simpson, S. Pearl and A. L. Zydney, *Biotechnol. Bioeng.*, 1997, **56**, 71–82.
70. C. Dennison and R. Lovrien, *Protein Expression Purif.*, 1997, **11**, 149–161.
71. K. W. Chew, T. C. Ling and P. L. Show, *Sep. Purif. Rev.*, 2019, **48**, 52–64.
72. S. Liu, Z. Li, B. Yu, S. Wang, Y. Shen and H. Cong, *Adv. Colloid Interface Sci.*, 2020, **284**, 102254.
73. A. T. Phillips and M. W. Signs, *Curr. Protoc. Protein Sci.*, 2004, **38**, 4.
74. W. W. Ward and G. Swiatek, *Curr. Anal. Chem.*, 2009, **5**, 85–105.
75. R. Bibi, H. Mokrane, K. Khaladi and H. Amoura, *J. Microbiol. Biotechnol. Food Sci.*, 2021, **11**, e3674–e3674.
76. S. Singh, M. Habib, D. Mondal, Y. Kumar, S. Jan, K. Bashir and K. Jan, *Int. J. Biol. Macromol.*, 2025, **318**, 145039.

CHAPTER 8

Bioactive Properties of Millet Protein

ANTHONY O. OBILANA

Department of Food Science and Technology, Cape Peninsula University of Technology, Bellville Campus, Cape Town, RSA, South Africa
Email: obilanaA@cput.ac.za

8.1 Introduction

8.1.1 Overview of Millet

8.1.1.1 Definition of Millet

Millet is a collective term for a group of small-seeded grasses cultivated primarily for food and fodder. These grains are valued for their resilience to harsh environmental conditions, including drought, poor soil fertility, and extreme temperatures. Millets belong to the *Poaceae* family and are widely grown in Africa, Asia, and parts of Europe and the Americas. Unlike conventional staple cereals, such as wheat and rice, millets are gluten free and nutritionally superior, offering a range of health benefits.[1]

8.1.1.2 History and Origin

Millets are among the oldest cultivated crops, with evidence of their domestication dating back over 10 000 years in Asia and Africa. Foxtail millet (*Setaria italica*) remains have been traced to northern China, dating back more than 4000 years, while pearl millet (*Pennisetum glaucum*) is believed to have been domesticated in Africa.[1] Due to their adaptability to various

Food Chemistry, Function and Analysis No. 48
Millet Proteins: Nutritional, Structural and Functional Attributes
Edited by Shweta Suri, Anuradha Dutta and Tarun Belwal

Published by the Royal Society of Chemistry, www.rsc.org

Table 8.1 Types of millets and their key characteristics.[1,3–5]

Millet type	Key characteristics
Pearl Millet (*Pennisetum glaucum*)	Most widely grown in Africa and India, high in protein and iron
Finger Millet (*Eleusine coracana*)	Rich in calcium and dietary fibre, widely consumed in India and Africa
Foxtail Millet (*Setaria italica*)	High-yielding, drought-resistant, commonly cultivated in China and India
Proso Millet (*Panicum miliaceum*)	Popular in Europe and the US, used in human diets and bird feed
Barnyard Millet (*Echinochloa spp.*)	Low glycaemic index, beneficial for diabetics

agro-climatic conditions, millets have historically played a crucial role in food security, especially in semi-arid and arid regions.[2]

8.1.1.3 *Types of Millet*

Table 8.1 categorises the five most cultivated types of millet, highlighting their key characteristics and referencing relevant sources. Each type of millet has distinct nutritional and agronomic properties that make it suitable for specific dietary and agricultural applications.

1. Pearl Millet (*Pennisetum glaucum*)
    - **Key Characteristics:** pearl millet is the most widely grown millet, particularly in **Africa and India.** It is known for its **high protein and iron content**, making it a crucial dietary staple in regions with high malnutrition rates.
    - **Significance:** due to its high nutritional value, it plays a role in improving **iron deficiency** and overall **protein intake**, especially in rural and semi-arid regions where it is extensively cultivated.[6–8]
2. Finger Millet (*Eleusine coracana*)
    - **Key Characteristics:** finger millet is rich in **calcium and dietary fibre**, making it a **valuable food for bone health and digestion.** It is widely consumed in **India and parts of Africa.**
    - **Significance:** its high calcium content makes it beneficial for **preventing osteoporosis and promoting bone strength**, while the fibre content supports **gut health and weight management.**[3,9–11]
3. Foxtail Millet (*Setaria italica*)
    - **Key Characteristics:** known for being **high-yielding and drought-resistant**, foxtail millet is commonly cultivated in **China and India.**
    - **Significance:** its ability to **thrive in harsh environments** makes it an essential crop for food security, especially in areas prone to **climate change and water scarcity.**[12–14]
4. Proso Millet (*Panicum miliaceum*)
    - **Key Characteristics:** popular in **Europe and the United States**, proso millet is used in both **human diets and bird feed.**

- **Significance:** it is often incorporated into **gluten-free diets** and is also a preferred grain for **poultry and livestock feed** due to its **high energy content.**[15–17]

5. Barnyard Millet (*Echinochloa spp.*)
   - **Key Characteristics:** known for its **low glycaemic index**, barnyard millet is particularly beneficial for **diabetics.**
   - **Significance:** because it leads to a **gradual increase in blood sugar levels**, it is recommended for individuals managing **diabetes and metabolic disorders.** Additionally, it is **gluten free**, making it a good alternative for those with **celiac disease.**[18–21]

8.1.1.4 Nutritional Composition

Millets are rich in macronutrients and micronutrients, making them a valuable dietary component (Table 8.2). Their nutritional composition varies slightly among different millet types.

Millets are recognised for their bioactive compounds, which have significant health benefits.

- **Phenolic compounds and flavonoids** act as natural antioxidants that combat oxidative stress and reduce inflammation.[26,27]
- **Antimicrobial peptides** help inhibit the growth of harmful bacteria, potentially enhancing gut health.[13,28,29]
- **Tannins and phytosterols** are known for their cholesterol-lowering and cardiovascular protective effects.[2,30]
- **Saponins and alkaloids** contribute to anti-inflammatory and immune-boosting properties (Sharma and Sharma 2022).

The presence of these bioactive compounds makes millet a promising candidate for functional foods and nutraceutical applications.

Table 8.2 Summary of nutrients in millets.

Carbohydrates	Millets contain 60–70% carbohydrates, primarily in the form of slow-digesting starches, which provide a sustained release of energy.
Protein	The protein content in millets ranges from 7% to 12%, depending on the variety.
Dietary fibre	Millets are high in dietary fibre, including both soluble and insoluble fibres, which promote gut health and reduce the risk of chronic diseases.
Micronutrients	Millets are rich in essential minerals such as iron, zinc, calcium, phosphorus, and magnesium.
Bioactive compounds	Millets contain phenolic acids, flavonoids, and tannins, contributing to their antioxidant, anti-inflammatory, and anti-diabetic properties.

8.1.2 Importance of Millet Protein

Millets provide a good source of plant-based protein, with content ranging from 7% to 12%. Although the overall protein level is comparable to wheat and rice, the protein quality varies due to millet's unique amino acid composition.[31–33] According to Taylor and Emmambux (2007), Abah *et al.* (2020) and Saleh *et al.* (2013), unlike many cereal grains, millets contain higher levels of sulfur-containing amino acids such as methionine and cysteine, which are crucial for various metabolic functions. The amino acid composition of millet protein is of considerable nutritional interest. Millet proteins are rich in: (1) **methionine and cysteine**, which are important for protein synthesis, antioxidant defence, and immune function; (2) **leucine and isoleucine**, which are essential branched-chain amino acids that support muscle protein synthesis and energy metabolism; and (3) **phenylalanine and tyrosine**, which are precursors for neurotransmitter production. However, millet proteins are generally low in lysine, an essential amino acid often abundant in legumes. This makes it beneficial to combine millet with leguminous crops to achieve a balanced amino acid profile.[27,35–38]

8.2 Bioactive Properties of Millet Protein

Millet protein exhibits several bioactive properties, making it a valuable component for functional foods and nutraceuticals. These properties are attributed to bioactive compounds such as phenolic acids, flavonoids, peptides, and phytochemicals, which contribute to antioxidant, antimicrobial, antihypertensive, antidiabetic, and anti-inflammatory effects.[2,30,39,40]

8.2.1 Antioxidant Activity

Millet proteins and their derived peptides possess strong antioxidant properties due to the presence of phenolic compounds, flavonoids, and other bioactive molecules. These compounds contribute to scavenging free radicals and reducing oxidative stress, which play a crucial role in preventing chronic diseases such as cardiovascular disorders, cancer, and neurodegenerative diseases.[17,41]

8.2.1.1 Phenolic Compounds and Flavonoids

Phenolic acids, flavonoids, and tannins present in millet grains act as primary antioxidant compounds. They function by donating hydrogen atoms or electrons, neutralizing free radicals, and preventing lipid peroxidation.[42] The presence of hydroxycinnamic acids, such as ferulic and *p*-coumaric acids, further enhances their antioxidant potential.[40]

8.2.1.2 Mechanisms of Action

The antioxidant activity of millet protein is attributed to multiple mechanisms:

- **Direct scavenging of free radicals.** Phenolic compounds neutralise reactive oxygen species (ROS) and prevent oxidative stress damage to cells.[17]
- **Chelation of metal ions.** Flavonoids and phenolic acids bind to metal ions, reducing their ability to catalyse oxidative reactions.[39,42]
- **Enzyme modulation.** Some millet bioactives upregulate endogenous antioxidant enzymes, such as superoxide dismutase (SOD) and catalase, further strengthening the body's defence against oxidative stress.[40]

Comparative studies of antioxidant potential of millet protein with other plant proteins suggest that millet-derived peptides exhibit similar or superior antioxidant activities to soy and wheat proteins. Processing techniques such as germination and fermentation have been found to enhance the bioavailability of phenolic compounds, further improving millet's antioxidant efficacy.

8.2.2 Antimicrobial Activity

Millet protein exhibits antimicrobial properties, primarily due to antimicrobial peptides (AMPs) and phenolic compounds that inhibit the growth of pathogenic bacteria, fungi, and viruses.[17]

8.2.2.1 Antimicrobial Peptides

Certain peptides derived from millet protein hydrolysates display antimicrobial properties by disrupting microbial cell membranes and inhibiting essential metabolic pathways in pathogens.[41] These peptides have shown effectiveness against foodborne pathogens such as *Escherichia coli*, *Staphylococcus aureus*, and *Salmonella spp.*[17]

8.2.2.2 Mechanisms and Evidence

- **Membrane disruption.** AMPs interact with bacterial membranes, causing pore formation and leakage of intracellular contents, leading to microbial cell death.[39,42]
- **Inhibition of enzymatic pathways.** Some antimicrobial peptides interfere with microbial enzyme functions, preventing their replication and survival.[41]
- **Synergistic effects with phenolics.** Phenolic compounds in millet enhance the antimicrobial action of peptides by destabilizing bacterial cell walls.[17]

8.2.2.3 *Applications*

Millet-derived antimicrobial peptides hold potential applications in:

- **Food preservation**. Natural preservatives in millet protein can extend the shelf life of food products.[39]
- **Functional foods**. Millet protein-enriched foods may help in gut health maintenance by reducing pathogenic bacterial load.[40]
- **Pharmaceutical and medical applications**. AMPs from millet protein could be explored as alternatives to synthetic antibiotics, particularly in combating antibiotic-resistant bacteria.[41]

8.2.3 Anti-hypertensive Effects

Millet protein hydrolysates contain bioactive peptides with antihypertensive effects, primarily due to their ability to inhibit angiotensin-converting enzyme (ACE), which plays a critical role in blood pressure regulation.[17]

8.2.3.1 *ACE Inhibitory Peptides*

Millet-derived peptides act as natural ACE inhibitors, reducing the conversion of angiotensin I to angiotensin II, thereby promoting vasodilation and lowering blood pressure.[41]

8.2.3.2 In Vitro *and* In Vivo *Studies*

- ***In Vitro* evidence**. Studies have demonstrated that millet protein hydrolysates show significant ACE inhibitory activity, with comparable efficacy to synthetic ACE inhibitors such as captopril.[40]
- **Animal and human studies**. Animal models consuming millet-based diets showed a reduction in systolic and diastolic blood pressure levels. Limited human trials also suggest that regular millet consumption is associated with improved cardiovascular health.[39,42]

8.2.4 Antidiabetic and Anti-inflammatory Properties

Millet protein has shown promising effects in regulating glucose metabolism and reducing inflammation, making it beneficial for managing diabetes and chronic inflammatory diseases.[40,43–45]

8.2.4.1 *Regulation of Glucose Metabolism*

Millet proteins influence carbohydrate metabolism through various mechanisms:

- **Slower starch digestion**. The complex carbohydrate matrix in millet delays starch hydrolysis, leading to a lower glycaemic response.[39]

- **Enhanced insulin sensitivity.** Millet-derived bioactive peptides stimulate glucose uptake in muscle cells, improving insulin sensitivity.[17,46]
- **Inhibition of α-amylase and α-glucosidase.** Certain peptides and polyphenols in millet inhibit digestive enzymes, reducing postprandial blood sugar spikes.[41,44]

8.2.4.2 Anti-inflammatory Effects

- **Modulation of inflammatory pathways.** Millet bioactives suppress pro-inflammatory cytokines such as TNF-α and IL-6, thereby reducing systemic inflammation.[40]
- **Reduction in oxidative stress.** By mitigating oxidative stress, millet compounds lower inflammation-driven diseases such as arthritis and cardiovascular disorders.[17,46]
- **Gut microbiome modulation.** Millet proteins may promote beneficial gut microbiota, contributing to reduced gut inflammation.[39,42]

Millet protein exhibits multiple bioactive properties, including antioxidant, antimicrobial, antihypertensive, antidiabetic, and anti-inflammatory effects. These properties are primarily attributed to bioactive peptides, phenolic compounds, and flavonoids. Millet-based functional foods and nutraceuticals hold potential for improving cardiovascular, metabolic, and immune health. However, further research, including clinical trials, is needed to fully understand the mechanisms and optimise the use of millet bioactives in health applications.

8.3 Processing Effects on Bioactive Properties

8.3.1 Impact of Processing

Processing methods significantly influence the bioactive properties of millet protein. Techniques such as boiling, roasting, steaming, and fermentation alter the bioavailability of essential nutrients, antioxidants, and peptides. While some methods enhance bioactive retention, others degrade essential compounds.

8.3.2 Effects of Selected Processing Methods

- **Boiling and steaming** reduce the levels of anti-nutrients such as phytic acid, thereby enhancing mineral bioavailability. However, excessive boiling can lead to a loss of water-soluble antioxidants such as phenolic acids.[2,14,26,47,48]
- **Roasting and popping** improve the sensory attributes and increase the availability of bioactive peptides by breaking down protein structures. However, prolonged exposure to high temperatures can reduce total polyphenol content.[2,49]

- **Fermentation** enhances the levels of bioactive peptides and antioxidants by breaking down complex proteins and releasing bound phenolic compounds.
- **Microwave processing** leads to an increase in flavonoid content but may reduce phenolic acids and tannins.[4,9,39,50]

8.3.3 Factors Affecting Protein Quality and Bioactivity

- **Thermal processing** treatments, such as extrusion and roasting, lead to protein denaturation, improving digestibility but potentially reducing bioactive peptides.[4,50]
- **Storage conditions**, such as extended storage and exposure to humidity, can lead to protein oxidation and degradation of bioactive peptides.[26,48]
- **Non-thermal processing** techniques, like ultrasound and cold plasma, preserve bioactive properties and improve functional characteristics of millet protein.[4,9,51]

8.4 Extraction and Characterisation of Millet Protein

8.4.1 Methods of Protein Extraction

8.4.1.1 Traditional Techniques

Traditional methods of extracting millet protein involve physical and aqueous techniques such as milling, dehulling, and fractionation. These methods have been widely used in the food industry due to their simplicity and cost-effectiveness. However, they often lead to significant protein loss and lower purity due to co-extraction of non-protein components such as starch and fibre.[52–54]

- Milling and dehulling are the preliminary steps that separate the protein-rich portion from the outer husk and bran. The efficiency of these methods depends on the type of millet and milling conditions.[52–55]
- Aqueous extraction involves soaking and grinding the millet grains in water, followed by centrifugation to separate the protein-rich fraction. This method is relatively inefficient as it extracts a lower yield of protein.[52–55]

8.4.1.2 Modern Techniques

To improve extraction efficiency, modern techniques such as enzymatic hydrolysis, solvent extraction, and ultrafiltration have been developed. These techniques optimize the protein yield and maintain its bioactivity.[52,56]

- Enzymatic hydrolysis uses specific proteases, such as alcalase and papain, to break down proteins into smaller peptides and amino acids, enhancing their bioavailability. Enzyme-assisted extraction also reduces the presence of anti-nutritional factors, improving the functional properties of millet protein.[52]

- Solvent Extraction utilizes organic or aqueous solvents to dissolve proteins selectively, followed by precipitation using acid or salt solutions. While effective, this method can lead to solvent residues that require removal before application in food products.
- Ultrafiltration and membrane technology use semi-permeable membranes to separate proteins based on molecular weight. Ultrafiltration enhances protein purity and functionality while minimizing the loss of essential amino acids.[52,56]

8.4.1.3 Optimization of Extraction Conditions

The efficiency of protein extraction is influenced by factors such as pH, temperature, solvent type, and processing time. Studies have shown that alkaline extraction at pH 9–10 followed by isoelectric precipitation at pH 4–5 results in higher protein yields and purity. Moreover, the combination of enzymatic hydrolysis with ultrafiltration has been found to improve the bioavailability of millet proteins.[53–56]

8.4.2 Characterisation of Millet Protein

8.4.2.1 Structural Properties

The characterisation of millet protein involves analysing its molecular structure, amino acid composition, and molecular weight distribution.[13,28,36]

- Amino acid composition. Millet proteins are rich in essential amino acids such as leucine, isoleucine, and valine, making them a valuable plant-based protein source. However, lysine is often the limiting amino acid, which may require supplementation.[7,57–59]
- Molecular weight distribution. The proteins in millet exist in various molecular weight ranges, with globulins and albumins being predominant. SDS-PAGE and mass spectrometry techniques are commonly used to determine protein subunits.[7,57–59]

8.4.2.2 Functional Properties

Millet protein exhibits several functional properties that influence its application in food and nutraceutical industries.

- Solubility. The solubility of millet protein is pH-dependent, with higher solubility observed in alkaline conditions. This property is crucial for developing protein-based beverages and emulsions.[23,48,55,60]
- Water-holding capacity. Millet proteins exhibit moderate water-holding capacity, which affects their textural properties in food applications such as bakery products and meat analogues.[19,28,43,61]

- Emulsifying and foaming properties. The ability of millet protein to stabilize emulsions and foams makes it useful in dairy alternatives and protein-enriched food formulations.[19,23,28,40,62]

8.4.3 Conclusion

Millet protein extraction and characterisation have advanced significantly with the adoption of modern techniques such as enzyme-assisted extraction and membrane filtration. These methods enhance protein yield, bioavailability, and functional properties, making millet protein a promising ingredient for functional foods and nutraceuticals. Further research is required to optimize these techniques for large-scale production and explore novel applications of millet protein in food systems.

8.5 Applications of Millet Protein in Functional Foods and Nutraceuticals

8.5.1 Functional Food Products

8.5.1.1 Incorporation into Diets

Millet protein is increasingly being integrated into gluten-free and vegetarian diets due to its high protein quality and digestibility. It serves as a suitable replacement for wheat and other cereals.[43,63]

8.5.1.2 Health Benefits

- Supports cardiovascular health through antihypertensive peptides.
- Aids in diabetes management by reducing glycaemic response.
- Improves gut health due to its high fibre and prebiotic content.[19,28,61,64]

8.5.1.3 Consumer Acceptance

The demand for plant-based protein sources is rising, increasing interest in millet-based products such as protein bars, fortified cereals, and dairy alternatives.[8]

8.6 Challenges and Future Perspectives

8.6.1 Technical Challenges in Extraction and Processing

8.6.1.1 Limitations and Challenges

- Inconsistent protein yield due to variation in millet varieties.[53,55,56]
- High processing costs for advanced extraction techniques.[52–54]
- Need for scalable and sustainable processing solutions.[52,53]

8.6.2 Future Research Directions

8.6.2.1 *Emerging Technologies*

- Development of enzymatic hydrolysis techniques to improve bioactive peptide yield.[39,42]
- Use of novel non-thermal processing methods such as high-pressure processing (HPP) and ultrasound-assisted extraction.

8.6.2.2 *Research Needs*

- Further *in vivo* studies to validate bioactive efficacy.
- Optimization of extraction techniques to retain functional properties of millet protein.[39,42]

8.7 Conclusion

Millet protein offers a unique combination of nutritional and bioactive properties. Its processing influences bioactive retention and digestibility, affecting its applications in functional foods and nutraceuticals. The inclusion of millet protein in diets can support metabolic health, cardiovascular function, and overall nutritional security. Further research and technological advancements will drive the adoption of millet-based products in global food systems, positioning it as a key sustainable protein source.

References

1. R. Sharma and S. Sharma, Anti-nutrient & bioactive profile, *in vitro* nutrient digestibility, techno-functionality, molecular and structural interactions of foxtail millet (*Setaria italica* L.) as influenced by biological processing techniques, *Food Chem.*, 2022, 368.
2. T. O. Akanbi, Y. Timilsena and S. Dhital, Bioactives from Millet: Properties and Effects of Processing on Bioavailability, in *Bioactive Factors and Processing Technology for Cereal Foods*, Springer Singapore; 2019, pp. 171–183.
3. D. Chandra, S. Chandra and A. K. Sharma, Review of Finger millet (*Eleusine coracana* (L.) Gaertn): A powerhouse of health benefiting nutrients, *Food Sci. Hum. Wellness*, 2016, **5**, 149–155.
4. N. Sharma, J. K. Sahu, V. Bansal, O. J. Esua, S. Rana and A. Bhardwaj, *et al.*, Trends in millet and pseudomillet proteins - Characterization, processing and food applications, *Food Res. Int.*, 2023, **164**, 112310.
5. S. Saleem, N. U. Mushtaq, W. H. Shah, A. Rasool, K. R. Hakeem and C. S. Seth, *et al.*, Millets as smart future food with essential phytonutrients for promoting health, *J. Food Compos. Anal.*, 2023, **124**, 105669.
6. J. R. N. Taylor, Millet Pearl: Overview, in *Encyclopedia of Food Grains*, 2nd edn, 2015, vol. 1–4, pp. 190–198.

7. N. G. Malleshi and C. F. Klopfenstein, Nutrient composition, amino acid and vitamin contents of malted sorghum, pearl millet, finger millet and their rootlets, *Int. J. Food Sci. Nutr.*, 1998, **49**(6), 415–422.
8. E. N. Uukule, W. Embashu, P. N. Kadhila, I. S. E. Ueitele and K. K. M. Nantanga, Climate smart, underutilised, healthful future cereal: Protein content, hydration properties, starch digestibility and consumer liking of pearl millet-based Oyster mushroom crackers, *Food Chem. Adv.*, 2023, **3**, 100467.
9. V. F. Abioye, G. O. Babarinde, G. O. Ogunlakin, J. A. Adejuyitan, S. J. Olatunde and A. O. Abioye, Varietal and processing influence on nutritional and phytochemical properties of finger millet: A review, *Heliyon*, 2022, **8**(12), 12310.
10. P. B. Devi, R. Vijayabharathi, S. Sathyabama, N. G. Malleshi and V. B. Priyadarisini, Health benefits of finger millet (*Eleusine coracana* L.) polyphenols and dietary fiber: a review, *J. Food Sci. Technol.*, 2014, **51**(6), 1021.
11. A. Kumar, M. Metwal, S. Kaur, A. K. Gupta, S. Puranik and S. Singh, *et al.*, Nutraceutical value of finger millet [*Eleusine coracana* (L.) Gaertn.], and their improvement using omics approaches, *Front. Plant Sci.*, 2016, **7**, 934.
12. P. Li, X. Cai, S. Li, W. Zhao, J. Liu and X. Zhang, *et al.*, Nutrient and metabolite characteristics of the husk, bran and millet isolated from the foxtail millet (*Setaria italica* L.) during polishing, *Food Chem.: X*, 2024, **23**, 101541.
13. I. Amadou, G. W. Le, T. Amza, J. Sun and Y. H. Shi, Purification and characterization of foxtail millet-derived peptides with antioxidant and antimicrobial activities, *Food Res. Int.*, 2013, **51**, 422–428.
14. V. D. Pawar and G. M. Machewad, Processing of Foxtail Millet for Improved Nutrient Availability, *J. Food Process. Preserv.*, 2006, **30**(3), 269–279.
15. D. D. Baltensperger, Progress with Proso, Pearl and Other Millets, in *Trends in new crops and new uses*, ed. J. Janick and A. Whipkey, ASHS Press, Alexandria, VA, 2002, pp. 100–103.
16. F. Akharume, D. Santra and A. Adedeji, Physicochemical and functional properties of proso millet storage protein fractions, *Food Hydrocolloids*, 2020, **108**, 105497.
17. P. Dixit-Bajpai and R. Ravichandra, The Potential of Millet Grains: A Comprehensive Review of Nutritional Value, Processing Technologies, and Future Prospects for Food Security and Health Promotion, *J. Food Technol. Nutr. Sci.*, 2024, **6**(1), 1–8.
18. D. Goswami, R. K. Gupta, D. Mridula, M. Sharma and S. K. Tyagi, Barnyard millet based muffins: Physical, textural and sensory properties, *LWT*, 2015, **64**(1), 374–380.
19. P. V. Bhavadharani and P. Gurumoorthi, Impact of germination on nutritional components, antinutritional, and functional properties of proso and barnyard millets, *Food Chem. Adv.*, 2025, **6**, 100896.
20. C. Charu, S. Vignesh, D. Chidanand, R. Mahendran and N. Baskaran, Impact of cold plasma on pearl and barnyard millets' microbial quality,

antioxidant status, and nutritional composition, *Food Humanity*, 2024, **2**, 100238.

21. P. Prasad and J. K. Sahu, Interplay of germination time, nutritional content, bioactive constituents, antioxidant activity, and *in-vitro* digestibility in kodo, little, and barnyard millets, *Food Humanity*, 2025, **4**, 100545.
22. B. Dayakar, R. K. Bhaskarachary, G. D. Arlene, C. G. Sudha, D. Vilas and A. Tonapi, *Nutritional and Health Benefits of Millets*, ed. V. Ramesh and V. Ravinder, ICAR_Indian Institute of Millets Research (IIMR), Hyderabad, India, 2017. Available from: www.millets.res.in
23. A. C. Karaca, M. Nickerson, C. Caggia, C. L. Randazzo, A. K. Balange and C. Carrillo, *et al.*, Nutritional and Functional Properties of Novel Protein Sources, *Food Rev. Int.*, 2023, **39**(9), 6045–6077.
24. A. Kumar, V. Tomer, A. Kaur, V. Kumar and K. Gupta, Millets: A solution to agrarian and nutritional challenges, *Agric. Food Secur.*, 2018, **7**(1), 1–15.
25. M. Samtiya, R. E. Aluko, N. Dhaka, T. Dhewa and A. K. Puniya, Nutritional and health-promoting attributes of millet: current and future perspectives, **81***Nutr. Rev.*, 2023, 684–704.
26. A. S. M. Saleh, Q. Zhang, J. Chen and Q. Shen, Millet grains: Nutritional quality, processing, and potential health benefits, *Compr. Rev. Food Sci. Food Saf.*, 2013, **12**(3), 281–295.
27. S. Liang and K. Liang, Millet grain as a candidate antioxidant food resource: a review, **22***Int. J. Food Prop.*, 2019, 1652–1661.
28. A. Majid and P. Priyadarshini, Millet derived bioactive peptides: A review on their functional properties and health benefits, **60***Crit. Rev. Food Sci. Nutr.*, 2020, 3342–3351.
29. R. Rani, C. Pandey and N. Kapoor, Nurturing health through millet derived nutraceuticals, *Food Humanity*, 2024, **2**, 100277.
30. F. Ansari and H. Pourjafar, *Influence of Fermentation and Germination on Food Bioactives*, 2022, 277–308.
31. S. Bleakley, Cereal Proteins: Potential Health Applications and Allergenicities, in *Novel Proteins for Food, Pharmaceuticals, and Agriculture*, 2018, pp. 77–100.
32. G. Kaur, S. Ahmadzadeh-Hashemi, S. Amir, Z. S. Khan, Z. Gulsunoglu-Konuskan and A. Karimidastjerd, *et al.*, Exploring sustainable novel millet protein: A look at the future foods through innovative processing, *Future Foods*, 2024, **9**, 100367.
33. S. Suri, K. S. Balasaheb, D. K. Yadav, S. Malakar, P. Choudhary and A. Mohapatra, *et al.*, Overview of Millet proteins: Quality characteristics, effect of thermal/ non-thermal processing and applications, *Food Biosci.*, 2024, **57**, 103434.
34. J. R. N. Taylor and M. N. Emmambux, Products containing other speciality grains: Sorghum, the millets and pseudocereals, *Technol. Funct. Cereal Prod.*, 2007, 281–335.
35. C. R. Abah, C. N. Ishiwu, J. E. Obiegbuna and A. A. Oladejo, Nutritional Composition, Functional Properties and Food Applications of Millet Grains, *Asian Food Sci. J.*, 2020, 9–19.

36. J. D. Hernández-Vega, E. G. Tovar-Pérez and I. Parola-Contreras, Biologically Active Peptides from Pearl Millet [*Pennisetum glaucum* (L.) R.Br.], Foxtail Millet [*Setaria italica* (L.) P.Beauv.] and Finger Millet [*Eleusine coracana* (L.) Gaertn] Grains, in *Potential Health Benefits of Biologically Active Peptides Derived from Underutilized Grains: Recent Advances in their Isolation, Identification, Bioactivity and Molecular Analysis*, 2023, pp. 76–93.
37. Z. M. Hassan, N. A. Sebola and M. Mabelebele, The nutritional use of millet grain for food and feed: a review, *Agric. Food Secur.*, 2021, **10**(1), 16.
38. F. Shahidi and A. Chandrasekara, Millet grain phenolics and their role in disease risk reduction and health promotion: A review, *J. Funct. Foods*, 2013, **5**(2), 570–581.
39. C. K. Sunil, N. A. N. Gowda, N. Nayak and A. Rawson, Unveiling the effect of processing on bioactive compounds in millets: Implications for health benefits and risks, *Process Biochem.*, 2024, **138**, 79–96.
40. A. Dubey and P. P. Tripathy, Ultrasound-mediated hydration of finger millet: Effects on antinutrients, techno-functional and bioactive properties, with evaluation of ANN-PSO and RSM optimization methods, *Food Chem.*, 2024, **435**, 137516.
41. D. Mehta, S. Vyas, D. Dudhagara, A. Patel and V. Parmar, Significance of Indian millets in enhancing global food security: A comprehensive review, *Trends Food Sci. Technol.*, 2024, **149**, 104527.
42. N. S. Verma, O. P. Prajapati, R. Prajapati, R. Lohare and A. Shukla, Unveiling the Nutritional Spectrum of Millet: An In-depth Review, *Int. J. Environ. Clim. Change*, 2023, **13**(11), 2546–2559.
43. T. J. Rao, A. Mounika and A. Shanmugam, Improved Bioactivity and Functional Qualities of Millet Protein Enzyme Hydrolysates by Ultrasonication, *Ind. Eng. Chem. Res.*, 2024, **63**(12), 5380–5396.
44. V. Singh, G. D. Lee, H. W. Son, S. Amani, M. Baunthiyal and J. H. Shin, Anti-diabetic prospects of dietary bio-actives of millets and the significance of the gut microbiota: A case of finger millet, *Front. Nutr.*, 2022, **9**, 1056445.
45. N. S. Rajasekaran, M. Nithya, C. Rose and T. S. Chandra, The effect of finger millet feeding on the early responses during the process of wound healing in diabetic rats, *Biochim. Biophys. Acta, Mol. Basis Dis.*, 2004, **1689**(3), 190–201.
46. K. Nagre, N. Singh, C. Ghoshal, G. Tandon, M. A. Iquebal and T. Nain, *et al.*, Probing the potential of bioactive compounds of millets as an inhibitor for lifestyle diseases: molecular docking and simulation-based approach, *Front. Nutr.*, 2023, **10**, 1228172.
47. I. Amadou, O. S. Gbadamosi and G. W. Le, Millet-based Traditional Processed Foods and Beverages-A Review, *Cereal Foods World*, 2011, **56**(3), 115.
48. A. N. A. Aryee, D. Agyei and C. C. Udenigwe, Impact of processing on the chemistry and functionality of food proteins, in *Proteins in Food Processing*, 2nd edn, 2017, pp. 27–45.

49. I. Amadou, M. E. Gounga and G. W. Le, Millets: Nutritional composition, some health benefits and processing - A review, *Emirates J. Food Agric.*, 2013, **25**(7), 501–508.
50. U. Dharmaraj and N. G. Malleshi, Changes in carbohydrates, proteins and lipids of finger millet after hydrothermal processing, *LWT*, 2011, **44**(7), 1636–1642.
51. C. K. Sunil, N. A. N. Gowda, N. Nayak and A. Rawson, Unveiling the effect of processing on bioactive compounds in millets: Implications for health benefits and risks, *Process Biochem.*, 2024, **138**, 79–96.
52. N. Nayak, R. R. Bhujle, N. A. Nanje-Gowda, S. Chakraborty, K. Siliveru and J. Subbiah, *et al.*, Advances in the novel and green-assisted techniques for extraction of bioactive compounds from millets: A comprehensive review, *Heliyon*, 2024, **10**(10), e30921.
53. M. del M. Contreras, A. Lama-Muñoz, J. Manuel Gutiérrez-Pérez, F. Espínola, M. Moya and E. Castro, Protein extraction from agri-food residues for integration in biorefinery: Potential techniques and current status, *Bioresour. Technol.*, 2019, **280**, 459–477.
54. A. Amin, I. L. Petersen, C. Malmberg and V. Orlien, Perspective on the Effect of Protein Extraction Method on the Antinutritional Factor (ANF) Content in Seeds, *ACS Food Sci. Technol.*, 2022, **2**(4), 604–612.
55. P. Singh, R. Singh, V. Bhadauria and H. Singh, The functionality and extraction of protein from sorghum, finger millet, and Kodo millet: a review, *Int. J. Food Sci. Technol.*, 2024, **59**(1), 512–521.
56. B. Zheng, Y. Yuan, J. Xiang, W. Jin, J. B. Johnson and Z. Li, *et al.*, Green extraction of phenolic compounds from foxtail millet bran by ultrasonic-assisted deep eutectic solvent extraction: Optimization, comparison and bioactivities, *LWT*, 2022, **154**, 112740.
57. S. Anitha, M. Govindaraj and J. Kane-Potaka, Balanced amino acid and higher micronutrients in millets complements legumes for improved human dietary nutrition, *Cereal Chem.*, 2020, **97**(1), 74–84.
58. S. Singh, D. Yadav, M. Beckmann, A. Naveen, P. I. Gangashetty and L. A. J. Mur, *et al.*, Variation in protein and amino acids in global collection of pearl millet (*Pennisetum glaucum*) germplasm, *J. Food Compos. Anal.*, 2024, **134**, 106557.
59. B. O. de Lumen, S. Thompson and W. J. Odegard, Sulfur Amino Acid-Rich Proteins in Acha (*Digitaria exilis*), a Promising Underutilized African Cereal, *J. Agric. Food Chem.*, 1993, **41**(7), 1045–1047.
60. J. R. N. Taylor, P. S. Belton, T. Beta and K. G. Duodu, Increasing the utilisation of sorghum, millets and pseudocereals: Developments in the science of their phenolic phytochemicals, biofortification and protein functionality, **59***J. Cereal Sci.*, 2014, 257–275.
61. V. S. Nambiar, N. Sareen, M. Daniel and E. B. Gallego, Flavonoids and phenolic acids from pearl millet (*Pennisetum glaucum*) based foods and their functional implications, *Funct. Foods Health Dis.*, 2012, **2**(7), 251–264.
62. M. Munshi and K. Dashora, Comparative study of physico-chemical composition, functional, morphological and pasting properties of major

and minor millet flours as a gluten free alternative to wheat flour, *Meas. Food*, 2024, **16**, 100202.
63. H. B. N. Itagi and V. Singh, Preparation, nutritional composition, functional properties and antioxidant activities of multigrain composite mixes, *J. Food Sci. Technol.*, 2012, **49**(1), 74.
64. Y. K. Muhammad, A. Bhasin, B. D. Maibam and M. Sharma, Review on Finger millet (*Eleusine coracana* L.): Nutritional composition, modification, its effect on physicochemical, structural, functional properties, and its applications, *J. Food Compos. Anal.*, 2024, 106623.

CHAPTER 9

Thermal and Non-thermal Ways of Processing Millet Protein

SENEM KAMILOGLU (ORCID 0000-0003-3902-4360)*[a] AND MERVE TOMAS[b]

[a] Department of Food Engineering, Faculty of Agriculture, Bursa Uludag University, Gorukle, 16059, Bursa, Turkey; [b] Department of Food Engineering, Faculty of Chemical and Metallurgical Engineering, Istanbul Technical University, 34469 Maslak, Istanbul, Türkiye
*Email: skamiloglu@uludag.edu.tr

9.1 Introduction

Growing consumer awareness of "healthy living" has driven increased demand for natural and functional food products. Foods rich in phytochemicals and high-quality proteins, particularly those derived from cost-effective plant sources, are gaining significant attention and undergoing extensive evaluation.[1] Current research and market trends are increasingly focused on identifying novel protein sources to meet dietary needs.[2] In recent years, there has been a growing interest in utilizing plant proteins as alternatives to animal proteins in food products. Millets have great potential as a plant-based protein source for human nutrition. Plant proteins are more eco-friendly and renewable, offering significant advantages such as reduced manufacturing costs, a lower carbon footprint, and decreased water usage throughout the supply chain. Additionally, they provide considerable health

Food Chemistry, Function and Analysis No. 48
Millet Proteins: Nutritional, Structural and Functional Attributes
Edited by Shweta Suri, Anuradha Dutta and Tarun Belwal

Published by the Royal Society of Chemistry, www.rsc.org

benefits, further enhancing their appeal. Millet, and its protein concentrates, is a highly nutritious cereal, particularly consumed in developing countries. It offers significant quantities of essential amino acids, including lysine, making it an excellent choice to enrich protein content in diets.[3] Thermal (soaking, boiling, roasting, extrusion, *etc.*) and non-thermal (ultrasound, cold plasma, high-pressure processing, pulsed electric fields, *etc.*) processing significantly impact the structure, functionality, and nutritional properties of millet proteins. In general, thermal processing negatively impacts protein quality by reducing solubility, digestibility, and overall protein content. It also causes structural denaturation, loss of essential amino acids, and an increase in β-sheet content, all of which can diminish the functionality and nutritional value of the proteins. On the other hand, non-thermal processing preserves and enhances millet protein properties. It improves digestibility, increases hydrophilicity, retains high protein content, and enhances the solubility of protein concentrates. Additionally, non-thermal methods support functional applications, such as in weaning foods, by improving foaming properties and causing minimal structural alteration.[4–6] Overall, non-thermal processing offers an advantageous approach for maintaining and enhancing the quality of millet proteins. This chapter aims to provide an overview of the effects of thermal and non-thermal treatments on millet protein.

9.2 Thermal Processing

Millets are widely consumed for their rich nutritional and phytochemical content. Various processing techniques, such as germination, soaking, boiling, roasting, drying, parboiling, baking, popping, and extrusion, are used individually or in combination to create value-added food products.[4,6] These methods aim to enhance nutrient bioavailability and improve grain digestibility. However, they can lead to significant loss of valuable components during heat processing. Thermal processing, in particular, induces changes in the functional and structural properties of proteins, including the formation of disulfide linkages and modifications to the secondary structure.[5]

The effects of thermal processing on millet protein are presented in Table 9.1. In a study, pearl millet grains underwent an initial hydro-treatment as a preparatory step. Following hydration, the wet grain flour was subjected to distinct and combined thermal processing techniques, including hydrothermal treatment and thermal exposure to near-infrared radiation. A significant reduction in *in vitro* protein digestibility was observed in the hydrothermally treated flour (87.9%) compared to the native flour (91.8%), hydrothermally treated flour exposed to near-infrared radiation (90.6%), and near-infrared radiation treated flour (89.16%). The reduction in *in vitro* protein digestibility may be attributed to protein denaturation induced by high-temperature processing, which hampers proteolytic hydrolysis.[7] Roasting resulted in a significant increase in non-essential amino acid

Table 9.1 Effect of thermal treatments on functional and structural properties of millet proteins.

Millet	Processing types	Processing conditions	Key findings	Ref.
Aohan millet	Steaming Stir-frying Puffing	95 °C for approximately 30 min 37 °C and then fried in a hot pan (300 W) for 30 min Dried at 37 °C and subjected to puffing	• The three processing methods led to a significant reduction in protein content (35–62%) in Aohan millet compared to the raw form. • Fried Aohan millet showed higher levels of sweet, fresh, and bitter amino acids. • Steamed Aohan millet had the lowest amount of bitter amino acids, indicating that steaming effectively reduced the bitter flavor.	12
Barnyard millet	Roasting	110 °C for 10 min	• The total amino acid content of the non-roasted sample was 116.76 ng, whereas the roasted sample exhibited a value of 123.51 ng.	40
Foxtail millet	Microwave and hot air-radiofrequency treatments	Microwave: rated power output (W) 575.71 W, 701.64 W, 784.30 W, 2 min, 4 min, 6 min Hot air-radiofrequency treatment at 180 mm of electrode distance with a conveyor speed of 2.5 m/h, 1 pass–14 min, 2 passes–28 min	• Microwave and high-frequency alternating radio frequency heating improved the functional properties of foxtail millet protein. • A reduction in the relative crystallinity of foxtail millet protein was observed. • Notable alterations were identified in the secondary structural elements of foxtail millet protein.	9

Indian barnyard millet flour	Soaking	25 °C for 12 h	• *In vitro* protein digestibility was significantly improved from 72.21 to 73.15 g hydrolyzed protein per 100 g.	41
Kodo millet	Conventional parboiling Ohmic heating parboiling	50 °C, 60 °C, and 70 °C for 5 h, 2.5 h, and 1.5 h, respectively	• Ohmic heating parboiling had a notable impact on the nutritional profile, retaining a higher protein content compared to conventional parboiling.	42
Pearl millet	Hydrothermal (HTh), and near-infrared (NIR) processing	HTh treatment at 100 °C for 5 min and NIR treatment at 40 °C for 5 min	• *In vitro* protein digestibility (IVPD) decreased. HT-HTh flour: 87.9%, native flour: 91.8%, HT-HTh-thNIR: 90.6%, and HT-thNIR: 89.16% treated flours.	7
Pearl millet	Hydrothermal treatment	Temperature: 35 °C, 45 °C, 55 °C, soaking time 2, 3, 4 h, and steaming time 5, 10, 15 min	• Protein content initially increased with soaking temperature but decreased at higher temperatures.	43
White finger millet (KMR-340)	Roasting	120 °C, 5 min	• Roasting resulted in a significant increase in non-essential amino acid content, while the essential amino acid content decreased following the pretreatment processes.	8

content, while the essential amino acid content decreased following the pretreatment processes.[8] More recently, microwave and hot air-assisted radiofrequency-treated foxtail millet protein exhibited improved emulsifying and oil/water absorption properties compared to the untreated sample, whereas foaming properties and protein purity were reduced. These changes are likely due to a decrease in relative crystallinity and alterations in secondary structural elements, such as an increase in β-turns and a reduction in α-helices and β-sheets, as evidenced by Fourier transform infrared spectroscopy (FTIR) and X-ray diffraction (XRD) analyses.[9] Prashanth *et al.*[10] investigated the effects of hydrothermal treatment on the quality characteristics of pearl millet. The untreated grain flour had a protein content of 10.23%. Protein content initially increased with soaking temperature but later decreased, likely due to higher temperatures causing protein molecules to separate and embed within the gelatinized starch, reducing their availability for extraction. Similarly, following hydrothermal treatment, a significant reduction in protein content was observed in the decorticated millet samples. This decrease is likely due to the leaching of soluble proteins, which appears to be associated with the drying temperature applied after treatment. Higher drying temperatures likely result in increased cellular disruption, causing greater protein loss as cellular contents are released into the soaking medium.[11] More recently, Cheng *et al.*[12] examined the impact of three thermal processing techniques—steaming, stir-frying, and puffing—on the amino acid profiles of Aohan millet. The three processing methods resulted in a significant reduction (35–62%) in the protein content of Aohan millet compared to the raw millet. The content of sweet amino acids was significantly higher in fried Aohan millet, alongside increased levels of fresh and bitter amino acids. In contrast, steamed Aohan millet contained the lowest amount of bitter amino acids, indicating that the steaming process effectively reduced the bitter flavor.

9.3 Non-thermal Processing

9.3.1 Ultrasonication

Ultrasound technology has emerged as a versatile tool for enhancing the functional, structural, and bioactive properties of millet proteins, offering promising applications in food science and nutraceuticals. Research has consistently demonstrated the efficacy of ultrasound treatment in improving solubility, emulsifying, and foaming properties, along with its ability to modify the structural and thermal characteristics of millet proteins (Table 9.2). These advancements are attributed to mechanisms such as protein molecular weight reduction, changes in zeta potential, and structural rearrangements that improve protein interactions and bioavailability.

Several studies have shown that ultrasound processing enhances the functional characteristics of millet proteins.[13–19] Nazari *et al.*[13] demonstrated that the solubility of proso millet protein concentrate increased

Table 9.2 Effect of ultrasound treatment on functional and structural properties of millet proteins.

Millet type	Conditions	Key findings	Ref.
Browntop millet (*Urochloa ramose* L.)	Power: 200 W Frequency: 26 kHz Durations: 0, 5, 10, 15, and 20 min	Ultrasound pretreatment significantly improved the degree of hydrolysis, solubility, emulsification, and foaming properties. Ultrasound-treated hydrolysates exhibited enhanced antidiabetic properties, with a 15-min treatment effectively inhibiting α-amylase and a 10-min treatment targeting α-glucosidase. 15-min ultrasound treatment exhibited high DPPH radical scavenging activity (73.43%), and also significantly inhibited cholesteryl esterase and lipase.	18
Commercial millet	Intensity: 23.47, 35.57 and 47.07 W cm^{-2} Solid : liquid ratio: 1 : 5, 1 : 10, and 1 : 15 pH: 7.0, 8.0, and 9.0	Ultrasound enhanced protein extraction yields by 165% compared to conventional techniques. The efficiency of protein extraction was primarily influenced by intensity, solid : liquid ratios, and pH levels, with intensity being the most significant factor. Intensity of 47.07 W cm^{-2} led to a 95% increase in the diffusion coefficient compared to lower intensities.	23
Proso millet (*Panicum miliaceum* L.)	Intensity: 18.4, 29.58, and 73.95 W cm^{-2} Duration: 5, 12.5, and 20 min	Optimal ultrasound treatment (73.95 W cm^{-2} for 12.5 min) significantly improved solubility, emulsification, and foaming properties of millet protein concentrate. Ultrasound-induced structural modifications contributed to improved functionality, making modified millet protein concentrate suitable for applications in food formulations such as emulsifiers or egg replacers. Longer ultrasound exposure (20 min) at high intensity led to slight aggregation, reducing functional benefits.	13
Proso millet (*Panicum miliaceum* L.)	Power: 400 W Amplitudes: 60%, 80%, and 100% Duration: 5, 12.5, and 20 min	High-intensity ultrasound effectively enhanced the nutritional profile of proso millet bran, increasing antioxidant activity and total phenolic content. Dietary fiber bioavailability improved significantly, with the best results seen at 100% amplitude for 20 min. Polyphenol oxidase activation led to increased enzymatic browning, requiring optimization of processing conditions. The optimized ultrasound treatment (100% amplitude for 9.3 min) provided the best balance between improved functional properties and controlled undesirable effects, making proso millet bran more suitable for bakery and cereal applications.	22

Table 9.2 (*Continued*)

Millet type	Conditions	Key findings	Ref.
Proso millet (*Panicum miliaceum*)	Power: 29.29, 38.08, and 52.72 W Amplitude: 50%, 75%, and 100% Frequency: 20 kHz Duration: 5 and 10 min	Ultrasound treatment significantly improved protein solubility and digestibility, making proso millet protein more bioavailable for human consumption. Structural changes, including particle size reduction and increased hydrophobicity, contributed to improved functionality, particularly in foaming and emulsification. Longer ultrasound exposure enhanced foaming properties but had mixed effects on emulsification, with a decline observed after 10 min.	15
Sorghum, pearl millet, and foxtail millet	Power: 100 W Frequency: 40 kHz Duration: 30 min Temperature: 20–30 °C	Ultrasonication effectively reduced millet protein particle size, leading to improved dispersion stability, solubility, and emulsifying/foaming properties. Structural modifications, such as reduced crystallinity and altered functional groups, enhanced protein functionality for food applications. Thermal stability decreased, indicating partial denaturation. Millet proteins treated with ultrasound showed potential as improved functional ingredients, particularly in emulsions, foams, and encapsulation applications.	14
Sorghum, pearl millet, and foxtail millet	Power: 100 W Frequency: 40 kHz Duration: 1 h	Foxtail millet protein nanoparticles exhibited the best encapsulation efficiency (83.78%), the smallest particle size, and the highest bioactivity retention. Encapsulated crocin retained its antimicrobial, anti-cancer, and anti-hypertensive properties, demonstrating the potential of millet proteins as bioactive carriers.	24
Finger millet (*Eleusine coracana* L.)	*Before inoculation*: 0.53–1.59 W mm^{-2} for 5–15 min *After inoculation*: 0.21–0.35 W mm^{-2} for 1–5 min	The application of ultrasound significantly reduced fermentation time, particularly after inoculation, where it was reduced by 2.6 times compared to control samples. Ultrasound treatment led to structural modifications in millet, which increased water-holding capacity and viscosity, thereby improving the beverage's texture and stability. Low-intensity ultrasound after inoculation enhanced the proliferation of *Lactobacillus rhamnosus*, leading to faster lactic acid production and improved fermentation kinetics.	25

Finger millet (*Eleusine coracana*)	Power: 100 W, 200 W, and 300 W Frequency: 20 kHz Duration: 12 min	Ultrasound treatment at 100–200 W significantly improved solubility, emulsification, and foaming properties. Ultrasound reduced particle size, increased surface hydrophobicity, and altered zeta potential, facilitating improved protein dispersion and stability. While moderate power levels enhanced protein functionality, high-intensity ultrasound (300 W) led to aggregation and reduced solubility.	19
Finger millet (*Eleusine coracana*) Pearl millet (*Pennisetum glaucum*) Barnyard millet (*Echinochloa* sp.) Sorghum (*Sorghum vulgare*) Kodo millet (*Paspalum scrobiculatum*)	*Germination*: 24 h *Ultrasound treatment*: Power: 180 W Duration: 20 min	Germination significantly improved antioxidant activity and reduced antinutrients, making the nutrients more bioavailable. Ultrasound treatment enhanced the structural and sensory attributes of millet milk, improving its physicochemical stability. Antinutrient levels increased with ultrasound treatment alone, but the combined approach (ultrasound + germination) successfully reduced them.	20
Foxtail millet (*Setaria italica*)	Amplitude: 5% and 10% Duration: 5, 10, and 20 min	High-intensity ultrasound significantly improved the solubility, emulsifying properties, foaming capacity, and thermal stability of foxtail millet protein. Different responses of prolamin and glutelin fractions to high-intensity ultrasound treatment were observed, suggesting the potential for customized applications based on protein type.	16
Foxtail millet (*Setaria italica* (L.) Beauv.)	Power: 0, 60, 150, 250, 350, 450 and 600 W Frequency: 20 kHz Duration: 30 min	Ultrasound treatment, especially at 450 W, induced structural modifications in foxtail millet bran protein hydrolysates, leading to improved solubility, emulsification, and foaming properties. Ultrasound-treated foxtail millet bran protein hydrolysates exhibited enhanced DPPH, ABTS, hydroxyl radical scavenging, Fe^{2+} chelating, and reducing power, with the best results observed at 450 W. Peptidomics analysis revealed that ultrasound increased the number and relative abundance of bioactive peptides, particularly at 450 W, making the hydrolysates more functionally relevant for food and nutraceutical applications.	17

across all ultrasound-treated samples, with the highest solubility observed at 73.95 $W cm^{-2}$ for 12.5 min. Furthermore, the effect on foaming capacity varied with treatment intensity; low-intensity ultrasound reduced foaming capacity, whereas high-intensity treatments enhanced it, reaching maximum levels again at 73.95 $W cm^{-2}$ for 12.5 min. Similarly, both the emulsion activity index and emulsion stability improved following ultrasound treatment, with optimal results at the same high-intensity condition. Another study by Jhan *et al.*[14] investigating the effect of nanoreduction *via* ultrasonication on the functional properties of millet proteins extracted from sorghum, pearl millet, and foxtail millet also found enhanced solubility, emulsifying and foaming capacities, and increased hydrophobicity, while turbidity decreased.

Akharume and Adedeji[15] investigated the effects of high-power ultrasound treatment on the functional properties of prolamin and glutelin protein fractions extracted from two proso millet cultivars (Dawn and Plateau). Their results showed that ultrasound treatment significantly improved protein solubility, with increases of 169%, 191%, and 235% for Dawn prolamin at power levels of 29.29, 38.08, and 52.72 W, respectively, after 5 min of treatment. Functional properties such as foaming capacity increased significantly with longer ultrasound exposure, while the emulsifying activity index improved after 5 min but declined with extended treatment (10 min). The study conducted by Sharma *et al.*[16] focused on the functionalization of foxtail millet protein through high-intensity ultrasound processing. Notably, treatments at higher ultrasound intensities (5% and 10% for durations of 5, 10, and 20 min) resulted in improved solubility, emulsifying properties, and foaming capacity compared to untreated samples. Another study examining the effects of ultrasound at different power levels (0–600 W) on the functionality of foxtail millet bran protein hydrolysates found that ultrasound treatment, particularly at 450 W, improved solubility, emulsifying, and foaming properties.[17]

Rao *et al.*[18] explored the impact of ultrasound pretreatment followed by enzymatic hydrolysis using bromelain on the functional properties of browntop millet protein hydrolysates. The findings revealed that emulsifying and foaming properties were enhanced, with ultrasound pretreatment at 200 W for 15 min showing the best emulsification activity (20.26 $m^2 g^{-1}$). Additionally, ultrasound pretreatment at 200 W for 10 min displayed the highest solubility (88.02%) and foaming capacity (18.52%). A separate study investigated the effects of high-intensity ultrasound treatment at 100 W, 200 W, and 300 W for 12 min on the functional properties of white finger millet protein fractions—albumin, globulin, and glutelin. The study demonstrated that ultrasound treatment enhanced solubility, emulsifying, and foaming properties, particularly at moderate power levels (100–200 W). However, at higher power (300 W), the solubility and emulsifying properties decreased, suggesting possible protein aggregation.[19]

Ultrasound treatment also induced changes in the structural and thermal properties of millet protein.[13,14,17,19] Nazari *et al.*[13] reported that ultrasound treatment of proso millet protein concentrate led to a reduction in molecular weight, while zeta potential analysis indicated a 36% increase in negative surface charge, promoting better dispersion stability. Moreover, FTIR analysis revealed alterations in the protein's secondary structure, including an increase in β-sheet content and positional shifts in the amide regions. Additionally, differential scanning calorimetry (DSC) analysis showed a reduction in enthalpy, suggesting a weakening of intermolecular interactions. Another study on the impact of nanoreduction *via* ultrasonication on the structural and thermal properties of sorghum, pearl millet, and foxtail millet proteins indicated that nanoreduction significantly reduced the particle size of the proteins, with mean sizes of 335.40 nm for pearl millet, 587.98 nm for foxtail millet, and 801.42 nm for sorghum. This process increased the zeta potential (from −11.25 to −15.87), improving dispersion stability. Sodium dodecyl sulfate-polyacrylamide gel electrophoresis (SDS-PAGE) analysis showed a molecular weight reduction (27–10 kDa), indicating the breaking of intermolecular bonds. Structural changes were confirmed by scanning electron microscopy (SEM) (disoriented microstructure), FTIR (alterations in functional groups), and XRD (reduced crystallinity). Thermal analysis with DSC showed a significant reduction in denaturation temperatures (50–65 °C) and enthalpy (5.5–7.0 $J\,g^{-1}$), suggesting lower thermal stability.[14]

Peng *et al.*[17] demonstrated that ultrasound treatment significantly altered the secondary and tertiary structures of foxtail millet bran protein hydrolysates, promoting the transformation of α-helix and β-sheet structures into more flexible random coils and β-turns. This led to greater exposure of hydrophobic and sulfhydryl groups, a decrease in particle size, and an increase in the absolute value of the zeta potential. In another study, Shylla *et al.*[19] reported that ultrasound also induced structural changes by reducing particle size, increasing surface hydrophobicity, and altering the zeta potential. FTIR analysis revealed minor structural modifications, with a slight increase in the intensity of the amide A band, suggesting protein unfolding. SDS-PAGE analysis showed a reduction in molecular weight, particularly in albumin and glutelin at 300 W, indicating partial hydrolysis or fragmentation. SEM further confirmed that ultrasound-treated protein fractions exhibited more irregular microstructures, suggesting alterations in their morphology.

Ultrasound treatment has been demonstrated to enhance the bioactive properties of millet. Peng *et al.*[17] reported that ultrasound processing significantly increased the antioxidant capacity of foxtail millet bran protein hydrolysates, as evidenced by elevated 2,2-diphenyl-1-picrylhydrazyl (DPPH), 2,2′-azino-bis(3-ethylbenzothiazoline-6-sulfonic acid) (ABTS), and hydroxyl radical scavenging activities. Peptidomics analysis further revealed that ultrasound treatment boosted the number of active peptides, particularly at 450 W, where the relative abundance of 17 active peptides was markedly enhanced. Similarly, Saxena *et al.*[20] investigated the effects of ultrasound, germination, and their combined application on the bioactive properties of millet-based milk. Germination led to a 92% increase in total phenolic content and a 33% improvement in antioxidant activity, enhancing the bioavailability of bioactive compounds. Ultrasound treatment contributed to higher total dissolved solids, electrical conductivity, and oil adsorption capacity, refining the structural and sensory attributes of the extracted milk. However, ultrasound alone resulted in a 32% rise in antinutrient levels, while germination effectively reduced them. Notably, the combined application of ultrasound and germination lowered antinutrient concentration by 23%, achieving an optimal balance between enhanced functionality and reduced undesirable compounds.

Rao *et al.*[18] demonstrated that the *in vitro* antidiabetic activity of browntop millet protein hydrolysates was significantly enhanced when pretreated with ultrasound followed by enzymatic hydrolysis. Hydrolysates subjected to ultrasound for 15 min and 10 min effectively inhibited α-amylase (0.47 $mg\,mL^{-1}$) and α-glucosidase (0.68 $mg\,mL^{-1}$). Furthermore, 15-min ultrasound treatment exhibited strong antilipidemic activity, as indicated by the inhibition of cholesteryl esterase (0.34 $mg\,mL^{-1}$) and lipase (0.51 $mg\,mL^{-1}$). Additionally, antioxidant activity, measured through DPPH radical scavenging capacity (73.43%), was significantly improved.

The digestibility of plant proteins is a crucial factor as it directly influences their nutritional value and health benefits.[21] Akharume and Adedeji[15] reported that the application of high-power ultrasound treatment enhanced the *in vitro* digestibility of proso millet protein by 28–58%, indicating improved bioavailability. Conversely, Mustac *et al.*[22] observed a slight reduction in the *in vitro* digestibility of proso millet protein following high-intensity ultrasound treatment, possibly due to polyphenol–protein interactions, which may hinder enzymatic hydrolysis. Therefore, the impact of ultrasound on the *in vitro* digestibility of millet protein depends on several factors, including the intensity and duration of ultrasound treatment, and the presence of bioactive compounds, such as polyphenols, which may interact with proteins and influence enzymatic accessibility.

Other studies have investigated the application of ultrasound treatment to enhance millet protein extraction efficiency. Ma *et al.*[23] demonstrated that ultrasound can increase protein extraction yields by 165% compared to conventional methods. The key parameters influencing this process included acoustic energy density, solid : liquid ratios, and pH levels, with acoustic energy density emerging as the most critical factor affecting the effective diffusion coefficient of proteins. Notably, an acoustic energy density of 47.07 $W\,cm^{-2}$ led to a 95% increase in the diffusion coefficient compared to lower energy levels. Additionally, the study utilized an adaptive neuro-fuzzy inference system to model and predict protein yields, achieving a high coefficient of determination ($R^2 > 0.97$). Similarly, Sharma *et al.*[16] reported that ultrasound treatment resulted in a protein recovery yield of 70.79%, with prolamin and glutelin contributing 31.10% and 34.69% of the total protein content, respectively.

Jhan *et al.*[24] explored the use of ultrasonication-assisted nanoencapsulation to enhance the stability and bioavailability of crocin using millet-derived protein nanoparticles. Proteins extracted from sorghum, pearl millet, and foxtail millet served as encapsulation carriers, forming crocin-loaded nanocomplexes. The particle sizes of these nanocomplexes were successfully reduced, with foxtail millet (387 nm) exhibiting the smallest size, followed by sorghum (487 nm) and pearl millet (665 nm). The encapsulation efficiency was highest for foxtail millet (84%), followed by sorghum (79%) and pearl millet (70%). FTIR spectroscopy confirmed successful encapsulation through molecular interactions. *In vitro* release studies demonstrated a controlled release of crocin under simulated gastrointestinal conditions, ensuring stability and targeted intestinal delivery. Antimicrobial assays showed activity against *E. coli, S. aureus, and Fusarium oxysporum*, with foxtail millet nanocomplexes exhibiting the strongest inhibition. Furthermore, the encapsulated crocin retained its anti-cancer and anti-hypertensive properties, highlighting the potential of millet protein nanoparticles as effective carriers for protecting and delivering functional bioactives in nutraceutical applications.

Meena *et al.*[25] investigated the impact of ultrasound pre-treatment on the fermentation kinetics and quality attributes of millet-based probiotic beverages, using white finger millet and *Lactobacillus rhamnosus*. The study evaluated two ultrasound treatment approaches: before and after inoculation. Results showed that ultrasound treatment significantly reduced fermentation time, with the post-inoculation treatment decreasing fermentation time by 2.6-fold compared to untreated samples. Additionally, ultrasound application modified the structural properties of millet, influencing the beverage's water-holding capacity and viscosity. Fermentation kinetics followed a Weibull model, with ultrasound-treated samples exhibiting enhanced lactic acid production. Notably, low-intensity ultrasound after inoculation promoted microbial proliferation, further optimizing fermentation efficiency.

Overall, ultrasound treatment has proven to be a transformative technique for optimizing the functional, bioactive, and structural properties of millet proteins. Studies highlight its ability to enhance protein solubility, emulsifying and foaming properties, antioxidant capacity, and bioavailability, while also reducing antinutrient content and promoting controlled release of bioactives. Additionally, ultrasound has shown significant potential in improving protein digestibility and extraction yields, paving the way for its application in sustainable food systems. As research progresses, a deeper understanding of the interactions between ultrasound parameters, protein structure, and bioactive compounds will enable more precise and effective utilization of ultrasound technology, driving innovation in functional foods and nutraceuticals.

9.3.2 Cold Plasma

Millets, recognized for their rich nutrient profile and resilience in diverse agro-climatic conditions, have gained considerable interest in food science research. However, their utilization has been hindered by the presence of antinutritional factors, suboptimal functional properties, and microbial contamination, limiting their application in modern food formulations.

Cold plasma, a non-thermal processing technology, has emerged as a promising approach for enhancing the functional, nutritional, and bioactive properties of millets without causing significant heat-induced degradation.[26] A series of studies have explored the effects of cold plasma treatment on different millet varieties, each revealing unique improvements in hydration properties, protein functionality, antinutrient reduction, and bioactive enhancement, reinforcing the potential of this novel technique for millet-based food applications (Table 9.3).

Studies on barnyard millet (*Echinochloa* spp.), foxtail millet (*Setaria italica* L.), little millet (*Panicum miliare*), and pearl millet (*Pennisetum glaucum*) have shown that cold plasma treatment significantly enhances their hydration properties, solubility, and oil absorption capacity.[27–33] Lokeswari *et al.*[27] demonstrated that atmospheric pressure cold plasma treatment improved the swelling capacity, water absorption, and pasting properties of pearl millet, making it more suitable for thickening and hydration-dependent food formulations. Similarly, Jaddu *et al.*[28] reported that cold plasma-treated little millet flour exhibited higher water solubility, swelling capacity, and oil absorption, attributed to starch depolymerization and a reduction in crystallinity. Structural modifications observed through XRD and SEM indicated that the breakdown of starch granules contributed to these functional improvements.

Building on these findings, Sharma *et al.*[29] investigated multipin atmospheric cold plasma treatment of barnyard millet flour, revealing notable increases in water solubility index (0.83–3.92 $g\,g^{-1}$), emulsifying capacity (37.40–44.51%), and foaming capacity (20.61–24.99%). Sarkar *et al.*[30] similarly found that cold plasma treatment of pearl millet flour significantly enhanced emulsification and foaming properties, key attributes for bakery and functional food formulations. These studies confirm that cold plasma improves hydration and dispersion properties, facilitating the incorporation of millet-based ingredients in modern food products, including gluten-free baked goods and plant-based formulations.

Beyond flour processing, cold plasma has demonstrated significant benefits in seed germination and protein modification. Monica *et al.*[31] evaluated cold plasma treatment on foxtail millet seeds and found that it accelerated germination rate, sprout length, and seed vigor indices, with 2 kV for 3 minutes yielding the highest germination percentage (84%). The increase in soluble protein and γ-aminobutyric acid (GABA), coupled with a reduction in phytic acid and tannins, suggests that cold plasma enhances nutrient bioavailability and seed metabolic activity, making it a viable method for improving millet sprouting efficiency in food and feed applications.

Further extending its application, Monica *et al.*[32] explored the effects of cold plasma on sprouted foxtail millet protein concentrate. The treatment improved protein solubility, emulsification, and foaming properties, critical for meat alternatives and functional food formulations. These enhancements were linked to surface roughness, protein oxidation, and increased hydrophobicity, as observed through SEM and FTIR. This highlights cold

Table 9.3 Effect of cold plasma on functional and structural properties of millets.

Millet type	Conditions	Key findings	Ref.
Barnyard millet (*Echinochloa* spp. var. DHBM 93-3)	Gas: air Voltage: 10 kV, 20 kV, and 30 kV Duration: 10, 20, and 30 min Electrode spacing: 7 cm	• Cold plasma treatment significantly enhanced the functional properties of barnyard millet flour, improving its solubility, absorption capacities, and emulsifying and foaming abilities. • The treatment increased bioactive compound content, including total phenols and flavonoids, leading to improved antioxidant potential and potential health benefits. • Structural and morphological modifications, including surface roughness and changes in crystallinity, were observed, contributing to improved hydration and dispersion properties. • Cold plasma effectively reduced antinutritional factors (phytic acid and tannins), improving bioavailability of nutrients.	29
Foxtail millet (*Setaria italica* L. var. ATL 1)	Gas: air Voltage: 1 kV (57 W) and 2 kV (114 W) Duration: 1, 3, and 5 min Low-pressure range: 66.66–79.99 kPa	• Cold plasma significantly enhanced germination efficiency by increasing germination percentage, germination speed, sprout length, and seed vigor indices, with 2 kV for 3 min being the most effective condition. • Nutrient composition of foxtail millet is improved, as cold plasma treatment leads to higher soluble protein and GABA content while reducing anti-nutritional factors like phytic acid and tannins. • Surface modifications by reactive species (ROS and RNS) contributed to seed coat etching, increased water absorption, and enhanced germination without significant damage to seed integrity.	31

Table 9.3 (*Continued*)

Millet type	Conditions	Key findings	Ref.
Foxtail millet (*Setaria italica* L.)	Voltage: 1 kV and 2 kV Duration: 5, 15, and 25 min Plasma type: dielectric barrier discharge (DBD) Low-pressure: 2 mbar	• Cold plasma treatment enhanced the functional properties of foxtail millet protein concentrate, significantly improving solubility, emulsification, and foaming capacities, making it a suitable plant-based protein alternative. • Structural modifications, including surface roughness, oxidation, and increased surface hydrophobicity, suggested that cold plasma alters protein interactions, leading to improved functional properties. • Plasma treatment improved water and oil absorption capacities, contributing to better textural properties in food formulations, particularly in plant-based meat and bakery applications.	32
Little millet (*Panicum miliare* var. DHLM 14-1)	Gas: air Power: 13 W and 24 W Duration: 10, 20, and 30 min	• Cold plasma treatment significantly enhanced the functional properties of little millet flour by increasing its water absorption, swelling capacity, and solubility, making it more suitable for food processing applications. • Structural changes, including starch depolymerization and reduced crystallinity, were observed due to the action of reactive plasma species, which contributed to improved hydration and viscosity properties. • The treatment had minimal impact on physical properties such as bulk and tapped densities, while slight surface damage to starch granules was evident in SEM images.	28
Pearl millet (*Pennisetum glaucum*)	Gas: air Voltage: 40 kV and 45 kV Duration: 5, 10, and 15 min	• Cold plasma treatment significantly enhanced hydration and pasting properties, making pearl millet more suitable for food formulations such as bakery products, soups, and thickening agents. • Nutritional modifications were observed, with increased crude fat content but decreased crude fiber, protein, and moisture levels, potentially influencing the functional properties of millet-based products. • The physical properties of millet remained largely unaffected, except for minor color changes.	27

Pearl millet (*Pennisetum glaucum*)	Gas: air Voltage: 20 kV, 25 kV, and 30 kV Duration: 10, 15, and 20 min Inter-electrode spacing: 7 cm	• Cold plasma treatment significantly reduced antinutritional factors (tannin and phytic acid), improving the bioavailability of nutrients in pearl millet flour. • Functional properties such as water absorption, oil absorption, emulsification, and foaming capacity are enhanced, making the flour more suitable for various food applications. • Structural modifications in starch granules and reduced crystallinity improve hydration properties and viscosity, which can enhance the textural properties of processed foods.	30
Pearl millet (*Pennisetum glaucum* var. Dhanashakti) Barnyard millet (Echinochloa spp. var. VL-172)	Gas: air Voltage: 2 kV Duration: 5 min Electrode distance: 3 cm	• Cold plasma treatment significantly improved microbial safety, reducing bacterial and yeast counts, which enhances millet shelf-life. • Nutritional enhancements included increased fat and vitamin B-complex content, which contribute to improved bioavailability. • Antinutritional factors (phytic acid and tannins) are significantly reduced, improving mineral absorption and overall digestibility. • Despite a decrease in total phenolic and flavonoid content, cold plasma treatment presents an effective processing technology for enhancing the functional, nutritional, and safety attributes of millets for food applications.	33

plasma's ability to modify protein structure, expanding its potential use as a plant-based emulsifier and stabilizer in diverse food applications.

One of the most significant challenges in millet utilization is the presence of antinutritional factors such as phytic acid and tannins, which limit mineral bioavailability. Cold plasma treatment has been highly effective in reducing these compounds. Sarkar *et al.*[30] reported that cold plasma treatment of pearl millet flour reduced tannins and phytic acid, with the most significant reduction observed at 30 kV for 20 minutes. Similarly, Charu *et al.*[33] found that cold plasma treatment significantly decreased phytic acid (57% in pearl millet and 61% in barnyard millet) and tannins (18% in pearl millet and 24% in barnyard millet), improving nutrient accessibility.

Additionally, cold plasma treatment contributes to microbial safety and shelf-life extension. Charu *et al.*[33] demonstrated that 2 kV for 5 min effectively reduced bacterial and yeast populations, enhancing the microbial stability of pearl millet and barnyard millet. This effect is particularly valuable for minimizing foodborne risks and extending storage stability, making cold plasma a viable alternative to traditional decontamination techniques.

Collectively, these studies highlight the versatility of cold plasma as a non-thermal food processing technology, with applications ranging from flour modification to seed enhancement and protein functionality improvement. The observed benefits, including improved hydration properties, bioactive compound retention, antinutrient reduction, and microbial safety, reinforce its potential for value addition in millet-based food products. Future research should focus on scaling up cold plasma technology, optimizing treatment conditions for different millet varieties, and evaluating its long-term effects on storage and product quality. As an eco-friendly and energy-efficient alternative to conventional processing methods, cold plasma paves the way for the sustainable utilization of underutilized millets, expanding their application in functional foods, gluten-free formulations, and plant-based nutrition.

9.3.3 High-pressure Processing

Millets, valued for their nutritional density and functional versatility, are gaining increased attention in food processing. However, their structural rigidity, strong hydrophobic nature, and antinutritional factors limit their use in various food formulations. Recent studies have demonstrated that high hydrostatic pressure processing treatments significantly enhance the functional, structural, and bioactive properties of millets, making them more suitable for bakery, protein-based, and functional food applications (Table 9.4). These advancements include improving starch hydration, enhancing protein solubility, modifying gluten-like structures, and increasing antioxidant potential, positioning millets as sustainable and nutritious alternatives to conventional grains.

One of the primary applications of high hydrostatic pressure in millet processing is the modification of starch properties to enhance water absorption, gelatinization, and solubility. Sherin *et al.*[34] investigated the effects of high hydrostatic pressure on white finger millet starch, demonstrating that pressure levels up to 600 MPa significantly increased amylose content, swelling power, and solubility, particularly at high temperatures. Similarly, Sharma *et al.*[35] studied the effects of high-pressure soaking on foxtail millet grains, showing that hydration rates, starch gelatinization (65%), and diffusion coefficients improved significantly under pressures of up to 200 MPa at 60 °C. These studies suggest that high-pressure techniques enhance the water-binding capacity of millet starch, making it more suitable for applications in porridges, baked goods, and sauces where hydration properties are critical.

Additionally, composite flour formulations incorporating high-pressure-treated millets show promising results. Angioloni and Collar[36] found that high hydrostatic pressure-treated finger millet, oat, and sorghum flours (350 MPa, 10 min) improved bread textural properties and antioxidant activity, although some volume loss and faster staling were observed in millet-containing breads. These findings indicate that high hydrostatic pressure can improve the sensory and nutritional quality of millet-based baked products, offering potential gluten-free alternatives.

Beyond starch modifications, high hydrostatic pressure processing has profound effects on millet protein functionality. Zhang *et al.*[37] demonstrated that high hydrostatic pressure treatment of millet gliadin altered its secondary structure, reduced hydrophobic interactions, and improved its solubility and stability. These changes increased protein particle size and thermal denaturation temperature, mitigating the hydrophobic nature of millet gliadin, which often limits its direct use in food formulations.

Building on these structural modifications, Zhang *et al.*[38] explored the antioxidant potential of high hydrostatic pressure-treated millet gliadin hydrolysates. High hydrostatic pressure treatment at 300 MPa, followed by enzymatic hydrolysis (pepsin, 50 min), resulted in bioactive peptides with strong DPPH, hydroxyl radical, and superoxide anion scavenging properties. Seven key antioxidant peptides were identified, containing amino acids such as glycine, leucine, phenylalanine, and proline, which contribute to lipid peroxidation inhibition. These results highlight the potential of high hydrostatic pressure-treated millet proteins as natural antioxidant ingredients, reinforcing their role in functional and nutraceutical food formulations.

High-pressure techniques offer multiple advantages beyond hydration and protein modification. Sharma *et al.*[35] and Zhang *et al.*[38] reported a notable reduction in antinutritional factors such as phytic acid and tannins, leading to improved mineral bioavailability and enhanced functional properties. Angioloni and Collar[36] further confirmed that high hydrostatic pressure-treated millet-containing breads retained a superior dietary fiber profile and higher antioxidant capacity, reinforcing the nutritional benefits of high-pressure processing in improving millet-based products.

Despite these benefits, some challenges remain, such as increased staling rates in millet-based bread formulations[36] and decreased paste clarity in

Table 9.4 Effect of high-pressure processing on functional and structural properties of millets.

Millet type	Conditions	Key findings	Ref.
Finger millet (*Eleusine coracana*), oat and sorghum	Pressure: 350 MPa Duration: 10 min	• Finger millet, oat, and sorghum-based breads benefit from high hydrostatic pressure processing in terms of better dietary fiber content and polyphenol retention, supporting their use in functional food applications. • Despite improvements in texture and nutritional properties, high hydrostatic pressure treatment resulted in volume reduction and faster staling in millet- and sorghum-containing breads, requiring further optimization.	36
Finger millet (*Eleusine coracana* var. KMR 340)	Pressure: 200, 400, and 600 MPa Duration: 10 min	• High-pressure processing enhanced the functional properties of finger millet starch, increasing amylose content, water absorption, and solubility, making it more adaptable for food applications. • Structural modifications due to high-pressure processing enhanced the functional properties of finger millet starch, increasing amylose led to partial gelatinization and changes in crystallinity, with 600 MPa-treated starch resembling a B-type crystalline structure, affecting pasting behavior. • Improved hydration properties (swelling and solubility) at higher temperatures suggest the potential use of high-pressure processing-treated starch in heat-stable formulations like bakery products, sauces, and thickeners.	34
Foxtail millet (*Setaria italica* L.)	Pressure: 200, 400, and 600 MPa Temperature: 20, 40, 60, and 80 °C Duration: 30, 60, 90, and 120 min	• High-pressure soaking significantly improved water absorption and starch gelatinization, making foxtail millet more suitable for food applications requiring pre-gelatinized starch (*e.g.*, porridges and bakery products). • Germinated foxtail millet exhibited higher diffusion rates and gelatinization compared to non-germinated grains, highlighting the benefits of combining germination and high-pressure processing for improved functionality. • Antinutritional factors such as phytic acid and tannins were reduced, enhancing the bioavailability of nutrients, particularly minerals. • Despite minimal changes in protein content, the treatment increased total phenolic content and antioxidant activity.	35

Millet gliadin	Pressure: 100, 300, and 500 Mpa Duration: 10 min	• High hydrostatic pressure treatment increased the hydrophilicity of millet gliadin, reducing its strong hydrophobic nature, which enhances its potential application in food formulations. • Protein aggregation and structural changes were observed, including an increase in particle size, surface charge redistribution, and the formation of disulfide bonds, leading to improved protein stability. • High hydrostatic pressure treatment resulted in a more ordered protein structure, increasing the denaturation temperature, which enhanced its functional properties for use in food applications.	37
Millet gliadin	Pressure: 100, 300, and 500 Mpa Enzymatic hydrolysis: pepsin, trypsin, and alkaline protease	• 300 MPa treatment significantly improved enzymatic hydrolysis, increasing the yield of antioxidant peptides by modifying gliadin's conformation and increasing its solubility. • Seven bioactive peptides were isolated, showing strong DPPH, hydroxyl radical, and superoxide anion scavenging abilities, making them potential natural antioxidants. • High hydrostatic pressure altered the secondary structure of gliadin, increasing its reactivity with pepsin and enhancing the efficiency of bioactive peptide release.	38

high hydrostatic pressure-treated starch.[34] These factors indicate the need for further optimization of processing conditions to balance functional improvements with product stability.

Overall, high-pressure processing has demonstrated significant potential in enhancing the functional, structural, and nutritional properties of millets, making them more suitable for diverse food applications. These treatments significantly improve starch hydration properties, increasing solubility, swelling power, and water absorption capacity, which enhances their usability in gluten-free baking, porridges, and sauces. Additionally, protein modifications under high pressure enhance solubility, aggregation behavior, and structural stability, enabling millet proteins to function as emulsifiers, stabilizers, and antioxidant-rich ingredients in functional foods. Moreover, high-pressure processing reduces antinutritional factors such as phytic acid and tannins, improving mineral bioavailability and increasing the release of bioactive compounds, including antioxidant peptides, making millets more suitable for nutraceutical and health-oriented food formulations. Despite some challenges, such as increased staling rates in bread and decreased paste clarity in starch-based applications, high-pressure techniques remain a promising non-thermal processing method, offering a sustainable, clean-label alternative to conventional food processing. Integrating high-pressure processing into millet-based food production can enhance nutritional quality, functional properties, and consumer acceptability, reinforcing the role of millets as a key ingredient in sustainable and health-promoting diets.

9.3.4 Pulsed Electric Field

Zhang *et al.*[39] investigated the effects of moderate-intensity pulsed electric field on the structure and physicochemical properties of foxtail millet (*Setaria italica*) prolamin. The study found that moderate-intensity pulsed electric field altered the spatial conformation of millet prolamin, increasing its hydrophobicity and solubility in alcohol solutions. However, the treatment did not change the subunit composition distribution. Moderate-intensity pulsed electric field treatment also promoted the formation of soluble aggregates, leading to an increase in protein particle size. Changes in secondary structure were observed, with a decrease in α-helix content and an increase in β-sheet content, indicating a more orderly protein structure post-treatment. Additionally, moderate-intensity pulsed electric field increased the number of disulfide bonds, which contributed to protein aggregation and improved structural stability. Spectroscopic analyses, including UV spectroscopy and FTIR, confirmed that moderate-intensity pulsed electric field expanded the protein's tertiary structure, leading to greater exposure of tryptophan and tyrosine residues. Surface hydrophobicity initially increased with moderate-intensity pulsed electric field treatment but decreased at higher electric field intensities, likely due to changes in electrostatic interactions and protein aggregation. The study also revealed that moderate-intensity pulsed electric field treatment affected sulfhydryl content, reducing total sulfhydryl concentration due to the formation of disulfide bonds, which contributed to the structural reorganization of the protein. Furthermore, particle size analysis and zeta potential measurements showed that moderate-intensity pulsed electric field promoted protein aggregation and rearrangement, leading to an increase in overall protein size and stability.

9.4 Conclusion

Millet proteins have garnered significant attention in recent years due to their nutritional benefits, sustainability, and functional applications in food processing. As consumer demand for plant-based proteins rises, understanding the impact of thermal and non-thermal processing on millet proteins becomes increasingly crucial. This chapter has comprehensively explored various processing techniques, highlighting their effects on the structural, functional, and nutritional properties of millet proteins.

Thermal processing, including boiling, roasting, steaming, and extrusion, has long been employed to improve the digestibility and bioavailability of millet proteins. However, it has been found to induce protein denaturation, reduce solubility, and lower essential amino acid content, ultimately affecting the overall protein functionality. While some techniques, such as microwave-assisted treatments, have shown improvements in emulsifying and oil absorption properties, conventional thermal processing often leads to undesired structural changes, such as increased β-sheet content and reduced crystallinity, which may negatively impact protein stability and digestibility. Therefore, while thermal treatments enhance digestibility to some extent, they also pose challenges in terms of preserving protein integrity and bioactivity.

On the other hand, non-thermal processing techniques, including ultrasound, high-pressure processing, cold plasma, and pulsed electric field, have emerged as promising alternatives to enhance the functionality of millet proteins while minimizing structural damage. Ultrasound processing, for instance, has been shown to increase protein solubility, emulsification, and foaming capacity by modifying protein molecular interactions and reducing particle size. High-pressure processing has demonstrated significant potential in enhancing starch hydration, improving protein solubility, and increasing antioxidant potential without the drawbacks associated with excessive heat exposure. Cold plasma technology has been shown to be effective in reducing antinutritional factors such as phytic acid and tannins, thereby enhancing mineral bioavailability and functional applications of millet proteins in bakery, weaning foods, and gluten-free formulations. Additionally, pulsed electric field treatment has been shown to enhance the solubility and stability of millet prolamin proteins, making them more suitable for emulsification and structuring applications in plant-based food formulations.

Beyond functionality, non-thermal processing methods have also been recognized for their ability to preserve the bioactive properties of millet proteins. Several studies have demonstrated that these techniques increase the yield of antioxidant peptides, improve phenolic content, and enhance protein digestibility, reinforcing their potential role in functional food and nutraceutical applications. The application of high-pressure processing and ultrasound in combination with enzymatic hydrolysis has further been shown to improve the release of bioactive peptides, contributing to enhanced antioxidant and lipid-lowering properties.

Despite these advancements, challenges remain in optimizing processing conditions to achieve a balance between improved functionality and minimal structural degradation. The variability in millet varieties, processing parameters, and protein compositions suggests that further research is needed to standardize processing techniques and explore synergistic approaches that combine different non-thermal methods for maximizing protein functionality and nutritional value.

Overall, this chapter highlights the potential of non-thermal processing technologies as superior alternatives to conventional thermal methods for enhancing the functional, structural, and bioactive properties of millet proteins. As research progresses, integrating these techniques into commercial food production could lead to the development of high-quality, protein-enriched, and sustainable millet-based food products, contributing to the growing market for plant-based and functional foods. The adoption of non-thermal processing in millet protein applications represents a significant step toward sustainable food innovation, leading to broader utilization of millet proteins in health-conscious and eco-friendly diets.

References

1. O. D. Medina Martinez, R. C. Lopes Toledo, V. A. Vieira Queiroz, M. R. Pirozi, H. S. Duarte Martino and F. A. Ribeiro de Barros, *LWT*, 2020, **129**, 109597.
2. S. Mahalaxmi, P. Himashree, B. Malini and C. K. Sunil, *Food Chem. Adv.*, 2022, **1**, 100147.
3. A. G. A. Sá, Y. M. F. Moreno and B. A. M. Carciofi, *Trends Food Sci. Technol.*, 2020, **97**, 170–184.
4. S. P. Bangar, S. Suri, S. Malakar, N. Sharma and W. S. Whiteside, *J. Food Process. Preserv.*, 2022, **46**, e17042.
5. S. Suri, K. S. Balasaheb, D. K. Yadav, S. Malakar, P. Choudhary, A. Mohapatra and P. Dhurve, *Food Biosci.*, 2024, **57**, 103434.
6. N. Sharma, J. K. Sahu, V. Bansal, O. J. Esua, S. Rana, A. Bhardwaj, S. Punia Bangar and A. A. Adedeji, *Food Res. Int.*, 2023, **164**, 112310.
7. T. Vinutha, D. Kumar, N. Bansal, V. Krishnan, S. Goswami, R. R. Kumar, A. Kundu, V. Poondia, S. G. Rudra, V. Muthusamy, G. Rama Prashat, P. Venkatesh, S. Kumari, P. Jaiswal, A. Singh, A. Sachdev, S. P. Singh, T. Satyavathi, S. V. Ramesh and S. Praveen, *Int. J. Biol. Macromol.*, 2022, **195**, 207–216.

8. N. Navyashree, A. Singh Sengar, C. K. Sunil and N. Venkatachalapathy, *Food Chem.*, 2022, **374**, 131665.
9. M. V. Rao, C. K. Sunil and N. Venkatachalapathy, *J. Cereal Sci.*, 2023, **114**, 103774.
10. P. Prashanth, T. J. Joshi, S. M. Singh and P. S. Rao, *J. Food Meas. Charact.*, 2025, **19**, 1064–1078.
11. R. Iralepatil, R. S. Priyadharshini, D. Srenuja, V. Hema, V. R. Sinija and R. Vidyalakshmi, *Food Nutr. Chem.*, 2024, **2**, 130.
12. L. Cheng, S. Qu, Y. Yun, Y. Ren, F. Guo, Y. Zhang and G. Wang, *Food Sci. Nutr.*, 2024, **12**, 9007–9024.
13. B. Nazari, M. A. Mohammadifar, S. Shojaee-Aliabadi, E. Feizollahi and L. Mirmoghtadaie, *Ultrason. Sonochem.*, 2018, **41**, 382–388.
14. F. Jhan, A. Gani, N. Noor and A. Shah, *ACS Food Sci. Technol.*, 2021, **1**, 1418–1427.
15. F. Akharume and A. Adedeji, *J. Food Meas. Charact.*, 2023, **17**, 178–186.
16. N. Sharma, J. K. Sahu, A. Choudhary, M. Meenu and V. Bansal, *Food Hydrocolloids*, 2023, **134**, 108083.
17. Z. Peng, F. Wang, L. Yu, B. Jiang, J. Cao, Z. Sun and J. Cheng, *Ultrason. Sonochem.*, 2024, **110**, 107044.
18. T. J. Rao, A. Mounika and A. Shanmugam, *Ind. Eng. Chem. Res.*, 2024, **63**, 5380–5396.
19. E. Shylla, C. K. Sunil, A. Rawson and N. Venkatachalapathy, *J. Food Sci.*, 2024, **89**, 8999–9014.
20. S. Saxena, H. Vasudevan, S. Saini and S. Sasmal, *ACS Food Sci. Technol.*, 2023, **3**, 600–607.
21. S. Kamiloglu, M. Tomas, G. Ozkan, T. Ozdal and E. Capanoglu, *Curr. Opin. Food Sci.*, 2024, **57**, 101148.
22. N. Mustač, B. Voučko, D. Novotni, S. Drakula, A. Gudelj, F. Dujmić and D. Ćurić, *Food Technol. Biotechnol.*, 2019, **57**, 183–190.
23. Y. Ma, Y. Xu, W. Guo, Y. Shi, Y. Wu and Z. Chen, *Ultrason. Sonochem.*, 2024, **111**, 107153.
24. F. Jhan, A. Gani, N. Noor, B. Ahmad Malla and B. A. Ashwar, *Ultrason. Sonochem.*, 2022, **86**, 106006.
25. L. Meena, B. Malini, T. Byresh, C. Sunil, A. Rawson and N. Venkatachalapathy, *Food Chem. Adv.*, 2024, **4**, 100631.
26. S. D. Balendran, R. G. T. Kalaivendan and R. Theagarajan, *J. Food Process Eng.*, 2024, **47**, e14409.
27. R. Lokeswari, P. S. Sharanyakanth, S. Jaspin and R. Mahendran, *IEEE Trans. Plasma Sci.*, 2021, **49**, 1745–1751.
28. S. Jaddu, R. C. Pradhan and M. Dwivedi, *Innovative Food Sci. Emerging Technol.*, 2022, **77**, 102957.
29. V. Sharma, M. S. Tomar, S. Sahoo and R. C. Pradhan, *J. Food Process Eng.*, 2024, **47**, e14667.
30. A. Sarkar, T. Niranjan, G. Patel, A. Kheto, B. K. Tiwari and M. Dwivedi, *J. Food Process Eng.*, 2023, **46**, e14317.
31. V. Monica, R. Anbarasan and R. Mahendran, *Plasma Chem. Plasma Process.*, 2023, **43**, 1843–1861.

32. V. Monica, R. Anbarasan and R. Mahendran, *Food Bioprocess Technol.*, 2025, **18**, 850–867.
33. C. Charu, S. Vignesh, D. V. Chidanand, R. Mahendran and N. Baskaran, *Food Humanity*, 2024, **2**, 100238.
34. A. J. Sherin, C. K. Sunil, D. V. Chidanand and N. Venkatachalapathy, *Int. J. Biol. Macromol.*, 2024, **261**, 129919.
35. N. Sharma, S. K. Goyal, T. Alam, S. Fatma, A. Chaoruangrit and K. Niranjan, *J. Cereal Sci.*, 2018, **83**, 162–170.
36. A. Angioloni and C. Collar, *J. Cereal Sci.*, 2012, **56**, 713–719.
37. Y. Zhang, X. Zhang, Z. Zhang, Z. Chen, X. Jing and X. Wang, *LWT*, 2022, **154**, 112755.
38. Y. Zhang, X. Jing, Z. Chen and X. Wang, *LWT*, 2022, **164**, 113654.
39. Y. Zhang, X. Jing, Z. Chen and X. Wang, *Cereal Chem.*, 2023, **100**, 360–370.
40. M. S. Mohd, N. H. Shafie, F. Ahmad and S. R. Shafie, *J. Med. Health Sci.*, 2023, 69–76.
41. S. Sharma, R. Sharma and B. Singh, *Int. J. Food Sci. Technol.*, 2022, **57**, 4744–4753.
42. K. K. Kabui and K. A. Athmaselvi, *J. Cereal Sci.*, 2024, **120**, 104049.
43. P. Prashanth, T. Jayasree Joshi, S. M. Singh and P. S. Rao, *Grain Oil Sci. Technol.*, 2024, **7**, 87–95.

CHAPTER 10

Uses of Millet Protein in Food and Non-food Sectors

SAPNA LANGYAN,*[a] NAND LAL MEENA,[a] SUNITI YADAV,[b] RENU SINGH[c] AND SAKSHI GOYAL[a]

[a] ICAR-National Bureau of Plant Genetic Resources, Pusa Campus, New Delhi-12, India; [b] IIMT University, Meerut, UP, India; [c] ICAR-Indian Agricultural Research Institute, Pusa Campus, New Delhi-12, India
*Emails: sapna82@icar.org.in; singh.sapna06@gmail.com

10.1 Introduction

10.1.1 Overview of Millet as a Crop

Millets are considered as superfoods and are an extremely varied ancestral crops that can be grown in a variety of arid and unfavorable climates. They have the potential to improve food security and, because of their high nutritional content, can help maintain a balanced diet. The term "millets," often known as "nutri-cereals," encompasses a diverse group of dryland cereals, including fonio, teff, Job's tears, foxtail, barnyard, little, kodo, finger, sorghum, proso, and pearl millet. Native to many regions of the world, millets are an essential component of traditional culinary traditions, particularly in India and sub-Saharan Africa, where they are produced in the greatest quantities. Millets can survive extreme environmental conditions like nutrient-depleted soils and erratic weather.[1]

Among cereal crops, millets have a low global carbon footprint. Wheat has the highest global warming potential among major cereals, emitting approximately 4 tons of CO_2 equivalent per hectare, followed by rice and maize, which each emit around 3.4 tons of CO_2 equivalent per hectare. Additionally,

Food Chemistry, Function and Analysis No. 48
Millet Proteins: Nutritional, Structural and Functional Attributes
Edited by Shweta Suri, Anuradha Dutta and Tarun Belwal

Published by the Royal Society of Chemistry, www.rsc.org

the carbon equivalent emissions for maize, rice, and wheat are significantly high—1000, 956, and 935 kg C ha^{-1}, respectively. Despite their higher emissions, these crops are widely cultivated and serve as primary food sources worldwide. In contrast, minor cereals like sorghum and millet have comparatively lower carbon footprints.[2] Various millet varieties are cultivated across different regions of the world.

Globally, pearl millet is the sixth highest produced crop, after maize, wheat, rice, barley, and sorghum. Pearl millet is a staple food for millions in Africa and Asia. The protein content ranges from 9.45 to 20.80 g per 100 g for genotypes, and 4.5 to 17.10 g per 100 g for varieties.[3]

In India, finger millet, also called *Eleusine coracana* L., ranks sixth among the major cereal grains, after wheat, rice, maize, sorghum, and bajra, and is cultivated in both Africa and India.[4] One study reported that the calcium level was consistently high in finger millet and teff regardless of varieties at 331.29 ± 10 mg per 100 g and 183.41 ± 29 mg per 100 g, respectively. Iron content was highest for finger millet at 12.21 ± 13.69 mg per 100 g, followed by teff at 11.09 ± 8.35 mg per 100 g.[3]

Proso millet originated in Central and Eastern Asia, and later spread to India, Russia, the Middle East and Europe. One of the fastest-growing cereal crops, proso millet is ideal for areas with short growing seasons since it matures in about 60 to 100 days. Protein content was highest in proso millet (12.42 ± 1.99 g per 100 g), followed by Job's tears (12.66 g per 100 g) and barnyard millet (12.05 ± 1.77 g per 100 g). Calcium content in proso millet showed wide variability (9.11–235.4 mg per 100 g), with an average of (107.7 ± 94.17 mg per 100 g) across evaluated genotypes.[3]

Foxtail millet seeds have been discovered across various regions of Europe, the Middle East, and Eastern and Central Asia, dating back to the Bronze and Neolithic periods. Known for its high photosynthetic efficiency and rapid ripening, foxtail millet is well-suited as a catch crop. It is also rich in nutrients and exhibits strong resistance to diseases and insects. Notably, this crop thrives with just a single pre-sowing rainfall. The average fat and protein content of foxtail millet genotypes has been reported as 3.52 ± 0.29 g per 100 g and 12.53 ± 2.19 g per 100g, respectively.[5]

Barnyard millet comes in two cultivated species: *Echniochloa utilis* and *E. frumentacea*. *E. frumentacea* is also known as Indian barnyard millet, sawa millet, and billion-dollar grass, and *Echniochloa utilis* is also known as Japanese barnyard millet. Grown extensively in India, China, Japan, Pakistan, Africa, and Nepal, this kind of millet is regarded as a minor grain. In India, barnyard millet comes in second to finger millet in terms of production (87 000 tons annually) and productivity (857 kg ha^{-1}). Barnyard millet cultivars were found to have an exceptionally high total dietary fiber content, ranging from 23.25 to 31.70 g per 100 g. The mineral content of barnyard millet varieties was reported in Korea and India, with variations between 2.28 and 22.98 mg per 100 g for iron, 0.44 and 5.92 mg per 100 g for zinc, and 5.81 and 36.13 mg per 100 g for calcium.[6]

Kodo millet originates from India. Tropical and sub-tropical climates are ideal for kodo millet cultivation. Kodo millet is thought to provide a decent yield and have the best tolerance to drought of the minor millets. Its growth season lasts between 80 and 135 days. The average protein content of kodo millet genotypes reported was 8.00 ± 0.57 g per 100 g, whereas pooled varieties had 8.92 g per 100 g.[3]

10.1.2 Nutritional Profile of Millet Protein

Millet stands out among commonly consumed staple grains due to its high protein content, making it an excellent source of essential amino acids. The protein derived from millet grains offers significant nutritional and functional benefits. Millets are a moderate to good source of protein, providing about 9%–12% protein content by weight, depending on the type of millet.[7] Millet protein contains higher levels of both essential and non-essential amino acids, including leucine, phenylalanine, proline, serine, tyrosine, aspartic acid, and glutamic acid.[8] However, deficiencies or low levels of certain amino acids—such as lysine, threonine, methionine, cysteine, and tryptophan—particularly in the prolamin fraction of millet flours, can impact the effectiveness of protein utilization for nutrition and other purposes.[9] Taylor and Taylor[10] confirmed that pearl millet and finger millet are rich in lysine, owing to the presence of albumin, glutelin, or globulin fractions, which are high in lysine. Additionally, the high germ-to-endosperm ratio in pearl millet may also play a role in its elevated lysine content. Millet proteins are seen as a nutritious substitute for traditional protein sources like nuts, meat, and dairy. They offer a promising solution to enhance balanced diets and improve the nutritional health of populations, especially those affected by malnutrition.[11] Millet protein provides numerous health benefits when complemented with legumes. When specific millets are combined with legumes in a 3 : 1 ratio, the nutritional value of the food is significantly improved, offering a balanced amino acid profile, better protein digestibility, and high levels of iron (7.58 mg) and zinc (4.96 mg) per 100 g of pearl millet, along with calcium (400.57 mg) per 100 g of finger millet.[12]

10.1.2.1 Millet Proteins: Isolates, Hydrolysates, Extracts, and Additives

Millet proteins, including isolates, hydrolysates, extracts, and additives, are gaining attention for their functional and bioactive properties, making them suitable for diverse applications in the food, pharmaceutical, and nutraceutical industries.[13] Millet protein isolates, obtained through alkaline solubilization, isoelectric precipitation, and ultrafiltration, have high protein content (>90%) and exhibit excellent emulsification, water-holding, and foaming properties, enhancing their use in protein-fortified foods and plant-based alternatives.[14] Hydrolysates, derived *via* enzymatic hydrolysis using proteolytic enzymes such as pepsin and alcalase, generate bioactive peptides with antioxidant, antihypertensive, and antimicrobial properties, contributing to their growing use in functional foods and medical nutrition. Millet protein extracts, obtained through aqueous and supercritical fluid extraction, retain essential amino acids, globulins, and albumins, playing a role in nutraceutical formulations and biodegradable packaging.[15] These proteins have also demonstrated superior digestibility and hypoallergenic properties compared to other cereal proteins, making them an ideal alternative for specialized dietary needs.[16,17]

As food additives, millet proteins function as emulsifiers, gelling agents, and foaming agents, improving texture and stability in bakery, dairy alternatives, and plant-based meat products.[18] Their role in enhancing the nutritional profile of gluten-free and allergen-free foods has positioned them as a sustainable alternative to soy and whey proteins.[19] The future of millet proteins depends on advancements in extraction and processing technologies, which can optimize yield, functionality, and large-scale industrial applications.[20] Further research on protein modification and bioactive peptide characterization will enhance their role in food security, personalized nutrition, and biodegradable biomaterials.[14] With increasing consumer preference for plant-based proteins, millet-derived proteins hold significant potential in addressing global protein demand while promoting sustainable agriculture and human health.

10.1.3 Millet Protein as a Gluten Replacement

Millet protein is emerging as a viable gluten replacement due to its unique structural and functional properties, making it suitable for gluten-free formulations in bakery and processed food industries.[18] Unlike wheat gluten, which provides elasticity and structure to baked goods, millet proteins, primarily composed of albumins, globulins, and prolamins, exhibit good water-binding and emulsification properties, contributing to dough stability and texture enhancement.[14] Enzymatic and thermal modifications of millet proteins have been explored to improve their viscoelasticity, mimicking gluten functionality in bread and pastry applications. Additionally, millet protein-based hydrocolloids and protein–polysaccharide complexes have demonstrated improved dough rheology and gas retention, enhancing the quality of gluten-free products.[19] Research suggests that millet proteins can complement other gluten-free alternatives, such as rice and sorghum proteins, to improve the nutritional profile and sensory attributes of gluten-free formulations.[15] Their high digestibility and hypoallergenic nature further position them as a superior alternative for individuals with gluten intolerance and celiac disease.[17]

Beyond bakery applications, millet proteins play a significant role in gluten-free pasta, snacks, and plant-based meat analogues, where they enhance texture and binding properties.[13] The inclusion of millet protein hydrolysates in gluten-free dough formulations has been shown to improve dough extensibility and moisture retention, preventing dryness and crumbliness in finished products.[21] Moreover, millet proteins contribute to the sustainability of gluten-free product development by offering a low-input, drought-resistant alternative to conventional gluten-containing grains, reducing the environmental impact of wheat production.[20] Future research should focus on optimizing millet protein processing techniques, such as enzymatic crosslinking and extrusion, to further enhance its gluten-mimicking properties for large-scale food production.[14] With increasing consumer demand for clean-label and allergen-free food products, millet protein is expected to play a crucial role in advancing the next generation of gluten-free and functional food innovations.

Millet proteins offer a wide array of nutritional and health benefits. The bioactive compounds in millets contribute to various biological activities, including antioxidant, anti-hyperglycemic, anti-cholesterol, and anti-hypertensive effects; improvements in anthropometric measures; and the regulation of gut microbiota composition.[22] Millets are an excellent food choice for managing diabetes. Their hypoglycemic effect is attributed to the presence of low-digestible starch fractions, the ability of phenolic compounds to limit carbohydrate digestion, their role in reducing reactive oxygen species (ROS), promoting probiotic bacteria growth, modulating enzyme activity, and regulating various signaling pathways.[23] In addition to primary nutrients, what makes millets therapeutic is the presence of phenolic compounds, which are the primary form of secondary metabolites.[7] In brief, millet protein is a nutritious and sustainable alternative to traditional protein sources. Rich in essential amino acids, fiber, vitamins, and minerals, it supports overall health. Its gluten-free nature and heart-healthy fats make it suitable for diverse diets, especially for populations facing malnutrition. Incorporating millet into diets can help address nutritional challenges, particularly in developing regions.

10.1.4 The Importance of Millet Protein in the Context of Global Food Security

The demand for millets as food and value-added goods has increased due to a sharp rise in knowledge of their nutritional advantages. Millets comprise approximately 65%–75% carbohydrates, 7%–12% proteins, 2%–5% fat, and 8%–15% fiber. They contain higher levels of essential amino acids compared to conventional cereals, and their prolamin content enhances protein digestibility. While the nutritional benefits of millets have gained recognition through recent knowledge dissemination, their therapeutic potential requires further exploration. Scientific evidence indicates that millet consumption can slow the progression of prediabetes, improve glycemic control, lower body mass index (BMI), and reduce the risk of atherosclerotic cardiovascular disease. The high resistant starch (RS) and slowly digestible starch (SDS) content in millets contribute to lower postprandial glucose and insulin levels.[7] Due to their exceptional nutritional profile, millets are increasingly being incorporated into weaning foods and health-conscious diets, driving market expansion. Millet has gained attention in the context of current issues such as overcrowding, water shortages, climate change, and sustainable food production. This has given farmers, dietitians, and producers of food and feed the chance to conduct studies to learn more about the nutritional and functional properties of millet grains.[24] Globally, the current agricultural systems are under tremendous strain. Millets may practically offer much-needed solutions to the present problems with global food and nutritional security.[25] Indian millets have gained recognition as a possible solution to global food security issues, although the nation has been struggling with food-related issues since the 1970s and has sought to maximize resource usage while increasing food production. The historic Green Revolution was one significant turning point that led to self-sufficiency in several food areas. Compared to frequently consumed staple grains like rice and wheat, millets provide superior nutritional, environmental, and financial advantages.[25,26]

The Food and Agriculture Organization of the United Nations (FAO) reports that millets are produced in more than 100 countries worldwide, with the top three producers being Nigeria, Niger, and India. In 2021, 30.1 million tons of millet were produced globally, with India accounting for 41% of the total production. India is the world's greatest producer of millet, accounting for 83% of Asian and 26.6% of worldwide millet production areas. About 12.7 million acres of land are utilized in India to produce 16.9 million tons of millets, which accounts for about 6% of the nation's total food grain output. In recognition of millet's nutritional benefits, the FAO declared 2023 to be the "International Year of Millets," with the goal of boosting millet production and productivity globally.[27]

Millets have great potential to provide food security worldwide. Millets show an amazing ability to sprout and flourish, even in conditions with high salt concentrations and despite the challenges presented by erratic weather patterns and continuous climatic swings. By exhibiting stimulatory effects on germination and development under seawater treatments, they have demonstrated their resistance to salt stress. Millets are desirable for intensive cropping systems and as a companion or intercrop with slower-maturing plants because of their hardiness, quick maturity, and possibility for year-round farming.[28] These qualities will enable millets to contribute eventually to food security in regions with difficult climates by taking the place of staples like rice and wheat. Nonetheless, millet is regarded as an underappreciated agro-biodiversity resource, despite its rich nutritional profile, climate resilience, and significant potential to contribute to food and nutritional security.

10.2 Millet Protein in Food Applications

10.2.1 Functional Properties of Millet Protein

Millet protein possesses unique functional properties that make it suitable for various food applications. It exhibits excellent water absorption, emulsification, and foaming properties, which contribute to texture enhancement and stability in food formulations. Additionally, millet protein has high digestibility and a well-balanced amino acid profile, making it a valuable alternative to traditional plant-based proteins.[14] These functional attributes allow millet protein to enhance the mouthfeel, structure, and overall quality of different food products.

10.2.2 Millet Protein in Bakery Products

10.2.2.1 Bread and Pastry Fortification

Millet protein can be incorporated into bread and pastry formulations to enhance their protein content while maintaining desirable texture and taste. Research indicates that millet flour and protein isolates improve the nutritional value of baked goods, making them ideal for consumers seeking protein-rich options.[21] When combined with other plant-based proteins, millet enhances the protein profile of baked products while retaining softness and shelf stability.

10.2.2.2 Gluten-free Products

Given its naturally gluten-free nature, millet protein is an excellent ingredient for developing gluten-free bakery products. It provides the necessary structure and texture to gluten-free bread, cakes, and cookies, offering a nutritious option for individuals with celiac disease or gluten intolerance. Studies have shown that millet flour can improve the texture and sensory attributes of gluten-free baked goods while increasing their protein content.[19]

10.2.3 Millet Protein in Dairy Alternatives

10.2.3.1 Plant-based Milks and Yogurts

Millet protein is increasingly being used as a key ingredient in plant-based dairy alternatives. Its emulsifying properties help stabilize formulations, while its amino acid profile enhances the nutritional benefits of plant-based milk and yogurt products. Millet-based dairy alternatives appeal to consumers looking for lactose-free, allergen-free, and non-GMO options.[18]

10.2.3.2 Cheese and Creamer Substitutes

Millet protein is also a promising candidate for plant-based cheese and creamer substitutes. Its ability to form stable emulsions and provide a creamy texture makes it ideal for non-dairy cheese products, such as spreads and hard cheeses. Innovations in fermentation and processing techniques have further enhanced the taste and functionality of millet-based cheese substitutes, increasing their acceptance in the growing plant-based dairy market.[20]

10.2.4 Millet Protein in Meat Analogues

10.2.4.1 Plant-based Meat Products

Millet protein is gaining recognition as an ingredient in plant-based meat alternatives due to its favorable texture and binding properties. When combined with other plant proteins such as pea and soy, millet protein enhances the firmness, chewiness, and overall mouthfeel of meat analogues like burgers, sausages, and nuggets.[17] Its ability to retain moisture and provide a fibrous texture makes it an attractive option for food manufacturers developing high-quality plant-based meats.

10.2.4.2 Textured Vegetable Protein (TVP)

Millet protein can also be processed into textured vegetable protein (TVP) for use in a variety of plant-based meat applications. TVP derived from millet protein offers a fibrous structure that closely mimics animal protein, providing a nutritious, plant-based option for consumers seeking sustainable and high-protein alternatives.[14] The demand for TVP-based products continues to grow as more consumers adopt plant-based diets for health and environmental reasons.

10.2.5 Millet Protein in Snacks and Cereals

10.2.5.1 Protein-enriched Snack Bars

Millet protein serves as an excellent ingredient in protein-enriched snack bars. It enhances the protein content while maintaining a desirable texture and taste. These snack bars cater to fitness enthusiasts, athletes, and individuals seeking convenient and nutritious snack options.[21] The ability of millet protein to blend seamlessly with other ingredients such as nuts, seeds, and fruit makes it an ideal addition to protein bars.

10.2.5.2 Ready-to-eat Cereals

Millet protein is also being incorporated into ready-to-eat cereals to boost their protein content while maintaining a crunchy texture. These cereals cater to health-conscious consumers seeking high-protein breakfast options that align with their dietary goals.[19] Additionally, millet-based cereals provide sustained energy release, making them a preferred choice for active individuals.

10.2.6 Millet Protein in Beverages

10.2.6.1 Protein Shakes and Smoothies

Millet protein is becoming a popular ingredient in protein shakes and smoothies as a plant-based alternative to whey and soy protein. It offers high digestibility and a neutral taste, making it an excellent option for fortifying beverages without altering their flavor profiles. Consumers looking for allergen-free and easily digestible protein sources find millet protein an appealing choice for their dietary needs.[18]

10.2.6.2 Functional Beverages

Functional beverages fortified with millet protein are gaining traction in the health and wellness industry. These include fortified teas, electrolyte drinks, and protein-enriched coffee blends, providing consumers with an easy way to incorporate plant-based protein into their daily routines. Millet protein's compatibility with various beverage formulations makes it a versatile ingredient for the expanding functional beverage market.[20]

The growing recognition of millet protein across multiple food sectors underscores its potential as a sustainable and nutritious alternative to traditional protein sources. As research and technological advancements continue to improve its extraction and application processes, millet protein is expected to play an increasingly vital role in shaping the future of plant-based food innovations.

A brief overview of millet protein applications in the food sector is given in the Table 10.1.

Table 10.1 Applications of millet protein in the food sector.

Application	Example products	Description	Ref.
Functional food ingredients	Protein bars, energy drinks, fortified cereals	Millet protein hydrolysates offer antioxidant, antihypertensive and other bioactive properties, making them ideal for functional foods.	84
Protein isolate production	Protein powders, dairy alternatives	Used to increase protein content in plant-based formulations.	79
Protein fortification	Millet-based flour, porridge mixes	Used to enhance the nutritional profile of staple foods, especially in undernourished populations.	53
Nutraceuticals and supplements	Protein capsules, health drinks	Rich in essential amino acids and bioavailable nutrients for dietary supplements.	71
Synbiotic foods	Paneer with probiotics, probiotic yogurt	Millet protein isolate aids in microencapsulation of *Lactiplantibacillus plantarum*, improving probiotic survival.	75
Emulsion and gelling agents	Whipped toppings, dairy-free spreads	Millet proteins modified *via* ultrasonication can improve foaming, emulsifying, and water-holding capacity.	83
Meat analogs/plant-based protein	Veggie patties, plant-based meat strips	Good digestibility and functionality for clean-label meat substitutes.	78

10.3 Millet Protein in Non-food Applications

Millet proteins are increasingly gaining attention for their potential in various non-food applications because of their functional properties and biodegradability.[29,30] In the pharmaceutical industry, millet protein-based biomaterials are being investigated for drug delivery systems and wound healing because of their biocompatibility and film-forming abilities.[31,32] In the textile sector, millet protein can be used as a sustainable fabric coating to enhance fabric strength and antimicrobial properties.[16] Millet protein-based bioplastics are emerging as eco-friendly alternatives to petroleum-based plastics, as they offer biodegradable packaging solutions.[19] Through recent advancements in science and protein engineering, millet proteins have great potential for diverse industrial applications beyond their traditional role in nutrition.[33]

10.3.1 Millet Protein in Bioplastics

Ecological concerns stemming from environmental pollution caused by petroleum-based plastics, due to their complex structure, are a significant cause of concern worldwide. Millet protein has emerged as a promising biopolymer for the development of sustainable bioplastics due to its biodegradability, film-forming ability, and mechanical strength.[34] As environmental concerns over petroleum-based plastics grow, researchers are increasingly exploring natural alternatives, and millet protein is a viable option owing to its renewable nature and functional properties. The key advantage of millet protein-based bioplastics is their eco-friendly composition.[35–37] Unlike conventional plastics, which persist in the environment for hundreds of years, millet protein bioplastics degrade naturally, reducing plastic waste accumulation. Moreover, they have potential applications in food packaging, agricultural films, and disposable cutlery, where biodegradability is a crucial factor. The processing of millet protein into bioplastics involves techniques such as extrusion, casting, and blending with plasticizers, such as glycerol or sorbitol, to improve flexibility and durability.[34] Millet proteins contain hydrophilic amino acids, which enhance their water-absorbing capacity but may reduce their moisture resistance. Therefore, researchers are working on modifying millet proteins through cross-linking, chemical modifications, and blending with other biopolymers, such as starch or polylactic acid (PLA), to enhance their properties.[38] In addition to their environmental benefits, millet protein-based bioplastics can also contribute to value addition in the agricultural sector. Utilizing millet byproducts for bioplastic production can provide an additional revenue stream for farmers and promote a circular economy.[39] Further advancements in processing techniques and functional enhancements will be key to commercializing millet protein bioplastics and replacing synthetic plastics in various industries.

Foxtail millet husk (FMH) can be transformed into a new natural filler for biodegradable polymers with potential applications in the furniture, automotive, home goods, and packaging sectors. FMH differs from other plant husk fillers in that it acts as a nucleating agent, enhancing the crystallinity and structural integrity in polymer matrices. Additionally, FMH offers a method for utilizing undervalued crop residue as reinforcement in thermoplastic composites. The strength provided by natural fillers, based on the cellulose-to-hemicellulose ratio, is vital; a lower ratio is associated with improved mechanical properties, which FMH successfully provides.[40]

10.3.1.1 Sustainable Packaging Solutions

In recent decades, the depletion of natural resources and growing environmental concern has increased the desire to find innovative methods to manufacture environmentally friendly solutions. Sustainable food packaging can be a suitable alternative to reduce agricultural waste from the food industry. These agricultural wastes are inevitable products of the millet processing industry. The byproducts are major sources of cellulose, lignin, pectin, starch, chitin, alginates, and hemicelluloses, which can be processed into food packaging.[41] Bioplastics appear to address the concerns of the food industry regarding many packaging-related issues and appear to play an important role in the sustainability of the food industry, ensuring high quality throughout the production and transportation stages.[42] Packaging materials rich in cellulose, pectin, zein, and starch can protect packaged foods against microorganisms and oxidation.

Ragi appeals to consumers because of its high-quality carbohydrates, essential amino acids, and low-fat content, as well as its mineral and micronutrient composition, which collectively contribute to its low glycemic index.[43] However, ragi grains contain a substantial amount of husk and bran, necessitating their removal before consumption.[44] These lignocellulosic residue materials hold considerable potential for application in the composite manufacturing industry. The composite industry is increasingly exploring the use of renewable materials, such as lignocellulosic materials, as reinforcements for polymer composites. This approach can effectively reduce product costs and enhance the value of lignocellulosic biomass residue. Consequently, the use of Ragi husks as a low-cost lignocellulosic substrate in microbial fermentation processing has garnered attention as a means of waste utilization.[43] This utilization can find applications in food packaging production, including the development of bioplastics and petrochemical plastic packaging.

Kodo millet-based edible cups incorporate guar gum (GG) and *hibiscus rosa sinensis* extract (HB) to enhance their physical, thermal, and functional properties, such as structural integrity, antioxidant activity, and water absorption.[45] Thaker and co-workers revealed that millet proteins, primarily prolamins, exhibit good tensile strength and elasticity, making them suitable for thin-film applications.[46] However, due to their hydrophilic nature, their moisture barrier properties need enhancement. Blending millet protein with other biopolymers, such as starch, cellulose, or polylactic acid (PLA), improves water resistance and mechanical stability.[46,47] However, natural additives, such as essential oils and antioxidants, can be incorporated to enhance antimicrobial properties and extend the shelf life of packaged products.[48] Millet-based packaging boosts the economy, reduces the use of non-renewable resources, and supports sustainability through bioplastics.

10.3.1.2 Biodegradable Films and Coatings

Natural biopolymers found in millets, known for their rich protein and polysaccharide contents, are emerging as promising raw materials for biodegradable films and coatings. Biodegradable polymers from plant-based sources such as millet, which is rich in starch, proteins, and fiber, improve the structural integrity, flexibility, and functionality of biodegradable films.[35] The ratio of amylose to amylopectin determines the crystallinity of starch granules, which can influence their properties and potential applications.[49] Amylose is a linear polymer with glucose units linked through α-1,4 bonds with excellent film forming ability, whereas amylopectin is a highly branched polymer of α-1,4 bonded glucose chains linked with α-1,6 glycosidic branching points.[50] Different millets, such as proso, pearl, finger, and foxtail millets, exhibit differences in the arrangement of the unit chain of amylopectin. The solubility and swelling power vary from 15.9%–18.7% and 14.7%–18.8%, respectively, depending on the leached amylose fraction in the heating process and water holding capacity of the amylopectin fraction.[34] Dhall observed that the solubility of starch is a desirable criterion for coatings of fresh produce, where it should be less soluble to achieve lower water vapor permeability.[51] Researchers have explored physical treatments, including dry heat, physical modification, and heat–moisture treatment, alone or with chemical modifications, to enhance the properties of finger, proso, and pearl millet starches.[52–54] Alteration of finger millet starch with the help of dry heat treatment has a great impact on the morphological, structural, and physicochemical properties of native finger millet starch.[55] Studies revealed that heat–moisture treatment of proso millet starch and pearl millet starch has a great effect on different physical properties like amylose content, pasting properties, mechanical and tensile properties, opacity and crystallinity, transmittance, *etc.* Millet-based films have been enhanced using glycerol as a plasticizer, or cellulose nanocrystals extracted from kudzu vine and clove bud oil, resulting in improved antimicrobial properties.[34] Kim and Kang modified adlay millet with benzalkonium chloride,[56] while Punia *et al.*[57] enhanced proso millet starch with κ-carrageenan for improved antioxidant activity. Pearl millet starch was modified *via* acetylation by adding acetic anhydride to the starch slurry. The resulting acetylated starch films degrade more rapidly and exhibit greater elongation than the native starch films. The use of biopolymers has increased owing to their biodegradability, environmental safety, and lower concern about their disposability. When used as edible films, biodegradable packaging can be used as a source of bioactive compounds for consumers. Millet-based biodegradable films have been extensively explored for food packaging because of their biocompatibility and barrier properties. These films act as protective coatings to extend the shelf life of perishable food products by reducing moisture loss and preventing microbial contamination.[58] Edible coatings derived from millet starch or proteins are also used to preserve fruits and vegetables and maintain their texture and nutritional quality. In addition to the food industry, millet-based films have potential applications in the pharmaceutical and agricultural industries.[34] In agriculture, these films can be utilized for seed coatings to enhance germination rates while minimizing the impact of synthetic materials on soil health.

10.3.2 Millet Protein in Cosmetics and Personal Care Products

Millet-based cosmetics are becoming increasingly popular because of the collagen-boosting and antioxidant-rich properties of millet extract, which provide effective skin protection. Millet is rich in amino acids, such as proline, glycine, and serine, which are directly involved in the creation and maintenance of collagen.[59] When incorporated into skincare, millet provides these amino acids, which support the skin's ability to produce collagen, helping the skin to stay firm, plump, and youthful.[60] Millets, rich in nutrients and high in fiber, provide essential vitamins (B1, B2, B3, B6, and folic acid) and minerals (iron, magnesium, phosphorus, potassium, zinc, and calcium).[61] Millets support heart health by lowering cholesterol levels and regulating blood pressure (BP). Their low glycemic index aids in diabetes management, and antioxidants combat oxidative stress. High protein levels promote muscle growth and overall development. They are rich in calcium and phosphorus and enhance bone health. Millets boost immunity, aid detoxification, and support skin and hair health.[59]

10.3.2.1 Skin Care: Lotions and Creams

Millet-based lotions and creams benefit skin by promoting collagen production and antioxidant properties. Millet milk and flour obtained from various millet species provide numerous nutritional and functional benefits, making them valuable ingredients in skincare lotions and creams. Collagen protein is the miracle component in millet that helps ensure skin elasticity and hydration and helps prevent aging by reducing the appearance of wrinkles.[62]

Millet have potential as exfoliants in the field of cosmetics. Jowar, mixed with rose petal powder and other natural products to form a loose facial powder and natural facial scrub, has been used in the cosmetics industry.[63]

However, the quantity of millet ingredients has been analyzed to ascertain their quality in the formulation of cosmetic products. The formulated products were assessed for their microbial properties, stability, and organoleptic properties, which are essential for demonstrating their potential use in the cosmetics industry. However, based on this general premise, millet extract has been used in certain formulations in combination with blood to enhance the efficacy of anti-toxin herbal preparations, and millets have also been suggested to exhibit anti-allergic properties These millets are sometimes used in the treatment of allergies and are also used topically in creams and lotions to clarify the skin, prevent skin aging, and reduce blemishes.[64]

10.3.2.2 Hair Care: Shampoos and Conditioners

Millets are increasingly being incorporated into hair care products, such as hair colors, shampoos, and conditioners, because of their rich nutritional profile. Packed with essential amino acids, vitamins, and minerals, millet provides nourishment to the scalp and hair, promoting overall hair health. Their natural properties make them a great ingredient in herbal and organic formulations, catering to the search for chemical-free alternatives.[65]

In hair color formulations, millets, particularly finger millet (ragi), contain natural pigments and bioactive compounds that help enhance hair color and prevent premature graying. Ayurvedic and herbal hair dyes incorporate millet extracts alongside henna and other plant-based ingredients to create a gentle yet effective coloring solution. Additionally, the antioxidants present in millets help combat oxidative stress, which is a key factor in hair aging and graying.[66] Therefore, millet-based shampoos are becoming increasingly popular for their mild yet effective cleansing properties. Free from harsh sulfates, they clean the scalp without stripping its natural oils.[67] Rich in iron, zinc, and magnesium, millets strengthen hair follicles, reducing hair fall and promoting healthy growth. Moreover, their antifungal properties make them beneficial in controlling dandruff and maintaining scalp health.

In conditioners, millet proteins act as natural moisturizers, helping to hydrate and soften the hair. They improve hair texture by strengthening the strands, reducing frizz, and repairing damage caused by heat and environmental stress. The presence of vitamin B and silica in millets enhances hair elasticity, making it more resilient to breakage. Fermented millet water is also used in scalp treatments, further nourishing the hair and improving overall hair quality. Millets have become a valuable ingredient in natural hair care products, offering a gentle and effective solution for maintaining strong, healthy, and vibrant hair in shampoos, conditioners, and hair colors.[68]

10.3.3 Millet Protein in Pharmaceuticals

Millets, a group of small-seeded grains, have been a staple in many traditional diets for centuries. While their nutritional benefits are widely recognized, recent research has revealed their potential applications in the pharmaceutical industry. Millet proteins, in particular, have attracted interest due to their bioactive properties, high digestibility, and functional benefits.[35] Their role in drug delivery, nutraceuticals, and treatments makes them a valuable natural resource in modern medicine.

Millet proteins have emerged as a promising natural ingredient in the pharmaceutical industry. Their bioactive properties, potential for drug delivery, antimicrobial effects, and metabolic benefits make them valuable for both preventive and therapeutic healthcare solutions. As research continues to explore new applications, millet-based pharmaceuticals are expected to play a significant role in the development of natural and sustainable medicines. With their ability to support human health while promoting environmental sustainability, millet proteins represent the future of plant-based pharmaceuticals.[69]

10.3.3.1 Drug Delivery Systems

The pharmaceutical industry is constantly searching for biocompatible and biodegradable materials for drug delivery. Millet-derived proteins offer a promising solution due to their ability to form nanoparticles, hydrogels, and biofilms that can serve as carriers for targeted drug delivery. These protein-based delivery systems ensure controlled release and enhanced bioavailability of therapeutic compounds, reducing side effects and improving patient outcomes. The use of millet proteins in drug delivery is an eco-friendly alternative to synthetic polymers, aligning with the growing trend of sustainable pharmaceuticals.[37]

10.3.3.2 Nutraceutical Formulations

Millet proteins possess outstanding nutraceutical qualities, making them highly valuable to the industry due to their rich nutritional profile and potent bioactive properties. Nutraceuticals, which bridge the gap between food and medicine, offer health benefits beyond basic nutrition. Millets, being an excellent source of essential amino acids, peptides, antioxidants, and dietary fiber, are emerging as a key ingredient in functional foods and dietary supplements. Their incorporation into nutraceutical formulations supports various health benefits, including improved metabolism, cardiovascular health, and immune function. Millet proteins have high digestibility and a balanced amino acid profile, which makes them an excellent plant-based alternative to animal proteins.[12] Unlike many other cereals, millets are gluten free, making them suitable for individuals with gluten intolerance and celiac disease. The presence of bioactive peptides in millet proteins contributes to their antioxidant, anti-inflammatory, and antimicrobial properties. These attributes help reduce oxidative stress, which is linked to chronic diseases such as diabetes, cardiovascular disorders, and neurodegenerative conditions.[30]

In nutraceutical formulations, millet proteins can enhance satiety and regulate blood sugar levels, making them ideal for weight management and diabetes control. Additionally, fermented millet extracts are utilized in probiotics to improve gut health and digestion. The presence of prebiotic fibers in millet proteins further enhances the growth of beneficial gut bacteria, supporting overall digestive well-being.[12] Their rich leucine content makes them beneficial for athletes and individuals looking to improve muscle mass and strength. The formulations containing millet protein hydrolysates are widely used in sports nutrition products to promote endurance and post-exercise recovery. Moreover, their natural ability to modulate cholesterol levels and improve lipid metabolism contributes to heart health, making them a valuable ingredient in cardioprotective nutraceuticals.[70]

With increasing consumers, millet proteins have emerged as a promising ingredient in nutraceutical formulations with multifunctional health benefits. Their hypoallergenic and environmentally friendly nature makes them an ideal choice for the future of functional foods and dietary supplements.[71] As research continues, millet-based nutraceuticals are expected to play a vital role in promoting long-term health and disease prevention.

10.3.4 Millet Protein in Textiles

The textile industry is continuously exploring eco-friendly and biodegradable materials to reduce environmental impact, and millet proteins have emerged as a promising resource in sustainable textile production. Derived from various millet grains, such as pearl millet, finger millet, and foxtail millet, these proteins offer unique properties that can be utilized in textile manufacturing. In the textile industry, millet protein can be applied to the development of protein-based fibers, similar to silk and wool proteins, offering softness, durability, and moisture-wicking properties. The incorporation of millet protein in textiles also improves the fabric's ability to retain dye, making it suitable for vibrant and long-lasting colors.[72] Millet protein coatings can enhance the antimicrobial and UV-resistant properties of fabrics, making them ideal for sportswear, medical textiles, and functional clothing. Researchers are also exploring millet protein films and nanofiber applications for creating advanced textiles with improved tensile strength and durability. As research progresses, millet-based textile materials could revolutionize the fashion industry by offering biodegradable, functional, and skin-friendly fabric options.[73]

10.3.4.1 Fiber Production and Blends

The demand for sustainable and eco-friendly fibers in the textile industry has led to the exploration of plant-based alternatives, and millets have emerged as a promising resource. Millets, such as pearl millet, foxtail millet, and finger millet, contain proteins and cellulose that can be processed into fibers for textile applications. The utilization of millet-based fibers offers an environmentally friendly alternative to synthetic fibers while maintaining strength, durability, and comfort.[72]

In fiber production, millet can be applied to the development of bio-based textile fibers. Millet husks and stalks, which are rich in cellulose, can be processed into biodegradable fibers, reducing the dependency on conventional cotton and synthetic materials. These fibers are lightweight, breathable, and highly absorbent, which makes them ideal for sustainable fabric production. Additionally, millet-derived protein fibers can be blended with other natural fibers, such as cotton, hemp, and bamboo, to enhance texture, softness, and durability.[72] Blending millet fibers with other plant-based or recycled materials enhances the overall fabric properties, while maintaining sustainability. Millet fiber blends improve moisture-wicking capabilities, providing comfort in activewear and breathable summer fabrics.[74] Moreover, the inclusion of millet fibers in textile blends reduces water consumption during fabric processing, making them an eco-friendly choice for the fashion industry.

The integration of millet-based fibers into mainstream textile manufacturing could revolutionize the industry, offering a balance between sustainability, durability, and innovation. The use of millet fibers in blends not only supports environmental conservation but also paves the way for the future of sustainable textiles.

10.3.4.2 Textile Coatings and Finishes

With the increasing global demand for eco-friendly and biodegradable materials, millet is emerging as a promising resource for textile coatings and finishes. Rich in proteins, polysaccharides, and bioactive compounds, millet-based extracts and formulations offer natural alternatives to synthetic coatings used in the textile industry. These coatings enhance fabric properties, providing improved durability, antimicrobial protection, water resistance, and UV protection while maintaining sustainability.[75] Millet protein-based coatings are particularly beneficial in creating functional textiles. When applied to fabrics, millet-derived proteins form a protective layer that enhances moisture-wicking, elasticity, and softness. Moreover, millet-based finishes improve dye absorption, allowing for vibrant and long-lasting colors while reducing the need for chemical dye fixatives.[76] Musere reported that textiles treated with millet protein extracts inhibited bacterial growth, making them suitable for hospital linens, sportswear, and everyday clothing. These natural antimicrobial finishes help reduce odor and extend fabric life without relying on synthetic chemicals that may cause skin irritation.[77]

Additionally, millet-based formulations have contributed to the development of water-resistant and UV-protective coatings. When combined with other natural polymers, millet extracts create breathable hydrophobic layers that enhance the water-repellent properties of outdoor and performance fabrics. The integration of millet-based coatings and finishes into textile production represents a significant step toward sustainable fashion and eco-friendly textile processing. Millet-derived coatings are expected to play a crucial role in the development of innovative and sustainable textile solutions.[78]

10.4 Technological Advances in Millet Protein Processing

Millet proteins have gained significant attention in recent years due to their nutritional benefits, functional properties, and potential applications in food, pharmaceuticals, and textiles. However, the efficient extraction, modification, and application of millet proteins require advanced processing technologies. Innovations in millet protein processing have led to improved protein yield, enhanced functionality, and expanded industrial applications.

10.4.1 Extraction and Isolation Techniques

Interest in millet proteins has increased substantially owing to their nutritional value, functional properties, and potential applications in food, pharmaceuticals, and biomaterials. However, efficient extraction and isolation techniques are essential to maximize the protein yield and preserve bioactivity. Various methods, including conventional and advanced technologies, have been developed to extract and isolate millet proteins while maintaining their structural integrity and functional properties.

Traditional methods of protein extraction from millet, such as alkaline and acid extraction, often lead to protein denaturation and reduced functionality.[33,35] Recent advancements in enzyme-assisted extraction (EAE) and ultrasound-assisted extraction (UAE) have improved protein recovery, while maintaining the structural integrity of millet proteins.[79] These techniques enhance the solubility and bioavailability of millet proteins, making them suitable for various applications. Additionally, supercritical fluid extraction (SFE) is being explored for extracting high-purity proteins without the use of harsh chemicals, ensuring a cleaner and eco-friendly process.[80] High-purity protein isolation and membrane filtration techniques, such as ultrafiltration and microfiltration, have been employed.[81] These methods separate proteins based on their molecular weights, thereby ensuring high-quality protein isolates. Additionally, chromatographic techniques, including ion exchange and gel filtration chromatography, provide selective purification of specific millet protein fractions with enhanced bioactivity.[82] A brief overview of millet protein applications in the non-food sector is given in Table 10.2.

Table 10.2 Applications of millet protein in the non-food sector.

Application	Example products	Description	Ref.
Biodegradable packaging films	Edible films, bio-based wraps	Millet protein blends and starch are used to make eco-friendly, biodegradable packaging with good barrier properties.	18
Cosmetics and skincare	Herbal creams, face masks, hair conditioners	Millet protein hydrolysates are incorporated into cosmetics for their moisturizing and anti-aging properties.	68
Nanotechnology and green synthesis	Silver nanoparticles for wastewater treatment	Millet husk extract is used in the synthesis of eco-friendly nanoparticles with antimicrobial and catalytic properties.	77
Textile processing	Eco-friendly dye binders and functional coatings	Protein-based compounds are studied for their film-forming and binding properties in textiles.	72
Agricultural applications	Biostimulants, slow-release fertilizers	Protein hydrolysates from millets can serve as organic sources of nitrogen and growth-promoting compounds.	63
Pharmaceutical delivery systems	Drug carriers, encapsulation agents	Millet proteins can serve as biocompatible carriers for drug delivery due to their low toxicity and stability.	88
Hair and skin health supplements	Nutraceutical tablets, topical gels	Millet proteins promote collagen synthesis and healthy hair/skin.	66

10.4.2 Enhancement of Functional Properties

To expand the usability of millet proteins, various modification techniques have been developed, for example, (a) hydrolysis using proteolytic enzymes improves digestibility and enhances bioactive peptide production, which is beneficial for nutraceutical applications; and (b) chemical and physical modifications, such as (i) acetylation, (ii) phosphorylation, and (iii) ultrasound treatment, have been used to improve solubility, emulsification, and foaming properties.[83] These modifications make millet proteins suitable for use in food formulations, including plant-based dairy alternatives and gluten-free baking products.

Nutrient-rich millets have profound emerging areas due to their health benefits and industrial applications.[84] However, their functional properties, such as solubility, digestibility, texture, and bioavailability, often require improvement to enhance their usability in food processing, pharmaceuticals, and other applications.[85]

Several other techniques, including physical, chemical, and biological modifications, have been developed to improve the functional properties.

(i) Physical processing methods, such as germination, roasting, extrusion, and high-pressure processing (HPP) have been widely used to enhance the functional properties of millets. Germination increases the bioavailability of essential amino acids, vitamins, and minerals while reducing anti-nutritional factors such as phytic acid and tannins. Roasting and extrusion improved the texture, digestibility, and flavor profiles of millets. HPP enhances protein functionality and starch digestibility while preserving the nutritional quality of millets.

(ii) Chemical processing methods such as acetylation, phosphorylation, and hydrolysis improve millet protein solubility, emulsification, and gelling properties. Enzymatic hydrolysis using proteolytic enzymes enhances protein digestibility and bioactive peptide production, thereby improving the role of millets in nutraceuticals and functional foods. These modifications also contributed to the improved foaming and water-holding capacities of millet-based ingredients.

(iii) Fermentation is an effective biological method for enhancing the functional properties of millets. It increases the availability of essential nutrients, improves flavor, and enhances the probiotic potential of millet-based foods. Fermentation reduces antinutritional factors, making millet more digestible and beneficial for gut health. Additionally, bioprocessing techniques, such as microbial fermentation and enzymatic treatment, can modify millet proteins to enhance their texture and nutritional value.

10.4.3 Novel Processing Methods

Millet has an excellent nutritional profile, bioactive properties, and potential applications in food, nutraceuticals, and biopolymers. However, conventional processing methods often limit functionality, solubility, and digestibility. To overcome these limitations, novel processing technologies have been developed to enhance the extraction efficiency, structural properties, and industrial applications of millet proteins.

(i) Enzyme-assisted extraction (EAE): This technique uses proteolytic enzymes, such as alcalase, pepsin, and trypsin, to break down cell walls and improve protein recovery and bioavailability. This method reduces the processing time and enhances the solubility and digestibility of millet proteins for food and pharmaceutical formulations.

(ii) High-pressure processing (HPP): This is a non-thermal technique in which hydrostatic pressure (100–600 MPa) is applied to modify the structural and functional properties of millet proteins. HPP enhances protein unfolding, increasing water-holding capacity, emulsification, and gelation properties.

(iii) Ultrasound-assisted processing (UAP): High-frequency sound waves are applied to break down protein aggregates and improve protein solubility. This technique enhances the extraction efficiency of millet proteins while preserving their native structures.

(iv) Fermentation and biofunctionalization: Fermentation using probiotic bacteria or fungi enhances the bioactivity of millet proteins by increasing the production of bioactive peptides with antioxidant, antihypertensive, and antimicrobial properties. Fermented millet protein hydrolysates are used as functional foods and nutraceuticals to improve digestibility and health benefits.

(v) Nanotechnology and protein structuring: This method modifies millet proteins for advanced applications in food packaging, drug delivery, and biodegradable films. Millet protein nanoparticles and nanofibers improve the stability of functional foods.

10.5 Sustainability and Economic Implications

10.5.1 Environmental Impact of Millet Protein Production

Millets are recognized for their resilience to harsh environmental conditions and their relatively low ecological footprint compared to major cereal crops such as wheat and rice. Millet protein production presents several environmental advantages, including reduced water consumption, lower greenhouse gas emissions, and improved soil health, making it a sustainable choice in modern agriculture.[86] One of the key environmental benefits of millet cultivation is its minimal water requirement. Millets are known to thrive in arid and semi-arid regions, requiring significantly less irrigation than conventional cereals. This contributes to water conservation efforts and reduces the strain on freshwater resources. Furthermore, millet cultivation requires fewer chemical inputs, such as synthetic fertilizers and pesticides, resulting in lower soil and water contamination.[87] Millet-based protein extraction processes are also being refined to minimize waste and energy consumption. Traditional wet and dry fractionation techniques are being optimized to improve protein yield while reducing carbon emissions associated with industrial processing. Additionally, byproducts from millet protein extraction, such as bran and fiber, can be repurposed into animal feed or biodegradable materials, contributing to a circular economy.[14] Despite these advantages, large-scale millet protein production still faces challenges regarding land use efficiency and scalability. Compared to soybean and other plant-based protein sources, millet protein yield per hectare is relatively lower, necessitating further research into crop improvement and processing efficiency. However, with advancements in sustainable agricultural practices and protein isolation technologies, millet protein has the potential to become a key player in environmentally responsible food and industrial applications.[88] Through sustainable farming, eco-friendly processing, and reduced resource consumption, millet protein production aligns with global efforts to mitigate climate change and promote food security. As scientific advancements continue to improve its efficiency, millet protein could play a crucial role in sustainable agriculture and industrial applications.[14]

10.5.2 Economic Viability and Market Trends of Millet Protein

Millet protein is gaining global recognition as an alternative plant-based protein source due to its nutritional benefits and environmental sustainability. The economic viability and market trends of millet protein are influenced by factors such as production costs, demand in various industries, and consumer preferences for sustainable food options.[89] Millet cultivation is economically viable due to its low input requirements, including minimal water and fertilizer use. Compared to high-input crops like wheat and soy, millet production is less costly and it can be grown in marginal soils, reducing land acquisition costs.[87] However, the processing of millet protein, including extraction and refinement, remains a challenge due to limited technological advancements, impacting overall profitability. The global plant-based protein market is expected to grow significantly, driven by the rising demand for sustainable and allergen-free protein sources. Millet protein is finding applications in food, pharmaceutical, and non-food industries, including biodegradable packaging and textiles. The Asia-Pacific region, particularly India and China, leads millet production, with increasing exports to Europe and North America.[19] Millet protein competes with established plant-based proteins such as soy, pea, and rice proteins. Unlike soy, millet does not pose major allergenic concerns and is free from genetically modified organism (GMO) controversies. Additionally, millet protein's unique amino acid profile and high digestibility make it an attractive option for specialized dietary needs.[14]

10.5.3 Role of Millet Protein in Sustainable Agriculture

Millet protein is emerging as a promising alternative in sustainable agriculture due to its nutritional value, environmental benefits, and economic viability. As global demand for plant-based proteins rises, millet serves as a crucial component in sustainable food systems, offering advantages over conventional protein sources in terms of resource efficiency and adaptability to climate change.[89] Millet is a highly resilient crop that thrives in arid and semi-arid regions, requiring significantly lower water inputs compared to traditional cereal crops such as wheat and rice.[87] Unlike high-input crops, millet cultivation relies on minimal fertilizers and pesticides, thereby reducing the environmental impact associated with chemical runoff and soil degradation. Additionally, millet contributes to biodiversity by serving as a rotation crop, enhancing soil health, and reducing monoculture dependency.[14] The cost-effectiveness of millet farming stems from its ability to grow in marginal soils with limited resources, making it an economically viable option for smallholder farmers. Compared to soy and wheat, which demand intensive agricultural inputs, millet production offers lower operational costs and higher resilience to climate variability.[19] However, despite these advantages, challenges persist in the large-scale processing and refinement of millet protein, limiting its market competitiveness. The global plant-based protein market is expanding rapidly, with millet protein gaining attention in the food, pharmaceutical, and industrial sectors. Applications of millet protein extend beyond human nutrition to biodegradable packaging and eco-friendly textiles, further reinforcing its role in sustainable industries.[20] The Asia-Pacific region, particularly India and China, leads in millet production, with growing export potential to Europe and North America. Millet protein is distinguished by its high digestibility and balanced amino acid profile, making it a viable alternative to traditional plant-based proteins such as soy and pea.[14] Unlike soy, millet is free from major allergens and GMO-related concerns, aligning with consumer preferences for clean-label and allergen-free protein sources. Moreover, its functional properties, including water retention and emulsification, make it suitable for diverse food formulations.[87]

10.6 Challenges and Future Prospects

As millet protein gains global recognition as a sustainable plant-based protein, several challenges must be addressed to enhance its market potential and scalability. These challenges range from technological barriers in protein extraction to regulatory frameworks and future research needs.

10.6.1 Overcoming Technical Barriers in Product Development

One of the main hurdles in the widespread adoption of millet protein is the inefficiency of current extraction and processing techniques. Unlike soy or pea protein, millet protein extraction faces difficulties due to its relatively lower protein content and complex biochemical composition. Traditional wet and dry fractionation methods yield lower protein recovery, impacting the economic feasibility of large-scale production. To overcome these technical barriers, advancements in enzymatic hydrolysis, membrane filtration, and novel extraction technologies, such as ultrasound-assisted or supercritical fluid extraction, are being explored. Additionally, improving the solubility, emulsification, and foaming properties of millet protein will enhance its functionality in food applications, making it more competitive with established plant proteins. Investment in research and infrastructure for millet protein refinement is crucial to scaling up production while maintaining nutritional integrity and sensory qualities.

10.6.2 Regulatory and Safety Considerations

The commercialization of millet protein requires adherence to food safety and regulatory standards. Unlike widely studied plant proteins such as soy and wheat, millet protein lacks well-established global regulatory frameworks, which can hinder its market penetration. Safety assessments regarding potential allergens, anti-nutritional factors, and heavy metal contamination are necessary to ensure compliance with food safety authorities like the FDA and EFSA. Furthermore, achieving certifications such as non-GMO, organic, and allergen-free status will enhance consumer confidence and marketability. Transparent labeling, traceability in supply chains, and standardized protein quality parameters must be developed to facilitate millet protein's entry into regulated markets in North America and Europe.[90]

10.6.3 Future Research Directions

Future research on millet protein should focus on optimizing extraction techniques to improve yield and bioavailability. Investigations into the structural and functional properties of millet proteins under different processing conditions will help develop innovative food applications. Additionally, breeding programs aimed at enhancing the protein content of millet varieties through conventional and biotechnological approaches could significantly improve its economic viability. Research into the environmental benefits of millet farming, such as its role in carbon sequestration and climate resilience, will further strengthen its position in sustainable agriculture. Collaborations between food scientists, agronomists, and industry stakeholders will be crucial in driving innovation and addressing knowledge gaps in millet protein research. The integration of millet protein into functional foods, sports nutrition, and alternative protein markets presents a promising avenue for future growth.

10.7 Conclusion

10.7.1 Summary of Millet Protein Applications

Millet protein has demonstrated potential applications across multiple sectors, primarily in the food industry as a plant-based protein alternative. Its high digestibility, unique amino acid composition, and non-allergenic properties make it suitable for health-conscious consumers seeking sustainable dietary options. Beyond food, millet protein is gaining traction in pharmaceuticals, biodegradable packaging, and eco-friendly textile industries, reinforcing its versatility and sustainability.

10.7.2 Potential Impact on Food and Non-food Sectors

In the food sector, millet protein presents a viable alternative to traditional proteins such as soy and pea, offering clean-label, non-GMO, and allergen-free benefits. Its integration into plant-based meat substitutes, protein supplements, and functional foods aligns with the growing consumer demand for sustainable nutrition. Additionally, in the non-food sector, millet protein's application in biodegradable materials and environmentally friendly textiles supports the global movement toward reducing plastic waste and promoting circular economies.

10.7.3 Final Thoughts on the Future of Millet Protein

The future of millet protein depends on continued research, technological advancements, and strategic policy support to enhance its market adoption. Addressing challenges related to extraction efficiency, regulatory compliance, and consumer awareness will be crucial for its mainstream acceptance. With increasing interest in sustainable agriculture and alternative proteins, millet protein has the potential to play a significant role in shaping resilient food systems and environmentally conscious industries. By leveraging its ecological benefits, economic viability, and expanding applications, millet protein can contribute to a more sustainable and diversified global protein supply.

References

1. R. R. Kumar, N. Singh, S. Singh, T. Vinutha, V. Krishnan, S. Goswami, B. Kumar, S. L. Jat, L. N. Yogeesh, S. P. Singh and G. P. Mishra, *J. Plant Biochem. Biotechnol.*, 2022, **31**(4), 837–852.
2. P. V. Prasad and S. A. Staggenborg, *Soils, Plant Growth and Crop Production*, EOLSS Publishers Co., Ltd., Oxford, UK, 2009, vol. 2.
3. S. Anitha, A. Rajendran, R. Botha, C. Baruah, P. Mer, J. Sebastian and J. Kane-Potaka, *Front. Sustainable Food Syst.*, 2024, **8**, 1324046.
4. P. B. Devi, R. Vijayabharathi, S. Sathyabama, N. G. Malleshi and V. B. Priyadarisini, *J. Food Sci. Technol.*, 2014, **51**, 1021–1040.
5. S. U. H. Dwivedi, S. Senthilvel, C. T. Hash, K. Fukunaga, X. Diao, D. B. D. Santra and M. Prasad, *Plant Breed. Rev.*, 2012, **35**, 246–375.
6. P. Panwar, A. Dubey and A. K. Verma, *J. Food Sci. Technol.*, 2016, **53**, 2779–2787.
7. J. Jacob, V. Krishnan, C. Antony, M. Bhavyasri, C. Aruna, K. Mishra and K. B. Visarada, *Front. Nutr.*, 2024, **11**, 1346869.
8. A. G. A. Sá, Y. M. F. Moreno and B. A. M. Carciofi, *Trends Food Sci. Technol.*, 2020, **97**, 170–184.
9. G. R. Abah, C. N. Ishiwu, J. E. Obiegbuna and A. A. Oladejo, *Asian Food Sci. J.*, 2020, **14**(2), 9–19.
10. J. R. Taylor and J. Taylor, in *Sustainable Protein Sources*, Academic Press, 2017, pp. 79–104.
11. J. Narayan, D. John and N. Ramadas, *J. Public Health Policy*, 2019, **40**, 126–141.
12. S. Anitha, M. Govindaraj and J. Kane-Potaka, *Cereal Chem.*, 2020, **97**(1), 74–84.
13. O. A. Adebo and S. A. Oyeyinka, *Front. Nutr.*, 2020, **7**, 217.

14. R. Singh, *et al.*, *Food Chem. Adv.*, 2023, **6**, 100312.
15. J. Chen, J. Rao, B. Yan and Y. Wang, *Eur. J. Nutr.*, 2017, **56**(6), 2315–2326.
16. S. Sarabhai, M. Kaur and A. Kaur, *LWT*, 2021, **137**, 110472.
17. M. Al-Hashimi, *et al.*, *Agroecol. Sustainable Food Syst.*, 2020, **44**(7), 923–938.
18. S. P. Bangar, *et al.*, *Food Sci. Technol.*, 2021, **56**(3), 405–418.
19. A. Johansson, *et al.*, *Agric. Econ. Policy*, 2018, **12**(4), 310–325.
20. H. Lee, *et al.*, *Environ. Res. Lett.*, 2022, **17**(1), 014005.
21. R. Kumar and P. Sharma, *J. Agric. Econ.*, 2020, **45**(3), 210–225.
22. A. A. Sabuz, M. R. Rana, T. Ahmed, M. M. Molla, N. Islam, H. H. Khan and Q. Shen, *Separations*, 2023, **10**(2), 80.
23. X. Ren, R. Yin, D. Hou, Y. Xue, M. Zhang, X. Diao and Q. Shen, *Nutrients*, 2018, **10**(10), 1509.
24. L. Gong, W. Cao, H. Chi, J. Wang, H. Zhang, J. Liu and B. Sun, *Food Res. Int.*, 2018, **103**, 84–102.
25. T. Bandyopadhyay, R. K. Singh, P. Ramesh, J. Maurya and M. Prasad, *J. Soil Sci. Plant Nutr.*, 2023, **23**(1), 628–637.
26. I. Amadou, G. Le and Y. H. Shi, *Food Res. Int.*, 2013, **51**, 422–428.
27. S. A. Kheya, S. K. Talukder, P. Datta, S. Yeasmin, M. H. Rashid and A. K. Hasan, *et al.*, *Heliyon*, 2023, **9**, e14832.
28. D. Mehta, S. Vyas, D. Dudhagara, A. Patel and V. Parmar, *Trends Food Sci. Technol.*, 2024, **149**, 104527.
29. A. Kumar, R. Singh and P. Sharma, *J. Pharm. Sci.*, 2017, **106**, 2532–2540.
30. J. Singh, P. Rasane, V. Nanda, S. Ercisli and H. Verma, in *Cereals and Nutraceuticals*, Springer Nature Singapore, Singapore, 2024, pp. 1–21.
31. H. Lee, *et al.*, *Environ. Res. Lett.*, 2021, **16**, 024003.
32. Y. Zhang, *et al.*, *Food Ind. J.*, 2023, **29**, 89–105.
33. P. Sachdev, *et al.*, *J. Food Sci. Innovation*, 2023, **78**, 1123–1139.
34. R. Majumder, S. Miatur, S. Roy, A. Saha, S. Manik and N. Biswas, *et al.*, *Trends Food Sci. Technol.*, 2024, **151**, 104618.
35. S. Bangar, PhD thesis, Clemson University, 2022.
36. P. Singh and B. K. Yadav, in *Nutri-Cereals*, CRC Press, 2023, pp. 289–304.
37. S. Hussain, R. Akhter and S. S. Maktedar, *Sustainable Food Technol.*, 2024, **2**, 1297–1364.
38. J. R. V. Matheus, R. R. Dalsasso, E. A. Rebelatto, K. S. Andrade, L. M. D. Andrade, C. J. D. Andrade and A. E. C. Fai, *Compr. Rev. Food Sci. Food Saf.*, 2023, **22**, 1148–1183.
39. S. Oyedeji, N. Patel, R. Krishnamurthy and P. O. Fatoba, *Advances in Biochemical Engineering/Biotechnology*, 2024, vol. **191**, pp. 215–248.
40. A. Babu, A. R. Kumar, N. R. Amrutha, S. Madhurya, H. P. Kumar and J. P. Reddy, *et al.*, *Int. J. Biol. Macromol.*, 2024, **282**, 137283.
41. C. Johansson, J. Bras, I. Mondragon, P. Nechita, D. Plackett and P. Simon, *et al.*, *BioResources*, 2012, **7**, 2506–2552.
42. E. Shlush and M. Davidovich-Pinhas, *Trends Food Sci. Technol.*, 2022, **125**, 66–80.
43. W. Rasri, S. R. Kumar and P. Khampratueng, in *Agro-Wastes for Packaging Applications*, CRC Press, 2024, pp. 83–104.

44. *Wealth out of Food Processing Waste: Ingredient Recovery and Valorization*, ed. M. Selvamuthukumaran, CRC Press, 2024.
45. B. P. Meshram, P. Jain and K. K. Gaikwad, *ACS Food Sci. Technol.*, 2025, **5**, 788–799.
46. S. Anitha, J. Kane-Potaka and T. W. Tsusaka, *et al.*, *Front. Nutr.*, 2021, **8**, 687428
47. R. Majumder, S. Miatur, S. Roy, A. Saha, S. Manik and N. Biswas, *et al.*, *Trends Food Sci. Technol.*, 2024, **151**, 104618.
48. B. K. Singh, A. K. Chaudhari, S. Das, S. Tiwari and N. K. Dubey, *Pestic. Biochem. Physiol.*, 2022, **187**, 105214.
49. S. Varghese, M. Awana, D. Mondal, M. H. Rubiya, K. Melethil, A. Singh *et al.*, in *Handbook of Biopolymers*, Springer Nature Singapore, Singapore, 2022, pp. 1–30.
50. J. Tang, F. Zou, L. Guo, N. Wang, H. Zhang, B. Cui and X. Liu, *Carbohydr. Polym.*, 2022, **279**, 119012.
51. R. K. Dhall, *Crit. Rev. Food Sci. Nutr.*, 2013, **53**, 435–450.
52. S. R. Ushakumari, PhD thesis, University of Mysore, 2009.
53. M. S. Amarnath, A. Muhammed, A. K. Antony, M. Yashini, A. J. Sherin and C. K. Sunil, *Food Humanit.*, 2024, **3**, 100356.
54. M. O. Olteo, PhD thesis, University of Nairobi, 2024.
55. M. S. Amarnath, A. Muhammed, A. K. Antony, B. Malini and C. K. Sunil, *Food Humanit.*, 2023, **1**, 599–606.
56. M. Kim and J. H. Kang, *Food Sci. Biotechnol.*, 2024, **33**, 355–362.
57. S. Punia Bangar, K. S. Sandhu, A. V. Rusu, P. Kaur, S. S. Purewal and M. Kaur, *et al.*, *Coatings*, 2021, **11**, 1167.
58. T. Fadiji, M. Rashvand, M. O. Daramola and S. A. Iwarere, *Processes*, 2023, **11**, 590.
59. K. Silva, PhD thesis, University of Guelph, 2018.
60. B. D. Rao, K. Bhaskarachary, G. D. Arlene Christina, G. Sudha Devi, A. T. Vilas and A. Tonapi, *Nutritional and Health Benefits of Millets*, ICAR-Indian Institute of Millets Research, Hyderabad, 2017.
61. R. Maurya, T. Boini, L. Misro, T. Radhakrishnan, A. P. Sreedharan and D. Gaidhani, *J. Drug Res. Ayurvedic Sci.*, 2023, **8**(Suppl 1), S82–S98.
62. M. Rawat, A. Varshney, M. Rai, A. Chikara, A. L. Pohty and A. Joshi, *et al.*, *J. Agric. Food Res.*, 2023, **12**, 100619.
63. R. S. Khillare and A. D. Chilkhalikar, *Agric. Food: e-Newsl.*, 2022, **8**, 7.
64. T. P. Mall, *Agric. Sci. Res. J.*, 2016, **6**, 227–234.
65. E. G. Obahiagbon and M. C. Ogwu, in *Food Safety and Quality in the Global South*, Springer Nature Singapore, Singapore, 2024, pp. 299–329.
66. R. M. Trüeb and R. M. Trüeb, in *Nutrition for Healthy Hair: Guide to Understanding and Proper Practice*, 2020, pp. 225–255.
67. R. E. l Hage, Y. Khalaf, S. Abou Fayssal, M. Hammoud, Z. E. l Sebaaly and Y. N. Sassine, in *Mushrooms: Agaricus bisporus*, CABI, Oxfordshire, 2021, pp. 357–425.
68. S. Roy, *Recent Trends Cosmet.*, 2024, **1**(2), 26–43.

69. S. Nithiyanantham, P. Kalaiselvi, M. F. Mahomoodally, G. Zengin, A. Abirami and G. Srinivasan, *J. Food Biochem.*, 2019, **43**, e12859.
70. J. Sharifi-Rad, C. F. Rodrigues, F. Sharopov, A. O. Docea, A. Can Karaca and M. Sharifi-Rad, *et al.*, *Int. J. Environ. Res. Public Health*, 2020, **17**, 2326.
71. V. Sharma and M. Thakur, in *Sustainable Food Systems (Volume II): Novel Sustainable Green Technologies, Circular Strategies, Food Safety & Diversity*, Springer Nature Switzerland, Cham, 2023, pp. 329–341.
72. S. Adeel, S. Kamal, T. Ahmad, I. Bibi, S. Rehman, A. Kamal and A. Saleem, in *Advances in Functional Finishing of Textiles*, 2020, pp. 85–114.
73. P. Jindal and P. Nikhanj, *J. Food Process Eng.*, 2023, **46**, e14419.
74. S. Eyupoglu, in *Sustainability in the Textile and Apparel Industries: Sourcing Natural Raw Materials*, 2020, pp. 27–48.
75. S. N. Sandhya, T. Karthika, V. Swetha and G. C. Jeevitha, *Int. J. Food Sci. Technol.*, 2025, vvaf037.
76. N. Sharma and K. Niranjan, *Food Rev. Int.*, 2018, **34**, 329–363.
77. P. S. F. Musere, PhD thesis, University of Namibia, 2019.
78. A. K. Rashwan, A. I. Osman, A. M. Abdelshafy, J. Mo and W. Chen, *Crit. Rev. Food Sci. Nutr.*, 2025, **65**, 667–694.
79. P. Bharti, M. Mohapatra, R. P. Sah, A. Pradhan, S. Priyadarsani and K. Rayaguru, *Int. J. Food Sci. Technol.*, 2024, **59**, 8045–8057.
80. A. Famakinwa, J. Ilo, O. Olubi, O. O. Oguntibeju, J. Van Wyk and A. Obilana, *Curr. Res. Nutr. Food Sci. J.*, 2023, **11**, 910–940.
81. A. Saxena, B. P. Tripathi, M. Kumar and V. K. Shahi, *Adv. Colloid Interface Sci.*, 2009, **145**, 1–22.
82. H. B. Costa, P. M. Fernandes, W. Romão and J. A. Ventura, *Ind. Crops Prod.*, 2014, **59**, 163–168.
83. T. J. Rao, A. Mounika and A. Shanmugam, *Ind. Eng. Chem. Res.*, 2024, **63**, 5380–5396.
84. B. Zhang, Y. Xu, C. Zhao, Y. Zhang, H. Lv and X. Ji, *et al.*, *Food Funct.*, 2022, **13**, 2594–2605.
85. O. K. Ozturk and H. Turasan, *Crit. Rev. Food Sci. Nutr.*, 2022, **62**, 4481–4503.
86. V. Patil, *et al.*, *Agric. Water Manage.*, 2021, **250**, 106804.
87. R. Kumar and P. Sharma, *J. Sustainable Agric.*, 2019, **34**, 123–135.
88. A. T. Rajendran and A. N. Vadakkepushpakath, *Foods*, 2024, **13**, 2812.
89. FAO, *The State of Food Security and Nutrition in the World*, 2021.
90. V. Patil, *et al.*, *Econ. Perspect. Agric.*, 2021, **39**, 78–95.

CHAPTER 11

Advances, Market Value, and Future Outlook of Millet Protein

P. NITHYA AND SARIKA YADAV*

Department of Food Business Management and Entrepreneurship Development, National Institute of Food Technology Entrepreneurship and Management (NIFTEM), Sonepat, Haryana 131028, India,
Email: nithyapn9@gmail.com
*Emails: sarikasatvik@gmail.com; sarika.yadav@niftem.ac.in

11.1 Introduction

Millets, often known as ancient grains, have drawn a lot of interest lately because of their remarkable nutritional profile and advantages for sustainability. In semi-arid and arid parts of the world, millets are frequently grown because they can survive in hot, humid climates where it is very unlikely that wheat and rice would be grown.[1] Millet seeds are vital to the ecosystem, work well in mixed and intercropping systems, are resilient to extreme weather conditions like drought and salt, and give farmers a means of subsistence and food security.[2,3] Minor or tiny millets include finger (*Eleusine coracana*), foxtail (*Setaria italica*), proso (*Panicum miliaceum*), barnyard (*Echinochloa utilis*), little (*Panicum sumantrense*) and kodo millet (*Paspalum scrobiculatum*). Major millets are sorghum (*Sorghum vulgare*) and pearl millet (*Pennisetum glaucum*).[4] Sorghum, foxtail millet, pearl millet, and other C4 model crops have qualities that make them climate resilient.[5] Millet is essential for regions suffering from poor soil or water scarcity because of its capacity to flourish in arid environments with little water. Millets require very few

Food Chemistry, Function and Analysis No. 48
Millet Proteins: Nutritional, Structural and Functional Attributes
Edited by Shweta Suri, Anuradha Dutta and Tarun Belwal

Published by the Royal Society of Chemistry, www.rsc.org

external inputs, which lowers production costs and avoids environmental contamination from synthetic pesticides, fertilizers, and insecticides, as well as harm to human health. Traditionally used to feed animals and birds, millets are now becoming more and more important as innovative therapeutic meals for humans. Compared to the production of protein from livestock, which is associated with the transmission of diseases to humans, overuse of natural resources such as land, water, and feed crops, as well as contributing to biodiversity decline and environmental impacts like greenhouse gas emissions, the cultivation of millets represents a more sustainable and environmentally responsible agricultural practice.[6]

In many parts of the world, including Asia, Africa, and some portions of Europe, millet is a staple food. The demand for millet has increased significantly as people throughout the world switch to plant-based diets and gluten-free substitutes for wheat and rice. Millet protein, with its essential ingredients, is particularly noteworthy as a plant-based protein source with enormous potential in the food sector. Millet protein shows promise as a component for a number of applications, from plant-based meat alternatives to functional meals, as consumer demand for alternative proteins rises. While urbanization in developing nations has shifted preferences towards processed foods, millets are carving out a niche due to their health benefits as a whole grain. This trend is further amplified by the increasing availability of processed millet products.[7]

Millets are vastly nutritious grains, offering a rich supply of vitamins and minerals. They serve as excellent sources of dietary fiber, slowly digestible starch, and impervious starch, which together contribute to a gradual discharge of glucose and enhance satiety. Compared to conventional cereals, millets provide higher amounts of protein and sulfur-containing amino acids such as methionine and cysteine, along with a more favorable fatty acid composition. However, they are relatively low in lysine and tryptophan, with concentrations varying among different cultivars.[8]

Millets are also abundant in vitamins E and B, and are rich in essential minerals like calcium, phosphorus, magnesium, manganese, potassium, and iron. Due to their impressive nutrient profile, millets offer numerous health benefits, including the potential to lower the risk of chronic conditions such as cancer, obesity, diabetes, cardiovascular disorders, digestive issues, migraines, and asthma. Their high dietary fiber and slow-digesting carbohydrate content make them particularly suitable for managing blood sugar levels, thereby benefiting individuals with diabetes.[9]

In modern nutrition, millets are increasingly recognized as vital sources of essential nutrients, particularly in low-income and developing regions. Despite their nutritional advantages, millet consumption in India remains limited due to several socio-economic and cultural factors. However, in recent years, these grains have begun to gain attention, sparking innovation in the food sector and contributing to the growth of nutrition-focused start-ups, while also generating employment opportunities. The development, market value, and prospects for millet protein are examined in this chapter.

It explores the advancements in science and technology that have improved its functioning, extraction, and use in the food and nutraceutical sectors. The chapter also assesses millet protein's economic potential, consumer acceptability, and current market trends in the worldwide protein sector. Finally, it talks about the future, covering new developments, difficulties, and how millet protein might help with issues of sustainability and food security. This chapter seeks to shed light on the changing millet protein landscape in order to offer insights into its importance in the contemporary food industry and its potential to transform the market for alternative proteins completely.

11.2 Properties of Millet Proteins

Millet protein, a viable substitute protein source, has a number of special nutritional, physiological, and bioactive qualities. These characteristics affect its digestion, use in food formulations, and general health advantages. Similar to wheat and rice, the whole protein content of different millet seed species varies from 7.52% to 12.1%.[10] The species, culture, anatomical distribution, growing circumstances, and agricultural practices, such as soil fertility, moisture levels, use of chemical inputs, and ambient growth temperatures, all influence the yield, composition, and overall quality of proteins in millets. Understanding these environmental conditions is essential for optimizing protein production.[11]

After starch, protein makes up over 10% of the total weight of millet, making it the second most prevalent nutrient.[12] In addition to having a greater protein, fat, and fiber content than wheat and rice seeds, millets also include a number of beneficial polyphenolic chemicals and a lower net carbohydrate content.[13] The gluten-free and hypoallergenic qualities of millet species make them an excellent source of dietary protein. Despite having a balanced amino acid composition, millet protein lacks tryptophan, lysine, and threonine. To satisfy human protein needs, the diet's nutritional quality can be enhanced by supplementing it with complementary proteins or the lacking amino acids.[14] The protein fractions from millet have favorable nutritional and appealing qualities.[15,16] In the treatment of chronic human illnesses, the protein from these seeds offers a wide variety of advantages.[12,15,17]

11.2.1 Protein Distribution and Composition

Since the endosperm portion is the biggest of the seed's structural components, it accounts for the majority of the millet's overall protein weight.[18] Amyloplasts are granular structures that contain starch, and the cytoplasmic proteinaceous matrix contains endospermic proteins in protein bodies. The overall protein composition of seeds is also accounted for by certain proteins that support metabolic functions, such as enzymes, primarily located in the bran and germ (embryo) portions of millet grains. Millet seed proteins are generally categorized into four types—glutelin, prolamin, globulin, and albumin—with their proportions varying depending on species, cultivar, and extraction conditions. This classification is primarily based on differences in solubility.[11] Another widely accepted classification system is Landry and Moureaux's method,[19] which employs reducing agents and detergents to extract polymeric, cross-linked proteins. Among the millet varieties, pearl, sorghum, barnyard, little, proso, and foxtail millets contain relatively high protein content, averaging around 12%, followed by kodo millet at 11%, while finger millet shows the lowest at approximately 9%, based on solubility. Pearl millet, in particular, has a high concentration of water-soluble proteins, followed by sorghum, foxtail, and proso millet. Prolamins are identified as proteins that are soluble in aqueous solutions of alcohol. It was discovered that foxtail millet and pearl millet have the highest levels of prolamin. Conversely, the cross-linked prolamin fraction is present in low amounts in pearl millet but is significantly higher in finger millet. Only a small portion of millet proteins could be extracted using sodium dodecyl sulfate (SDS) alone. The use of reducing agents was necessary to break down the polymeric cross-linked protein structures in millets. Analytical results indicate that foxtail millet contains the lowest levels of cross-linked protein fractions, while barnyard millet exhibits the highest concentration of these proteins. Prolamin and glutelin are indeed recognized as the predominant soluble fractions, with their solubility, crosslinking, and extraction dictated by the use of alcohol, alkali, and reducing agents. These proteins contribute nearly half of the total seed storage protein by supplying nitrogen during germination, thereby supporting seed growth and development. The outer seed layers house low molecular weight, complex proteins known as the albumin and globulin fractions. These proteins are nutritionally important as they provide essential amino acids and can be readily extracted using salt and water solutions.[20] It has recently been determined that the maturity stage of the seed also affects the protein content. In addition to mature millet seeds, young millet seeds may also serve as a valuable source of plant-derived proteins.

11.2.2 Amino Acid Profile and Digestibility

Millet protein contains high levels of both essential and non-essential amino acids, including leucine, phenylalanine, proline, serine, tyrosine, aspartic acid, and glutamic acid.[21] Beyond amino acid limitations, the protein efficiency of millet flour is compromised by its structural composition, specifically a high prolamin fraction. This reduces solubility and enzyme accessibility, while anti-nutrients further limit digestibility, particularly in whole-grain flours.[22] This is why prolamin content rises and amino acid shortages become more noticeable as protein concentration rises due to agricultural practices like applying nitrogen-rich fertilizers. This explains the tenable relationship between the prolamin fraction and the nutritional value and features of manufactured goods. Prolamin is a very hydrophobic substance with a high proline and glutamine concentration. Though it has a better circulation of essential amino acids, the albumin and globulin components only provide 25% of the seed's total nitrogen since they make up a tiny portion of millet seeds. Nonetheless, seeds that are inherently high in nutrients or that hold bran and germ proteins (globulin and albumin) after processing may be of high quality. In contrast to minor millets like foxtail and tiny millets (14.2 and 2.42 per kg of flour, respectively), pearl and finger millet contain greater levels of lysine (2.83 kg and 31.9 per kg of flour, respectively).[23] Therefore, it can be concluded that the levels of essential amino acids in millet proteins seem to change depending on the levels of protein fractions. Additionally, because high-protein seeds have more balanced protein fractions and, as a result, an even amount of important amino acids, especially lysine, they appear to have greater nutritional value than low-protein mature seeds. The presence of lysine in protein is important since it is necessary for human nutrition.[24] Maintaining the integrity of the bran and germ layers during millet seed processing methods like dehulling and milling enhances the nutritional quality of millet-derived proteins.[25] Germination may also increase the amount of lysine.[26] Additionally, lysine supplements or protein fractions rich in lysine can be combined with millet protein to make up for limiting amino acid deficiencies and bring millet protein closer to animal protein. Combining millet protein with complementary alternative plant-based protein sources boosts the intake of essential amino acids such as lysine, threonine, and tryptophan, which are important for supporting growth and overall health.[14] Relying just on millet protein-based meals to achieve an amino acid composition that is comparable to that of animal protein is insufficient since digestibility and bioavailability are also important aspects of protein nutritional quality. These proteins often have poor solubility due to their compact structure, cross-linked nature, and high concentration of hydrophobic amino acid residues.[27] Additionally, the degradation of proteins into free amino acids and tiny peptides is prevented by inhibiting access by digestive enzymes.

11.2.3 Functional Properties

The basic physical–chemical characteristics known as functional properties are those that describe the intricate relationship between the molecular conformation, composition, structure, and physical–chemical features of food ingredients, as well as the type of environment in which these features are related and measured. The quantity of water that flour can absorb to reach an appropriate consistency and produce a high-quality final product is known as its water absorption capacity (WAC). The WAC of several millet kinds was compared with that of wheat and soybean flour. The WAC for millet, wheat, and soybean flour were 74.08 to 76.83 ml per 100 g, 74.08 to 78.83 ml per 100 g, and 58.17 to 60.02 ml per 100 g, respectively.[28] Finger millet flour had an oil absorption capacity of 1.93 $g\,g^{-1}$, whereas pearl millet flour had a capacity of 1.60 $g\,g^{-1}$. Pearl millet and finger millet flours had bulk densities ranging from 0.67 to 0.54 $g\ ml^{-1}$.

The modest propensity of millet protein to absorb oil improves texture and mouthfeel, making it more useful in processed meals, meat substitutes, and baked goods. Furthermore, millet proteins have the ability to emulsify, which allows them to stabilize oil–water mixes in products such as plant-based dairy, sauces, and dressings. However, millet's foaming capacity and stability are relatively weak compared to egg and soy proteins, though enzymatic modifications can improve this property for applications in aerated foods like cakes and whipped toppings. Additionally, compared to soy or pea proteins, millet proteins have limited gelation characteristics, which means they are less successful at creating strong gels. Nevertheless, they can still improve the texture of plant-based and gluten-free compositions. Because of these qualities, millet is a useful ingredient in the creation of food products, especially in the expanding market for gluten-free and plant-based cuisine.

11.3 Extraction Techniques

The rising ultimatum for animal-derived proteins in the biomaterials industry presents an opportunity to replace them with cost-effective, plant-based alternatives like isolated millet proteins. To explore their commercial viability—particularly their applicability in food, pharmaceutical, or biomaterial products—it is crucial to first understand the intrinsic properties and functionalities of millet proteins. This necessitates their extraction and purification prior to designing appropriate application systems. Researchers have reported various conventional and modern methodologies for the efficient extraction and characterization of cereal proteins. The methods employed for isolating proteins from millet seed flour typically fall under "wet isolation techniques." These approaches can generally be divided into two main categories: an analytical sequential protein fractionation process, and a preparative multi-step process for producing millet protein isolates or concentrates.

11.3.1 Sequential Millet Protein Fractionation

Traditional fractionation methods aim to separate proteins into distinct fractions based on their solubility in different solvents, a technique well-documented in literature.[29] Fractionation not only aids in identifying the variety and nutritional content of proteins but also helps in assessing the effects of processing and storage on protein interactions.[30,31] This method offers early insights into solubility, which is essential for scaling up to commercial production. Additionally, specific protein fractions that may have allergenic potential can be isolated and analyzed for their immunological responses. The process typically begins with either defatted or non-defatted, protein-rich millet flour as the starting material. The main goal is to isolate, identify, and characterize the different protein types present, accounting for variability due to species, cultivar, and agricultural practices.

11.3.1.1 Osborne's Protein Fractionation Method

Osborne's classification categorizes proteins based on their solubility, grouping them into albumins, globulins, prolamins, and glutelins, which are soluble in water, dilute salt solutions, alcohol, and weak acids or alkalis, respectively.[32] A study by Akharume *et al.* used proso millet flour as the starting material.[33] The flour underwent defatting using hexane in a 1 : 6 ratio over six hours to remove lipids, followed by decolorization using a chloroform–methanol mixture (40 : 60) to eliminate pigments such as carotenes and xanthophyll. This treatment was repeated three times, then the flour was air-dried for 24 hours. Albumins and globulins were extracted together by using a 0.5 M NaCl solution, which was stirred for four hours. The mixture was then centrifuged at 5000 g for 30 minutes at 4 °C to separate the proteins. The supernatant was dialyzed using a 10 kDa molecular weight cut-off membrane for five days in cold conditions (~4 °C) with periodic water replacement. After a second round of centrifugation, the albumin remained in the supernatant while globulin was obtained from the precipitate. Both the albumin and globulin fractions were freeze-dried and stored at −20 °C. For the extraction of prolamin and glutelin, a mixture of 70% isopropyl alcohol and 0.05 M NaOH was used. These dispersions underwent dialysis, centrifugation, and lyophilization, and were also stored at −20 °C. The protein content was then determined using the Dumas combustion method. Despite being widely used, Osborne's method has limitations, including low protein recovery and the potential denaturation or association/dissociation of proteins during extraction, which can affect their true native forms.

11.3.1.2 Landry and Moureaux's Protein Fractionation Method

Given the complex and heterogeneous nature of millet proteins, Osborne's method may not fully extract all protein fractions. Landry and Moureaux's approach incorporates reducing agents, *viz.* 2-mercaptoethanol (2-ME) or dithiothreitol (DTT) in concentrations ranging from 0.5% to 2%, as well as detergents like sodium dodecyl sulfate and protein-denaturing agents like guanidinium chloride and urea. These components assist in disrupting protein polymers in alcohol and alkaline solutions, enabling a more complete extraction process.[19,34] This method divides millet proteins into five fractions: albumin/globulin (fraction I), prolamin (fraction II), prolamin-like proteins (fraction III), glutelin (fraction IV), and glutelin-like proteins (fraction V). Fractions III and IV encompass cross-linked prolamins and glutelins that were previously unextractable. Previous studies have largely employed this method for millet protein characterization, underlining its effectiveness and applicability.[35] Although it allows for near-complete protein extraction, the method has drawbacks, such as overlapping protein separations, the need for result validation, and impracticality for commercial-scale application due to its labor-intensive nature.

11.4 Advances

The food industry may use millet proteins for a variety of purposes due to its digestibility, emulsification qualities, solubility, and functional qualities. Proteins are frequently altered by denaturation or degradation processes. Treatments that use mechanical (like extrusion), thermal (like conventional heating, microwave heating), or electrical energy (like pulsed electric field), alone or in conjunction with one another, constitute physical alteration of millet proteins.[36] Using enzymes to break down the millets' pericarp is an alternative to conventional mechanical treatments that reduces milling loss and enhances quality.

To stabilize millet products while maintaining their nutritious qualities, many food processing methods have been investigated. In addition to non-thermal methods, such as high-pressure processing (HPP), cold plasma (CP), pulsed electric field (PEF), ultrasonication, irradiation, and ozonation, there are also thermal techniques like radio frequency (RF), microwave (MW), infrared (IR), and ohmic heating. These technologies can be useful for producing a wide variety of millet-based products, including pasta, noodles, baked goods, extruded products, breakfast cereals, beverages, functional meals, malting and weaning foods.[37]

Microwave treatment, as observed through structural changes analyzed by Fourier transform infrared spectroscopy (FTIR) and X-ray diffraction (XRD), has been used to modify the functional characteristics of foxtail millet proteins. This treatment enhanced emulsification and oil/water absorption capacities but resulted in a decrease in foaming properties and protein purity.[38] Pulsed electric field treatment has been shown to affect millet protein's structure, change how digestible the protein is, and change the millet-based products' qualitative characteristics. However, the precise effects may vary based on the processing parameters and variables.[39] It has been demonstrated that HPP affects the digestibility, qualitative features, structural attributes, and functional qualities of millet protein. HPP may enhance the water and oil retention properties of millet proteins, increasing their versatility in a range of culinary applications.[16]

According to one study, millet protein nanoparticles were created using ultrasound, which improved their solubility, decreased their turbidity, and improved their foaming and emulsifying qualities, particularly in acidic meals. Protein structure and conformation were enhanced by the treatment, even though denaturation resulted in decreased thermal stability. Because of these improvements, the proteins may be used in a range of culinary products, such as ice cream and baked goods.[40]

Modern processing and ready-to-eat (RTE) product development methods are necessary for the value addition of millets and millet-derived goods.

Hand pounding is the traditional way of millet dehulling, and it takes a lot of time and effort. These challenges could be resolved by using contemporary dehulling techniques, such as rice hullers, centrifugal shellers, and disc shellers.

Recent advancements in millet protein research and technology have significantly improved the protein's functional properties, food industry use opportunities, and extraction efficiency. Protein recovery has been enhanced while maintaining its nutritional integrity because of innovations including membrane filtration, ultrasound-assisted extraction, and enzymatic hydrolysis. Conventional protein extraction methods often resulted in poor solubility and low yields. These techniques have also reduced the quantity of anti-nutritional chemicals, improving digestibility and bioavailability.

Protein isolates and concentrates are commercially viable forms of pure protein, offering an innovative and cost-effective method for enriching food formulations with new and underutilized protein sources.[41] This approach helps develop novel, health-oriented foods that are typically abundant in certain regions of the world but may not be widely available globally. These protein sources can contribute to diversifying the global food market and addressing nutritional gaps. By employing biochemical extraction techniques, the underused millet seeds might be commercially vended as a protein concentrate with an appropriate amino acid profile, serving as an alternative protein supply.[12] Insulated millet proteins may improve their application in the biopolymer, pharmaceutical, and food sectors.

Making millet protein isolates and concentrates is necessary to increase the use of millet proteins in commercial settings. In fact, the extraction procedure has an impact on the isolates' pharmacological, sensory, nutritional, and culinary uses. Choosing the right extraction method requires understanding of the properties of millet protein and its processing parameters. In the production of micro- and nanomaterials, millet proteins are a crucial biomaterial.[20,39] For the functionalization of protein constituents in commercial settings, millet therefore offers an intriguing supply of proteins.

By improving the solubility, emulsification, and gelation qualities of millet protein, methods for protein modification such as cross-linking, nanoencapsulation, and enzymatic hydrolysis have been developed to increase its competitiveness with other plant-based proteins like soy and pea. Further addressing earlier restrictions associated with lysine deficit, biotechnological developments in millet breeding have produced high-protein millet cultivars with enhanced amino acid profiles.

Additionally, millet protein is becoming more popular in the creation of substitute food products, such as plant-derived dairy, meat alternatives, protein-enriched snacks, and sports nutrition. In response to the increased need for sustainable food solutions, efforts are underway to develop more environmentally friendly millet protein production methods using sustainable processing technology. Consequently, millet protein is becoming an important, useful, and sustainable source of protein, helping to accelerate the worldwide transition to plant-based diets.

11.5 Millet Market Size and Trends

The millet market is predicted to increase at a compound annual growth rate (CAGR) of 6.2% from 2024 to 2030, from an appraised USD 36.94 billion in 2023. Among the many health advantages of millets are their high protein, fiber, vitamin, and mineral content. Demand is driven by growing consumer knowledge of these advantages. Furthermore, there is a rising trend towards vegetarian and vegan diets, as well as a general increase in the consumption of plant-derived proteins among omnivores. Millet, being a versatile and high-protein legume, is a key component of many plant-derived diets.[43]

A combination of customer preferences and general market dynamics is responsible for the growing millet industry. The most important is the increased knowledge of millet's nutritional and health advantages. Because millet is gluten free and rich in vitamins, minerals, dietary fiber, and protein, it appeals to health-conscious individuals, especially for those who have celiac disease or gluten sensitivity. An increasing consumer interest in dietary diversity and the fact that millet offers a wholesome substitute for conventional grains like wheat and rice are two factors contributing to this health-driven demand. Concerns about sustainability also contribute significantly to millet's growing appeal. Resilient crops are an environmentally sustainable alternative since they use less water, which fits with consumers' growing need for environmentally favorable food options. A cultural and traditional renaissance in regions where millets were formerly a mainstay, especially in Asia and Africa, supports this tendency even more.

The expansion of the millet market has also been facilitated by government initiatives and assistance in a number of nations. Governments are promoting millet through research, subsidies, and marketing efforts because of its minimal environmental effect and nutritional advantages. These grains are now more widely available and enticing to a wide range of consumers thanks to the food industry's inventiveness in creating a diverse range of millet-derived goods, from baking components to morning cereals.

Global millet output increased to about 28.33 million metric tons in 2019 and 30.08 million metric tons in 2021, according to statistics from the Food and Agriculture Organization (FAO). Notably, with a substantial 43.0% market share for millet in 2021, India became the world's top producer. In India, a wide variety of millets are cultivated, including sorghum (jowar), pearl millet (bajra), and finger millet (ragi), along with several minor millets. Data from the Ministry of Agriculture and Farmers Welfare highlights a positive trend in millet cultivation, with production rising from 14.52 million metric tons in 2015–16 to 17.96 million metric tons in 2020–21, reflecting growing awareness and demand for these nutritious and climate-resilient crops.[43]

Additionally, millet is now more widely accepted due to the globalization of cuisines and customers' openness to trying new foods. Due to their high dietary fiber and low glycemic index, which help regulate blood sugar levels and promote prolonged satiety, millets are increasingly being incorporated into weight loss and fitness-oriented diets, gaining popularity among health-conscious consumers.

11.5.1 Current Market Trends

The small millet market has witnessed significant growth in recent years, spurred by a rise in consumer awareness regarding their nutritional value and eco-friendly cultivation. This positive shift is visible from grassroots levels to international trade spheres.

1. Increasing Health Awareness
   As consumers prioritize health and wellness, the demand for nutrient-rich foods is rising. Small millets, with their abundant fiber, protein, and minerals, are emerging as a preferred option for health-conscious individuals. Urban populations, particularly affected by lifestyle-related conditions such as diabetes and hypertension, are increasingly turning to millets. As a result, organic and health-focused retail outlets are expanding their range of millet-based products, including flours, snacks, and breakfast items.
2. Integration into Sustainable Food Practices
   Environmental sustainability is influencing food choices, pushing consumers to opt for ecologically responsible alternatives. Small millets are advantageous due to their low water requirement and adaptability to difficult climates, making them sustainable alternatives to conventional staples like rice and wheat. This awareness is growing not just among consumers but also among institutional buyers and policy makers advocating for climate-resilient farming practices.
3. Regional Consumption Trends
   Countries such as India, parts of Africa, and other Asian regions with a history of millet consumption are experiencing a revival in their use. National initiatives, like India's celebration of 2023 as the "International Year of Millets," have significantly amplified their visibility and market demand. Western nations are also seeing a steady uptick in millet demand, driven by the superfood trend.
4. Global Market Growth
   The international millet market is undergoing rapid expansion. Projections estimate a compound annual growth rate (CAGR) of approximately 4.5% over the coming decade. This momentum is largely attributed to increasing interest in gluten-free and organic food products.[44]

11.5.2 Economic Benefits

Small millets deliver notable economic advantages, particularly for small-scale farmers and rural entrepreneurs.

1. Income Stability Through Crop Diversification
   Cultivating millets allows small farms to diversify their crop portfolio, thereby reducing reliance on market-sensitive commodities. Millets require minimal inputs and are less vulnerable to extreme price swings. This provides a dependable income stream and reinforces economic stability for rural households.
2. Value-added Processing Potential
   Millets can be transformed into high-value products such as flours, snacks, breakfast mixes, and dietary supplements. These products typically fetch better market prices compared to raw grains. Encouraging local processing not only increases profitability for farmers but also generates employment and stimulates rural economic growth.
3. Opportunities in Export Markets
   The rising global appetite for nutritious, organic food opens export possibilities for millet growers. Nations with a solid agricultural infrastructure can benefit by investing in certification and strategic marketing to capture niche health-focused segments abroad.
4. Policy and Governmental Incentives
   Governments increasingly recognize the role of millets in enhancing food security and fostering sustainable farming. Supportive policies, including subsidies, technical guidance, and improved market access, are strengthening the commercial viability of millet cultivation. Incentives for seed purchase and millet-based enterprises are becoming more common as part of these efforts.

11.5.3 Market Opportunities

Capitalizing on emerging market segments is essential to fully realize the potential of small millets.

1. Health and Wellness Sector
   The health and wellness sector is a key growth driver for millet-based products. Innovations such as millet energy bars, nutritional drinks, and breakfast cereals are in demand. Engaging this audience through wellness stores, fitness centers, and digital platforms can result in robust sales growth.
2. Gluten-free Product Development
   With the increasing incidence of gluten-related health issues, the market for gluten-free foods is expanding. As naturally gluten-free grains, millets are ideally suited for this category. Developing an assortment of gluten-free products like flours, snacks, and convenience meals can cater to this specific but growing consumer base.
3. Organic Food Market Penetration
   Consumers seeking organic and minimally processed options represent a high-value market. Marketing millets as organic and non-GMO enhances their appeal in this premium sector. Certified organic branding can attract higher pricing and loyal customer segments.

4. Institutional Procurement Channels
   Institutions like schools, hospitals, and government-run nutritional programs offer large-scale opportunities. Promoting millet-based options in institutional catering can drive significant volume consumption. Advocacy for millet inclusion in public nutrition schemes ensures a stable and consistent demand stream.
5. Expansion in Export Markets
   Countries with active health and wellness markets—such as the U.S., Canada, and European nations—offer strong export potential. By ensuring compliance with international quality standards and building trade relationships, producers can effectively tap into these high-value markets.

11.5.4 Market Concentration and Characteristics

Innovative product options in the millet industry include ready-to-eat meals, morning cereals, snacks, and flour made from millet. These goods satisfy the convenience and variety demands of contemporary consumers. ITC Foods introduced its first millet biscuits under the Sunfeast Farmlite brand in May 2023. This launch aligns with ITC Food's Mission Millets effort, which seeks to increase consumer knowledge of millet's advantages, encourage sustainable cultivation, and offer a variety of millet-based goods. There are two varieties of these cookies: multi-millet and choco-chip multi-millet.

ADM, Wilmar International Limited, and Cargill are among the market participants engaged in merger and acquisition operations. These businesses can enter new markets and broaden their geographic reach through mergers and acquistions (M&A) activities.

International trade in millet can be hindered by import/export regulations, tariffs, and non-tariff barriers, which may reduce access to millet products in regions lacking local production and in those with slow market growth. Disparities in regulations or quality standards between countries can lead to confusion and logistical challenges for producers and exporters; therefore, harmonizing these standards is essential for facilitating smoother trade and supporting the expansion of the millet market.

A broad category of small-seeded grasses, millet is cultivated for its edible grains. Although millet has special qualities and is prized for its toughness and nutritional worth, there are several possible alternatives to millet in a variety of culinary applications, including rice, wheat, barley, oats, maize, and so on.[45]

11.5.5 Millet Market Dynamics

11.5.5.1 Market Drivers

According to the 2023 National Family Health Survey (NFHS) report, 59.1% of women are anemic, 35.6% of children experience stunted growth, and 32.2% are underweight. In light of these health challenges, small millets are gaining attention for their superior nutritional profile. They are non-acid forming; hypoallergenic; rich in protein, antioxidants and dietary fiber; and easy to digest. The recognition of these benefits has led to a growing demand for millets. Following a proposal submitted to the United Nations General Assembly, the year 2023 was officially designated as the *International Year of Millets*, a move that has further amplified global interest.

Millets are known to support weight management, boost hemoglobin levels, and mitigate conditions like anemia, diabetes, and heart disease. With these health advantages, production and productivity of millets have increased globally. Over 70 countries have committed to mobilizing stakeholders to enhance millet adoption.

A noticeable trend has emerged, especially in the Asia-Pacific region—primarily China and India—where farmers are shifting from traditional crops like wheat and rice to millet cultivation due to rising consumer demand. Urban consumers, in particular, are gravitating towards millet-based superfoods to boost immunity as an alternative to processed and unhealthy foods. This change in food preferences is expected to drive consistent demand throughout the forecast period. Furthermore, evolving lifestyles and a growing focus on organic and healthy diets are fueling the global appetite for organic millets. Their rich nutrient content is instrumental in managing obesity, diabetes, and cardiovascular issues, making them an attractive choice among health-aware consumers.

11.5.5.2 Market Restraints

Despite the increasing demand, certain barriers hinder the widespread consumption and production of millets. One of the most significant limitations is the limited availability of high-quality seeds. Additionally, a lack of basic processing infrastructure at the village level and long distances between farms and processing units contribute to elevated production costs for processed millet products.

Moreover, private entrepreneurs often remain unaware of existing government schemes and support mechanisms that provide financial aid, technology, and credit access. This gap in knowledge and resources restricts the development of value-added millet products and limits their market reach.

11.5.5.3 Market Opportunities

Government-led promotional campaigns are playing a pivotal role in expanding millet adoption across many regions. Authorities have introduced a range of support mechanisms—including funding, infrastructure, and technological assistance—for stakeholders across the supply chain.

The announcement of Minimum Support Prices (MSPs) for major millet varieties in most Indian states has incentivized millet cultivation. Processing units are increasingly being established near production centers, enabling efficient primary processing and quality checks. Secondary processing, packaging, and branding are being centralized on a larger scale, contributing to reduced factory gate prices and improved profitability for farmers. This integrated supply chain has also opened new avenues for rural employment and income generation.

11.5.6 Millet Market Segment Analysis

11.5.6.1 By Product Type

In countries across Asia and Africa, the most commonly consumed millets include sorghum, pearl millet, and a range of small millets. India leads the way in both production and consumption, with major varieties like finger millet, pearl millet, foxtail millet, barnyard millet, kodo millet, proso millet, and little millet being cultured extensively. India also ranks sixth globally in sorghum production. Research conducted in India and South Africa highlights growing public alertness regarding the health advantages of finger millet and other millet types.

11.5.6.2 By Application

The millet market can be segmented based on use, into food for livestock (fodder), baked goods, alcoholic and non-alcoholic beverages, breakfast cereals, and infant nutrition. Millet-based infant products, such as porridges, are gaining popularity for their role in combating early childhood malnutrition. In the Asia-Pacific region, processed millet foods—especially baked goods like cookies—are becoming increasingly common in modern retail channels like supermarkets, hypermarkets, and e-commerce platforms.

11.5.6.3 By Distribution Channel

Millets reach consumers through various channels, including supermarkets, hypermarkets, grocers, online platforms, and direct sales. While traditional retail outlets still dominate millet sales, online platforms are expected to see the fastest growth over the coming years. Distribution is also carried out through wholesalers, specialized stores, and direct links between producers and food processing companies or end consumers. Although specialty stores contribute less in terms of revenue, they offer premium options. Millet producers are increasingly leaning towards mainstream retail outlets where physical visibility of products builds consumer confidence.

11.5.7 Millet Market Regional Insights

As of 2021, millets ranked as the 3501st most traded product globally, with a total trade value of $205 million. Global millet exports saw an impressive 45.8% growth from 2020 to 2021. While India leads in millet production, the United States is the top exporter, accounting for nearly 30% of total global millet exports.

About half of the global millet exports go to countries such as Indonesia, Iran, Germany, Belgium, and South Korea, positioning the Asia-Pacific, Middle East, and European Union as high-demand regions.

The Asia-Pacific region, particularly, is emerging as a major hub for millet trade. Pearl millet is especially prominent due to its substantial contribution to global trade volumes. India alone contributes around 20% of the total global millet export value, reinforcing its strategic importance in meeting worldwide demand. While China continues to boost millet yields, other nations like Kenya and Oman—though smaller markets—have shown accelerated growth in millet imports from India in recent years.

11.6 Recent Developments

The Indo-US Millets Initiative is a partnership between the United States and India that seeks to increase the production, consumption, and knowledge of nutrient-dense millets. The Daugherty Water for Food Global Institute from the University of Nebraska, India's national India Millet Program, and non-profits Sorghum United and the North American Millets Alliance (NAMA) are among the important millet industry participants in this initiative.

In May 2023, the Bonn Group of Industries, a leading fast-moving consumer goods (FMCG) food manufacturer in India, introduced a new line of millet-based baked products including Millet Burger, Millet Bread, and Millet Pizza. These items are made using nutrient-rich grains such as bajra (pearl millet), jowar (sorghum), and ragi (finger millet), offering consumers a wholesome superfood alternative. Launched under the campaign slogan "Go Healthy with Millets," this initiative not only promotes healthier eating habits among Indians but also aligns with the Indian government's Millet Mission and the observance of the International Year of Millets.

Sunfeast Farmlite Super Millets Cookies were introduced by ITC in May 2023 as part of their goal to "Help India Eat Better." Crafted from a nutritious blend of millets such as jowar and ragi, these cookies are available in multi-millet and choco-chip multi-millet variants and are made entirely without maida (refined wheat flour), offering a healthier snacking alternative. They balance taste and nutrition by providing fiber, iron, and protein. These cookies are a tasty and healthy snack choice, and the chocolate-chip variety adds a little decadence. ITC's Mission Millets project is supported by this launch.

The India–Africa International Millet Conference, held in Nairobi, Kenya in July 2023, aimed to increase public awareness of prospects for South-South cooperation and exchange. Additionally, it highlighted the health and nutritional advantages of millets, as well as their adaptability to difficult growing circumstances made worse by climate change.

Among 61 crops, including millets like sorghum, pearl millet, and finger millet, the Indian Prime Minister introduced 109 new and hybrid varieties in August 2024. The goal of these hybrids is to boost quality and yields so that farmers may get higher prices and earn more money.

In September 2023, Nestlé India launched its Nestlé *a+ Masala* millet-based porridges, crafted with locally sourced ingredients, including bajra (pearl millet), barley, vegetables, and a variety of spices. Available in two flavors—Tangy Tomato and Veggie Masala—these quick-cook porridges are ready in just 2 minutes, offering a convenient and nutritious meal option.[7]

These porridges suit local dietary habits since they are produced with ingredients that are acquired locally, such as vegetables, barley, bajra (pearl millet), and a unique blend of ten spices. As a result, the market is expanding due to the increased popularity of millet, new product introductions, and an increase in the consumption of processed foods.

With an average reported production cost of approximately 1700 INR per quintal and a significant rise in global millet production from 26.7 to 30.0 million metric tonnes between 2012 and 2022, millet seeds are emerging as a readily accessible, eco-friendly, and sustainable source of plant-based proteins.[42]

Additionally, the introduction of novel millet varieties tailored to certain regions is driving market growth. For instance, in 2023, 'jirani' and 'chakti' varieties were launched in Nigeria through a joint initiative involving HarvestPlus, the International Crops Research Institute for the Semi-Arid Tropics (ICRISAT), and the Lake Chad Research Institute. These are the first pearl millets to be biofortified with iron. These naturally nutrient-dense, climate-smart genotypes are intended to increase farmer production and household nutrition in the face of global problems.

11.7 Future Outlook

The growing need for sustainable, plant-based protein sources and growing knowledge of the nutritional advantages of millets bode well for millet protein's future. As worries about climate change, food security, and health-conscious eating habits increase, millet protein is becoming more popular as a substitute for conventional protein sources, including wheat, soy, and animal-based proteins. A significant benefit of millet protein is its abundant nutritional profile, which includes important amino acids, especially methionine, which is sometimes lacking in other plant proteins. It is also hypoallergenic, gluten free, and quickly digested, which makes it a great option for anyone with dietary limitations, such as those who have allergies or gluten intolerance.

The market for functional foods and drinks is growing, and millet protein may be used to make a variety of food products, such as energy bars, morning cereals, plant-based dairy substitutes, protein shakes, and baked goods with added protein. It is also becoming more popular in weight control, sports nutrition, and senior nutrition due to its high fiber and protein content. Additionally, because millet is a climate-resilient crop that grows in poor soils, uses little water, and thrives in drought-prone areas, its production is extremely sustainable.

To improve global food security and promote environmental sustainability, governments, food corporations, and researchers are aggressively pushing millet-based goods. It is anticipated that technological developments in food processing and protein extraction would enhance millet protein's flavor, texture, and functioning even further, positioning it as a competitive substitute in the market for plant-based proteins. With sustained investment in research, product development, and consumer education, millet protein has the potential to be a major player in the development of functional foods and sustainable nutrition in the future.

11.8 Conclusion

Millet protein's high nutritional profile, sustainability, and health advantages have made it a prospective substitute in the nutraceutical and functional food sectors. Technological developments in processing, including fermentation, enzymatic hydrolysis, and nanoencapsulation, have improved its bioavailability and prospective uses. The market value of millet has grown dramatically as a result of consumers' increasing preference for plant-based proteins and growing knowledge of its advantages.

The need for high-protein, gluten-free, and environmentally friendly dietary alternatives is predicted to fuel the continuous rise of the millet protein market worldwide. Its growth is further supported by government programs.

The millet protein market is at a critical growth juncture, driven by shifting consumer preferences, the rise of functional foods, and the global push for sustainable nutrition. Trends such as clean-label products, personalized nutrition and plant-based innovation are expected to further shape the market landscape, presenting significant opportunities for stakeholders across the value chain. Strategic alignment with these evolving trends, alongside efforts to overcome infrastructural and awareness barriers, will determine the long-term success and scalability of millet protein in the global food system.

Future developments in millet protein extraction, fortification, and product development will be greatly influenced by research and innovation. Millet protein has the potential to become a widely used protein source with sustained investment and clever marketing. Millet protein supports sustainable nutrition and global food security, encouraging millet production and the growing popularity of functional foods. For it to be widely adopted, nevertheless, issues including low processing infrastructure, low consumer awareness, and competitive price must be resolved.

References

1. P. Shah, A. Dhir, R. Joshi and N. Tripathy, Opportunities and challenges in food entrepreneurship: In-depth qualitative investigation of millet entrepreneurs, *J. Bus. Res.*, 2023, **155**, 113372.
2. S. Maitra, Potential horizon of brown-top millet cultivation in drylands: A review, *Crop Res.*, 2020, **55**(1&2), 57–63.

3. M. Tomar, R. Bhardwaj, R. Verma, S. P. Singh, A. Dahuja, V. Krishnan, R. Kansal, V. K. Yadav, S. Praveen and A. Sachdev, Interactome of millet-based food matrices: A review, *Food Chem.*, 2022, **385**, 132636.
4. A. Kaur, K. Kumar and H. S. Dhaliwal, Physico-chemical characterization and utilization of finger millet (Eleusine coracana L.) *cultivars for the preparation of biscuits, J. Food Process. Preserv.*, 2020, **44**(9), e1467.
5. H. M. Wani, P. Sharma, K. Gul, J. P. Bishnoi, I. A. Wani, S. L. Kothari and A. A. Wani, Influence of γ-irradiation and genotype on the structural and techno-functional properties and microbial quality of Millet flour, *Appl. Food Res.*, 2022, **2**(2), 100181.
6. M. Henchion, M. Hayes, A. Mullen, M. Fenelon and B. Tiwari, Future protein supply and demand: strategies and factors influencing a sustainable equilibrium, *Foods*, 2017, **6**(7), 53.
7. https://www.mordorintelligence.com/industry-reports/millets-market.
8. D. B. Rao, N. G. Malleshi, G. A. Annor and J. V. Patil, *Millets Value Chain for Nutritional Security: A Replicable Success Model from India*, Indian Institute of Millets Research (IIMR), Hyderabad, India, Nutritional and health benefits of millets, 2017, p. 112.
9. S. Birania, P. Rohilla, R. Kumar and N. Kumar, Post-harvest processing of millets: A review on value added products, *Int. J. Chem. Stud.*, 2020, **8**, 1824–1829.
10. B. Sharma and H. S. Gujral, Influence of nutritional and antinutritional components on dough rheology and *in vitro* protein & starch digestibility of minor millets, *Food Chem.*, 2019, **299**, 12115.
11. K. S. Poutanen, A. O. Karlund, C. Gomez-Gallego, D. P. Johansson, N. M. Scheers, I. M. Marklinder, A. K. Eriksen, P. C. Silventoinen, E. Nordlund, N. Sozer, K. J. Hanhineva, M. Kolehmainen and R. Landberg, Grains – a major source of sustainable protein for health, *Nutr. Rev.*, 2022, **80**(6), 1648–1663.
12. N. Sachdev, S. Goomer and L. R. Singh, Foxtail millet: A potential crop to meet future demand scenario for alternative sustainable protein, *J. Sci. Food Agric.*, 2021, **101**(3), 831–842.
13. M. Tomar, R. Bhardwaj, M. Kumar, S. P. Singh, V. Krishnan, R. Kansal, R. Verma, V. K. Yadav, A. Dahuja, S. P. Ahlawat, J. Chand Rana, C. T. Satyavathi, S. Praveen and A. Sachdev, Development of NIR spectroscopy based prediction models for nutritional profiling of pearl millet (Pennisetum glaucum (L.)) R.Br: A chemometrics approach, *LWT*, 2021, **149**, 111813.
14. S. Anitha, M. Govindaraj and J. Kane-Potaka, Balanced amino acid and higher micronutrients in millets complements legumes for improved human dietary nutrition, *Cereal Chem.*, 2020, **97**(1), 74–84.
15. Y. Zhang, M. Xu, X. Zhang, Y. Hu and G. Luan, Application of zein in gluten free foods: A comprehensive review, *Food Res. Int.*, 2022, **160**, 111722.
16. A. Sharma, S. Garg, I. Sheikh, P. Vyas and H. S. Dhaliwal, Effect of wheat grain protein composition on end-use quality, *J. Food Sci. Technol.*, 2020, **57**(8), 2771–2785.

17. Y. Zheng, X. Wang, M. Guo, X. Yan, Y. Zhuang, Y. Sun and J. Li, Two novel anti-hypertensive peptides identified in millet bran glutelin-2 hydrolysates: Purification, *in silico* characterization, molecular docking with ACE and stability in various food processing conditions, *Foods*, 2022, **11**(9), 1355.
18. M. Ortiz-Martinez, J. T. Otero-Pappatheodorou, S. O. Serna-Saldívar and S. García-Lara, Antioxidant activity and characterization of protein fractions and hydrolysates from normal and quality protein maize kernels, *J. Cereal Sci.*, 2017, **76**, 85–91.
19. G. Gowthamraj, M. Raasmika and S. Narayanasamy, Efficacy of fermentation parameters on protein quality and microstructural properties of processed finger millet flour, *J. Food Sci. Technol.*, 2020, **58**(8), 3223–3234.
20. J. A. Tapia-Hernández, C. L. Del-Toro-Sánchez, F. J. Cinco-Moroyoqui, J. E. Juárez-Onofre, S. Ruiz-Cruz, E. Carvajal-Millan, G. A. López-Ahumada, D. D. Castro-Enriquez, C. G. Barreras-Urbina and F. Rodríguez-Felix, Prolamins from cereal by-products: Classification, extraction, characterization and its applications in micro- and nanofabrication, *Trends Food Sci. Technol.*, 2019, **90**, 111–132.
21. A. S. M. Saleh, Q. Zhang, J. Chen and Q. Shen, Millet grains: Nutritional quality, processing, and potential health benefits, *Compr. Rev. Food Sci. Food Saf.*, 2013, **12**(3), 281–295.
22. C. R. Abah, C. N. Ishiwu, J. E. Obiegbuna and A. A. Oladejo, Nutritional composition, functional properties and food applications of millet grains, *Asian Food Sci. J.*, 2020, **14**(2), 9–19.
23. T. Longvah, I. Anantan and K. Bhaskarachary, *Indian food composition tables*, 2017.
24. I. Leinonen, P. P. M. Iannetta, R. M. Rees, W. Russell, C. Watson and A. P. Barnes, Lysine supply is a critical factor in achieving sustainable global protein economy, *Front. Sustainable Food Syst.*, 2019, **3**, 1–11.
25. R. Devisetti, S. N. Yadahally and S. Bhattacharya, Nutrients and antinutrients in foxtail and proso millet milled fractions: Evaluation of their flour functionality, *LWT – Food Sci. Technol.*, 2014, **59**(2P1), 889–895.
26. T. Tamilselvan and A. Kushwaha, Effect of traditional processing methods on the nutritional composition of sorghum (Sorghum bicolour L. Moench) Flour, *Eur. J. Nutr. Food Saf.*, 2020, **12**(7), 69–77.
27. I. Joye, Protein digestibility of cereal products, *Foods*, 2019, **8**(6), 199.
28. T. Thilagavathi, S. Kanchana, P. Banumathi, G. Hemalatha, C. Vanniarajani and M. Sundar Ilamaran, Physico-chemical andfunctional characteristics of selected millets and pulses, *Indian J. Sci. Technol.*, 2015, **8**(S7), 147–155.
29. N. Ghanghas, S. Sharma and P. K. Prabhakar, Classification, composition, extraction, functional modification and application of rice (Oryza sativa) seed protein: A comprehensive review, *Food Rev. Int.*, 2020, 1–30.
30. D. Ramchandran, M. P. Hojilla-Evangelista, S. P. Moose, K. D. Rausch, M. E. Tumbleson and V. Singh, Changes in corn protein content during

storage and their relationship with dry grind ethanol production, *J. Am. Oil Chem. Soc.*, 2018, **95**(8), 923–932.
31. G. Jayaprakash, A. Bains, P. Chawla, M. Fogarasi and S. Fogarasi, A narrative review on rice proteins: Current scenario and food industrial application, *Polymers*, 2022, **14**, 3003.
32. J. Bouchard, M. Malalgoda, J. Storsley, L. Malunga, T. Netticadan and S. J. Thandapilly, Health benefits of cereal grain- and pulse-derived proteins, *Molecules*, 2022, **27**(12), 3746.
33. F. Akharume, D. Santra and A. Adedeji, Physicochemical and functional properties of proso millet storage protein fractions, *Food Hydrocolloids*, 2020, **108**, 105497.
34. R. Katoch, Methods for nutritional quality evaluation of food materials, in *Analytical techniques in biochemistry and molecular biology*, Springer, New York, 2011, pp. 251–322.
35. A. E.-M. M. R. Afify, H. S. El-Beltagi, S. M. Abd El-Salam and A. A. Omran, Protein solubility, digestibility and fractionation after germination of Sorghum varieties, *PLoS One*, 2012, **7**(2), e31154.
36. K. D. Kaur, A. Jha, L. Sabikhi and A. K. Singh, Significance of coarse cereals inhealth and nutrition: A review, *J. Food Sci. Technol.*, 2014, **51**(8), 1429–1441.
37. A. Kate, A. Singh, Processing technology for value addition in millets, *Millets and Millet Technology*, 2021, pp. 239-254.
38. M. V. Rao, C. Sunil and N. Venkatachalapathy, Effect of microwave and hot air radiofrequency treatments on physicochemical and functional properties of foxtail millet flour and its protein isolate, *J. Cereal Sci.*, 2023, 25–29.
39. S. Suri, K. S. Balasaheb, D. K. Yadav, S. Malakar, P. Choudhary, A. Mohapatra and P. Dhurve, Overview of Millet proteins: Quality characteristics, effect of thermal/ non-thermal processing and applications, *Food Biosci.*, 2024, **57**, 103434.
40. F. Jhan, A. Gani, N. Noor and A. Shah, Nanoreduction of millet proteins: Effect on structural and functional properties, *ACS Food Sci. Technol.*, 2021, **1**(8), 1418–1427.
41. A. Amin, I. L. Petersen, C. Malmberg and V. Orlien, Perspective on the Effect of Protein Extraction Method on the Antinutritional Factor (ANF) Content in Seeds, *ACS Food Sci. Technol.*, 2022, **2**(4), 604–612.
42. S. Rajashekar and S. Raju, *Introduction of Millets in PDS-Lessons from Karnataka*, M. S. Swaminathan Research Foundation, Chennai, 2017.
43. Anonymous, Millet Market Size, Share & Trends Analysis Report, 2024.
44. (Market Trends: The Growing Market for Small Millets).
45. (https://dhan.org/smallmillet_newsletter/market_for_small_millets.php)

Subject Index